Linear Systems Theory

LINEAR
SYSTEMS
THEORY

JOÃO P. HESPANHA

PRINCETON UNIVERSITY PRESS ~ PRINCETON AND OXFORD

In the United Kingdom: Princeton University Press, 6 Oxford Street,
Woodstock, Oxfordshire OX20 1TW

Library of Congress Cataloging-in-Publication Data

Hespanha, João P.
    Linear systems theory / João P. Hespanha.
       p. cm.
    ISBN 978-0-691-14021-6 (hardcover : alk. paper) 1. Linear systems—
Textbooks. 2. System analysis—Textbooks. I. Title
    QA402.H47 2009
    003′.74—dc22

                                                    2009006371

British Library Cataloging-in-Publication Data is available

This book has been composed in Minion
Printed on acid-free paper.

press.princeton.edu

Printed in the United States of America

10 9 8 7

TO MY WIFE STACY AND TO OUR SON RUI

CONTENTS

PREAMBLE

Linear systems theory is the cornerstone of control theory and a prerequisite for essentially all graduate courses in this area. It is a well-established discipline that focuses on linear differential equations from the perspective of control and estimation.

CONTENT

The first set of lectures (1–17) covers the key topics in linear systems theory: system representation, stability, controllability and state feedback, observability and state estimation, and realization theory. The main goal of these chapters is to provide the background needed for advanced control design techniques. Feedback linearization and the LQR problem are also briefly introduced to increase the design component of this set of lectures. The preview of optimal LQR control facilitates the introduction of notions such as controllability and observability, but is pursued in much greater detail in the second set of lectures.

Three advanced foundational topics are covered in a second set of lectures (18–25): poles and zeros for MIMO systems, LQG/LQR control, and control design based on the Q parameterization of stabilizing controllers (Q design). The main goal of these chapters is to introduce advanced supporting material for modern control design techniques. Although LQG/LQR is covered in some other linear systems books, it is generally not covered at the same level of detail (in particular the frequency domain properties of LQG/LQR, loop shaping, and loop transfer recovery). In fact, there are few textbooks in print that cover the same material, in spite of the fact that these are classical results and LQG/LQR is the most widely used form of state-space control. By covering the ARE in detail, I set the stage for H-2 and H-infinity.

In writing this book, it is assumed that the reader is familiar with linear algebra and ordinary differential equations at an undergraduate level. To profit most from this textbook, the reader would also have taken an undergraduate course in classical control, but these notes are basically self-contained regarding control concepts.

ORGANIZATION AND STYLE

This book was purposely designed as a textbook, and because it is not an adaptation of a reference text, the main emphasis is on presenting material in a fashion that makes it easy for students to understand. The material is organized in lectures, and it is divided so that on average each lecture can be covered in 2 hours of class time. The sequence in which the material appears was selected to emphasize continuity and motivate the need for new concepts as they are introduced.

In writing this manuscript there was a conscious effort to reduce verbosity. This is not to say that I did not attempt to motivate the concepts or discuss their significance (on the contrary), but the amount of text was kept to a minimum. Typically, discussion, remarks, and side comments are relegated to marginal notes so that the reader can easily follow the material presented without distraction and yet enjoy the

Attention! When a marginal note finishes with "▶ p. XXX," more information about that topic can be found on page XXX.

benefit of comments on the notation and terminology, or be made aware that there is a related MATLAB® command.

I have also not included a chapter or appendix that summarizes background material (e.g., a section on linear algebra or nonlinear differential equations). Linear algebra is a key prerequisite to this course, and it is my experience that referring a student who is weak on linear algebra to a brief chapter on the subject is useless (and sometime even counter-productive). I do review advanced concepts (e.g., singular values, matrix norms, and the Jordan normal form), but this is done at the points in the text where these concepts are needed. I also take this approach to referring the reader to MATLAB®, by introducing the commands only where the relevant concepts appear in the text.

LEARNING AND TEACHING USING THIS TEXTBOOK

Lectures 1–17 can be the basis for a one-quarter graduate course on linear systems theory. At the University of California at Santa Barbara I teach essentially all the material in these lectures in one quarter with about 40 hours of class time. In the interest of time, the material in the Additional Notes sections and some of the discrete-time proofs can be skipped. For a semester-long course, one could also include a selection of the advanced topics covered in the second part of the book (Lectures 18–25).

I have tailored the organization of the textbook to simplify the teaching and learning of the material. In particular, the sequence of the chapters emphasizes continuity, with each chapter appearing motivated and in logical sequence with the preceding ones. I always avoid introducing a concept in one chapter and using it again only many chapters later. It has been my experience that even if this may be economical in terms of space, it is pedagogically counterproductive. The chapters are balanced in length so that on average each can be covered in roughly 2 hours of lecture time. Not only does this greatly aid the instructor's planning, but it makes it easier for the students to review the materials taught in class.

As I have taught this material, I have noticed that some students arrive at graduate school without proper training in formal reasoning. In particular, many students come with limited understanding of the basic logical arguments behind mathematical proofs. A course in linear systems provides a superb opportunity to overcome this difficulty. To this effect, I have annotated several proofs with marginal notes that explain general techniques for constructing proofs: contradiction, contraposition, the difference between necessity and sufficiency, etc. (see, e.g., Note 9.2 on page 82). Throughout the manuscript, I have also structured the proofs to make them as intuitive as possible, rather than simply as short as possible. All mathematical derivations emphasize the aspects that give insight into the material presented and do not dwell on technical aspects of small consequence that merely bore the students. Often these technical details are relegated to marginal notes or exercises.

MATLAB®

Computational tools such as the MATLAB® software environment offer a significant step forward in teaching linear systems because they allow students to solve numerical problems without being exposed to a detailed treatment of numerical computations.

By systematically annotating the theoretical developments with marginal notes that discuss the relevant commands available in MATLAB®, this textbook helps students learn to use these tools. An example of this can be found, e.g., in MATLAB® Hint 9 in page 12, which is further expanded on page 51.

The commands discussed in the "MATLAB® Hints" assume that the reader has version R2007b of MATLAB® with Simulink®, the Symbolic Math Toolbox, and the Control System Toolbox. However, essentially all these commands have been fairly stable for several versions so they are likely to work with previous and subsequent versions for several years to come. Lecture 25 assumes that the reader has installed CVX version 1.2, which is a MATLAB® package for Disciplined Convex Programming, distributed under the GNU General Public License 2.0 [7].

MATLAB® and Simulink® are registered trademarks of The MathWorks Inc. and are used with permission. The MathWorks does not warrant the accuracy of the text or exercises in this book. This book's use or discussion of MATLAB®, Simulink®, or related products does not constitute an endorsement or sponsorship by The Math-Works of a particular pedagogical approach or particular use of the MATLAB® and Simulink® software.

Web

The reader is referred to the books's website at http://press.princeton.edu/titles/9102.html for corrections, updates on MATLAB® and CVX, and other supplemental material.

Acknowledgments

Several friends and colleagues have helped me improve this manuscript through their thoughtful constructive comments and suggestions. Among these, I owe special thanks to A. Pedro Aguiar, Karl Åström, Stephen Boyd, Bassam Bamieh, Maurice Heemels, Mustafa Khammash, Daniel Klein, Petar Kokotović, Kristi Morgansen, and Dale Seborg, as well as all the students that used early drafts of these notes and provided me with numerous comments and suggestions. I would also like to acknowledge the support of several organizations, including the National Science Foundation (NSF), the Army Research Office (ARO), the Air Force Office of Scientific Research (AFOSR), and the University of California at Santa Barbara.

Linear Systems I — Basic Concepts

PART I
SYSTEM REPRESENTATION

LECTURE 1

State-Space Linear Systems

Contents

This lecture introduces state-space linear systems, which are the main focus of this course.

1. State-Space Linear Systems
2. Block Diagrams
3. Exercises

1.1 STATE-SPACE LINEAR SYSTEMS

A *continuous-time state-space linear system* is defined by the following two equations:

$$\dot{x}(t) = A(t)x(t) + B(t)u(t), \qquad x \in \mathbb{R}^n, \ u \in \mathbb{R}^k, \qquad \text{(1.1a)}$$

$$y(t) = C(t)x(t) + D(t)u(t), \qquad y \in \mathbb{R}^m. \qquad \text{(1.1b)}$$

> **Notation.** A function of time (either continuous $t \in [0, \infty)$ or discrete $t \in \mathbb{N}$) is called a *signal.*

The signals

$$u : [0, \infty) \to \mathbb{R}^k, \qquad x : [0, \infty) \to \mathbb{R}^n, \qquad y : [0, \infty) \to \mathbb{R}^m,$$

are called the *input, state,* and *output* of the system. The first-order differential equation (1.1a) is called the *state equation* and (1.1b) is called the *output equation.*

> **Notation 1.** We write $u \overset{P}{\rightsquigarrow} y$ to mean that "y is one of the outputs that corresponds to u;" the (optional) label P specifies the system under consideration.

The equations (1.1) express an *input-output* relationship between the input signal $u(\cdot)$ and the output signal $y(\cdot)$. For a given input $u(\cdot)$, we need to solve the state equation to determine the state $x(\cdot)$ and then replace it in the output equation to obtain the output $y(\cdot)$.

Attention! For the same input $u(\cdot)$, different choices of the initial condition $x(0)$ on the state equation will result in different state trajectories $x(\cdot)$. Consequently, *one* input $u(\cdot)$ generally corresponds to *several* possible outputs $y(\cdot)$. □

1.1.1 TERMINOLOGY AND NOTATION

When the input signal u takes scalar values ($k = 1$), the system is called *single input (SI)*; otherwise, it is called *multiple input (MI)*. When the output signal y takes scalar values ($m = 1$) the system is called *single output (SO)*; otherwise, it is called *multiple output (MO)*.

When there is no state equation ($n = 0$) and we have simply

$$y(t) = D(t)u(t), \qquad\qquad u \in \mathbb{R}^k, \ y \in \mathbb{R}^m,$$

the system is called *memoryless*.

Note. The rationale behind this terminology is explained in Lecture 3.

When all the matrices $A(t), B(t), C(t), D(t)$ are constant $\forall t \geq 0$, the system (1.1) is called a *linear time-invariant (LTI)* system. In the general case, (1.1) is called a *linear time-varying (LTV)* system to emphasize that time invariance is not being assumed. For example, Lecture 3 discusses *impulse responses of LTV systems* and *transfer functions of LTI systems*. This terminology indicates that the *impulse response* concept applies to both LTV and LTI systems, but the *transfer function* concept is meaningful only for LTI systems.

To keep formulas short, in the following we abbreviate (1.1) to

$$\dot{x} = A(t)x + B(t)u, \quad y = C(t)x + D(t)u, \quad x \in \mathbb{R}^n, \ u \in \mathbb{R}^k, \ y \in \mathbb{R}^m \quad \text{(CLTV)}$$

and in the time-invariant case, we further shorten this to

MATLAB® **Hint 1.**
ss(A,B,C,D)
creates the
continuous-time LTI
state-space system
(CLTI). ▶ p. 6

$$\dot{x} = Ax + Bu, \qquad y = Cx + Du, \qquad x \in \mathbb{R}^n, \ u \in \mathbb{R}^k, \ y \in \mathbb{R}^m. \quad \text{(CLTI)}$$

Since these equations appear in the text numerous times, we use the special tags (CLTV) and (CLTI) to identify them.

1.1.2   DISCRETE-TIME CASE

A *discrete-time state-space linear system* is defined by the following two equations:

$$x(t+1) = A(t)x(t) + B(t)u(t), \qquad x \in \mathbb{R}^n, \ u \in \mathbb{R}^k, \qquad \text{(1.2a)}$$

$$y(t) = C(t)x(t) + D(t)u(t), \qquad y \in \mathbb{R}^m. \qquad \text{(1.2b)}$$

Attention! *One* input generally corresponds to *several* outputs, because one may consider several initial conditions for the state equation.

All the terminology introduced for continuous-time systems also applies to discrete time, except that now the domain of the signals is $\mathbb{N} := \{0, 1, 2, \dots\}$, instead of the interval $[0, \infty)$.

In discrete-time systems the state equation is a difference equation , instead of a first-order differential equation. However, the *input-output* relationship between input and output is analogous. For a given input $u(\cdot)$, we need to solve the state (difference) equation to determine the state $x(\cdot)$ and then replace it in the output equation to obtain the output $y(\cdot)$.

Note. Since this equation appears in the text numerous times, we use the special tag (DLTI) to identify it. The tag (DLTV) is used to identify the time-varying case in (1.2).

To keep formulas short, in the following we abbreviate the time-invariant case of (1.2) to

$$x^+ = Ax + Bu, \qquad y = Cx + Du, \qquad x \in \mathbb{R}^n, \ u \in \mathbb{R}^k, \ y \in \mathbb{R}^m. \quad \text{(DLTI)}$$

1.1.3   STATE-SPACE SYSTEMS IN MATLAB®

MATLAB® has several commands to create and manipulate LTI systems. The following basic command is used to create an LTI system.

Note. Initial conditions to LTI state-space MATLAB® systems are specified at simulation time.

MATLAB® **Hint 1** (ss). The command sys_ss=ss(A,B,C,D) assigns to sys_ss a continuous-time LTI state-space MATLAB® system of the form

$$\dot{x} = Ax + Bu, \qquad\qquad y = Cx + Du.$$

Note. It is common
practice to denote the
input and output
signals of a system by
u and y, respectively.
However, when
dealing with
interconnections, one
must use different
symbols for each
signal, so this
convention is
abandoned.

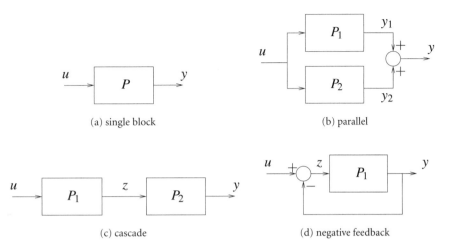

        (a) single block   (b) parallel

        (c) cascade   (d) negative feedback

Figure 1.1. Block diagrams.

Optionally, one can specify the names of the inputs, outputs, and state to be used in subsequent plots as follows:

```
sys_ss=ss(A,B,C,D,...
          'InputName', {'input1', 'input2',...},...
          'OutputName',{'output1','output2',...},...
          'StateName', {'input1', 'input2',...});
```

The number of elements in the bracketed lists must match the number of inputs, outputs, and state variables.

For discrete-time systems, one should instead use the command `sys_ss=ss(A, B,C,D,Ts)`, where `Ts` is the sampling time, or `-1` if one does not want to specify it.   □

1.2   BLOCK DIAGRAMS

It is convenient to represent systems by *block diagrams* as in Figure 1.1. These diagrams generally serve as compact representations for complex equations.

The two-port block in Figure 1.1(a) represents a system with input $u(\cdot)$ and output $y(\cdot)$, where the directions of the arrows specify which is which.

Although not explicitly represented in the diagram, one must keep in mind the existence of the state, which affects the output through the initial condition.

1.2.1   INTERCONNECTIONS

Interconnections of block diagrams are especially useful to highlight special structures in state-space equations. To understand what is meant by this, assume that the blocks P_1 and P_2 that appear in Figure 1.1 are the two LTI systems

$$P_1: \quad \dot{x}_1 = A_1 x_1 + B_1 u_1, \quad y_1 = C_1 x_1 + D_1 u_1, \quad x \in \mathbb{R}^{n_1}, \; u \in \mathbb{R}^{k_1}, \; y_1 \in \mathbb{R}^{m_1},$$

$$P_2: \quad \dot{x}_2 = A_2 x_2 + B_2 u_2, \quad y_2 = C_2 x_2 + D_2 u_2, \quad x \in \mathbb{R}^{n_2}, \ u \in \mathbb{R}^{k_2}, \ y_2 \in \mathbb{R}^{m_2}.$$

The general procedure to obtain the state-space for an interconnection consists of stacking the states of the individual subsystems in a tall vector x and computing $\dot{x}$ using the state and output equations of the individual blocks. The output equation is also obtained from the output equations of the subsystems.

In Figure 1.1(b) we have $u = u_1 = u_2$ and $y = y_1 + y_2$, which corresponds to a *parallel interconnection*. This figure represents the LTI system

$$\begin{bmatrix} \dot{x}_1 \\ \dot{x}_2 \end{bmatrix} = \begin{bmatrix} A_1 & 0 \\ 0 & A_2 \end{bmatrix} \begin{bmatrix} x_1 \\ x_2 \end{bmatrix} + \begin{bmatrix} B_1 \\ B_2 \end{bmatrix} u, \quad y = \begin{bmatrix} C_1 & C_2 \end{bmatrix} \begin{bmatrix} x_1 \\ x_2 \end{bmatrix} + (D_1 + D_2)u,$$

Notation. Given a vector (or matrix) x, we denote its transpose by x'.

with state $x := [x_1' \ x_2']' \in \mathbb{R}^{n_1 + n_2}$. The parallel structure is responsible for the block-diagonal structure in the matrix $\begin{bmatrix} A_1 & 0 \\ 0 & A_2 \end{bmatrix}$. A block-diagonal structure in this matrix indicates that the state-space system can be decomposed as the parallel of two state-space systems with smaller states.

In Figure 1.1(c) we have $u = u_1$, $y = y_2$, and $z = y_1 = u_2$, which corresponds to a *cascade interconnection*. This figure represents the LTI system

Note. How to arrive at equation (1.3)? *Hint: Start with the output equation.*

$$\begin{bmatrix} \dot{x}_1 \\ \dot{x}_2 \end{bmatrix} = \begin{bmatrix} A_1 & 0 \\ B_2 C_1 & A_2 \end{bmatrix} \begin{bmatrix} x_1 \\ x_2 \end{bmatrix} + \begin{bmatrix} B_1 \\ B_2 D_1 \end{bmatrix} u, \quad y = \begin{bmatrix} D_2 C_1 & C_2 \end{bmatrix} \begin{bmatrix} x_1 \\ x_2 \end{bmatrix} + D_2 D_1 u,$$

with state $x := [x_1' \ x_2']' \in \mathbb{R}^{n_1 + n_2}$. The cascade structure is responsible for the block-triangular structure in the matrix $\begin{bmatrix} A_1 & 0 \\ B_2 C_1 & A_2 \end{bmatrix}$, and, in fact, a block-triangular structure in this matrix indicates that the state-space system can be decomposed as a cascade of two state-space systems with smaller states.

In Figure 1.1(d) we have $u_1 = u - y_1$ and $y = y_1$, which corresponds to a *negative-feedback interconnection*. This figure represents the LTI system

MATLAB® Hint 2.
To avoid ill-posed feedback interconnections, MATLAB® warns about algebraic loops when one attempts to close feedback loops around systems like P_1 with nonzero D_1 matrices (even when $I + D_1$ is invertible).

$$\dot{x}_1 = \big(A_1 - B_1(I + D_1)^{-1} C_1\big)x_1 + B_1\big(I - (I + D_1)^{-1} D_1\big)u, \tag{1.3a}$$

$$y = (I + D_1)^{-1} C_1 x_1 + (I + D_1)^{-1} D_1 u, \tag{1.3b}$$

with state $x_1 \in \mathbb{R}^{n_1}$. Sometimes feedback interconnections are ill-posed. In this example, this would happen if the matrix $I + D_1$ was singular.

The basic interconnections in Figure 1.1 can be combined to form arbitrarily complex diagrams. The general procedure to obtain the final state-space system remains the same: Stack the states of all subsystems in a tall vector x and compute $\dot{x}$ using the state and output equations of the individual blocks.

1.2.2 SYSTEM DECOMPOSITION

MATLAB® Hint 3.
This type of decomposition is especially useful to build systems in Simulink®.

Block diagrams are also useful to represent complex systems as the interconnection of simple blocks. This can be seen through the following two examples:

1. The LTI system

$$\dot{x} = x + u, \qquad\qquad\qquad y = x$$

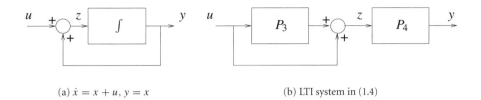

(a) $\dot{x} = x + u$, $y = x$ (b) LTI system in (1.4)

Figure 1.2. Block diagram representation systems.

can be viewed as a feedback connection in Figure 1.2(a), where the *integrator* system $\int$ maps each input z to the solutions y of

$$\dot{y} = z.$$

2. Consider the LTI system

$$\begin{bmatrix} \dot{x}_1 \\ \dot{x}_2 \end{bmatrix} = \begin{bmatrix} 1 & 1 \\ 0 & 3 \end{bmatrix} \begin{bmatrix} x_1 \\ x_2 \end{bmatrix} + \begin{bmatrix} 1 \\ 5 \end{bmatrix} u, \qquad y = \begin{bmatrix} 1 & 0 \end{bmatrix} \begin{bmatrix} x_1 \\ x_2 \end{bmatrix}. \qquad (1.4)$$

Writing these equations as

$$\dot{x}_2 = 3x_2 + 5u, \qquad\qquad y_2 = x_2 \qquad\qquad (1.5)$$

and

$$\dot{x}_1 = x_1 + z, \qquad\qquad y = x_1, \qquad\qquad (1.6)$$

Note. In general, this type of decomposition is not unique, since there may be many ways to represent a system as the interconnection of simpler systems.

where $z := x_2 + u$, we conclude that (1.4) can be viewed as the block diagram in Figure 1.2(b), where P_3 corresponds to the LTI system (1.5) with input u and output y_2 and P_4 corresponds to the LTI systems (1.6) with input z and output y.

1.2.3 SYSTEM INTERCONNECTIONS WITH MATLAB®

MATLAB® Hint 4 (series). The command sys=series(sys1,sys2) or, alternatively, sys=sys2*sys1 creates a system sys from the cascade connection of the system sys1 whose output is connected to the input of sys2.

Attention! Note the different order in which sys1 and sys2 appear in the two forms of this command.

For MIMO systems, one can use sys=series(sys1,sys2,outputs1, inputs2), where outputs1 and inputs2 are vectors that specify the outputs of sys1 and inputs of sys2, respectively, that should be connected. These two vectors should have the same size and contain integer indexes starting at 1. □

MATLAB® Hint 5 (parallel). The command sys=parallel(sys1,sys2) or, alternatively, sys=sys1+sys2 creates a system sys from the parallel connection of the systems sys1 and sys2.

For MIMO systems, one can use `sys=parallel(sys1,sys2,inputs1,`
`inputs2,outputs1,outputs2)`, where `inputs1` and `inputs2` specify
which inputs should be connected and `outputs1` and `outputs2` specify which
outputs should be added. All four vectors should contain integer indexes starting
at 1.   □

MATLAB® Hint 6 (`append`). The command `sys=append(sys1,sys2, ... ,`
`sysn)` creates a system `sys` whose inputs are the union of the inputs of all the systems `sys1, sys2, ... , sysn` and whose outputs are the union of the outputs of all
the same systems. The dynamics are maintained decoupled.   □

MATLAB® Hint 7 (`feedback`). The command `sys=feedback(sys1,sys2)`
creates a system `sys` from the negative feedback interconnection of the system `sys1`
in the forward loop, with the system `sys2` in the backward loop. A positive feedback
interconnection can be obtained using `sys=feedback(sys1,sys2,+1)`.
For MIMO systems, one can use `sys=feedback(sys1,sys2,feedinputs,`
`feedoutputs,sign)`, where `feedinputs` specify which inputs of the forward-
loop system `sys1` receive feedback from `sys2`, `feedoutputs` specify which outputs of the forward-loop system `sys1` are feedback to `sys2`, and `sign` $\in \{-1, +1\}$
specifies whether a negative or positive feedback configuration should be used. More
details can be obtained by using `help feedback`.   □

1.3 EXERCISES

1.1 (Block diagram decomposition). Consider a system P_1 that maps each input u
to the solutions y of

$$\begin{bmatrix} \dot{x}_1 \\ \dot{x}_2 \end{bmatrix} = \begin{bmatrix} 1 & 0 \\ -1 & 2 \end{bmatrix} \begin{bmatrix} x_1 \\ x_2 \end{bmatrix} + \begin{bmatrix} 4 \\ 1 \end{bmatrix} u, \qquad y = \begin{bmatrix} 1 & 3 \end{bmatrix} \begin{bmatrix} x_1 \\ x_2 \end{bmatrix}.$$

Represent this system in terms of a block diagram consisting only of

- *integrator* systems, represented by the symbol $\boxed{\int}$, that map their input $u(\cdot) \in$
  $\mathbb{R}$ to the solution $y(\cdot) \in \mathbb{R}$ of $\dot{y} = u$,

- *summation* blocks, represented by the symbol $\boxed{\sum}$, that map their input $u(\cdot) \in$
  $\mathbb{R}^k$ to the output $y(t) = \sum_{i=1}^{k} u_i(t), \forall t \geq 0$, and

- *gain* memoryless systems, represented by the symbol $\boxed{g}$, that map their input
  $u(\cdot) \in \mathbb{R}$ to the output $y(t) = gu(t) \in \mathbb{R}, \forall t \geq 0$ for some $g \in \mathbb{R}$.   □

LECTURE 2

Linearization

Contents

This lecture addresses how state-space linear systems arise in control.

1. State-Space Nonlinear Systems
2. Local Linearization around an Equilibrium Point
3. Local Linearization around a Trajectory
4. Feedback Linearization
5. Exercises

2.1 STATE-SPACE NONLINEAR SYSTEMS

Linear voltage versus current laws for resistors, force versus displacement laws for springs, force versus velocity laws for friction, etc., are only approximations to more complex nonlinear relationships. Since linear systems are the exception rather than the rule, a more reasonable class of systems to study appear to be those defined by nonlinear differential equations of the form

$$\dot{x} = f(x, u), \qquad y = g(x, u), \qquad x \in \mathbb{R}^n, \ u \in \mathbb{R}^k, \ y \in \mathbb{R}^m. \qquad (2.1)$$

It turns out that

1. one can establish properties of (2.1) by analyzing state-space linear systems that approximate it, and

2. one can design feedback controllers for (2.1) by reducing the problem to one of designing controllers for state-space linear systems.

2.2 LOCAL LINEARIZATION AROUND AN EQUILIBRIUM POINT

Definition 2.1 (Equilibrium). A pair $(x^{\mathrm{eq}}, u^{\mathrm{eq}}) \in \mathbb{R}^n \times \mathbb{R}^k$ is called an *equilibrium point* of (2.1) if $f(x^{\mathrm{eq}}, u^{\mathrm{eq}}) = 0$. In this case,

$$u(t) = u^{\mathrm{eq}}, \qquad x(t) = x^{\mathrm{eq}}, \qquad y(t) = y^{\mathrm{eq}} := g(x^{\mathrm{eq}}, u^{\mathrm{eq}}), \qquad \forall t \geq 0$$

is a solution to (2.1). $\qquad\qquad\square$

Suppose now that we apply to (2.1) an input

$$u(t) = u^{\text{eq}} + \delta u(t), \quad \forall t \geq 0$$

that is close but not equal to u^{eq} and that the initial condition

$$x(0) = x^{\text{eq}} + \delta x^{\text{eq}}$$

is close but not quite equal to x^{eq}. Then the corresponding output $y(t)$ to (2.1) will be close but not equal to $y^{\text{eq}} = g(x^{\text{eq}}, u^{\text{eq}})$. To investigate how much $x(t)$ and $y(t)$ are perturbed by $\delta u(\cdot)$ and δx^{eq}, we define

$$\delta x(t) := x(t) - x^{\text{eq}}, \qquad \delta y(t) := y(t) - y^{\text{eq}}, \qquad \forall t \geq 0$$

and use (2.1) to conclude that

$$\delta y = g(x, u) - y^{\text{eq}} = g(x^{\text{eq}} + \delta x, u^{\text{eq}} + \delta u) - g(x^{\text{eq}}, u^{\text{eq}}).$$

Expanding $g(\cdot)$ as a Taylor series around $(x^{\text{eq}}, u^{\text{eq}})$, we obtain

$$\delta y = \frac{\partial g(x^{\text{eq}}, u^{\text{eq}})}{\partial x} \delta x + \frac{\partial g(x^{\text{eq}}, u^{\text{eq}})}{\partial u} \delta u + O(\|\delta x\|^2) + O(\|\delta u\|^2), \tag{2.2}$$

where

$$\frac{\partial g(x, u)}{\partial x} := \left[\left(\frac{\partial g_i(x, u)}{\partial x_j} \right)_{ij} \right] \in \mathbb{R}^{m \times n}, \quad \frac{\partial g(x, u)}{\partial u} := \left[\left(\frac{\partial g_i(x, u)}{\partial u_j} \right)_{ij} \right] \in \mathbb{R}^{m \times k}.$$

To determine the evolution of δx, we take its time derivative

$$\dot{\delta x} = \dot{x} = f(x, u) = f(x^{\text{eq}} + \delta x, u^{\text{eq}} + \delta_u)$$

and also expand f as a Taylor series around $(x^{\text{eq}}, u^{\text{eq}})$, which yields

$$\dot{\delta x} = \frac{\partial f(x^{\text{eq}}, u^{\text{eq}})}{\partial x} \delta x + \frac{\partial f(x^{\text{eq}}, u^{\text{eq}})}{\partial u} \delta u + O(\|\delta x\|^2) + O(\|\delta u\|^2), \tag{2.3}$$

where

$$\frac{\partial f(x, u)}{\partial x} := \left[\left(\frac{\partial f_i(x, u)}{\partial x_j} \right)_{ij} \right] \in \mathbb{R}^{n \times n}, \quad \frac{\partial f(x, u)}{\partial u} := \left[\left(\frac{\partial f_i(x, u)}{\partial u_j} \right)_{ij} \right] \in \mathbb{R}^{n \times k}.$$

By dropping all but the first-order terms in (2.2) and (2.3), we obtain a *local linearization of (2.1) around an equilibrium point*.

Definition 2.2 (Local linearization around an equilibrium point). The LTI system

$$\dot{\delta x} = A \delta x + B \delta u, \qquad \delta y = C \delta x + D \delta u \tag{2.4}$$

defined by the following Jacobian matrices

$$A := \frac{\partial f(x^{\text{eq}}, u^{\text{eq}})}{\partial x}, \quad B := \frac{\partial f(x^{\text{eq}}, u^{\text{eq}})}{\partial u}, \quad C := \frac{\partial g(x^{\text{eq}}, u^{\text{eq}})}{\partial x}, \quad D := \frac{\partial g(x^{\text{eq}}, u^{\text{eq}})}{\partial u} \tag{2.5}$$

is called the *local linearization* of (2.1) around the equilibrium point $(x^{\text{eq}}, u^{\text{eq}})$. $\quad \square$

MATLAB® Hint 8. `jacobian(f,x)` can be used to symbolically compute the matrices in (2.5).
▶ p. 52

MATLAB® Hint 9. Symbolic computations in MATLAB® will be discussed in Lecture 6. ▶ p. 51

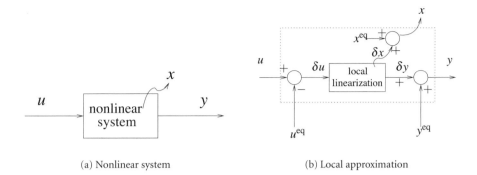

(a) Nonlinear system   (b) Local approximation

Figure 2.1. Nonlinear system (a) and its local approximation (b) obtained from a local linearization.

Attention! The local linearization (2.4) approximates (2.2) and (2.3) well *only while δx and δu remain small.* Therefore a key property that needs to be checked to make sure that a local linearization is valid is that δu and δx remain small along solutions to the linearized system. ☐

Attention! The local linearization (2.4) expresses only a relation between *perturbations* on the input, state, and output of the original system (2.1). Consequently, the input, output, and state of the local linearization system (2.4) are perturbation values with respect to u^{eq}, y^{eq}, and x^{eq}, respectively (cf. Figure 2.1). ☐

DISCRETE-TIME CASE

Consider the following discrete-time nonlinear system

$$x^+ = f(x, u), \qquad y = g(x, u), \qquad x \in \mathbb{R}^n, \ u \in \mathbb{R}^k, \ y \in \mathbb{R}^m. \qquad (2.6)$$

Definition 2.3 (Equilibrium). A pair $(x^{eq}, u^{eq}) \in \mathbb{R}^n \times \mathbb{R}^k$ is called an *equilibrium point* of (2.6) if $f(x^{eq}, u^{eq}) = x^{eq}$. In this case,

$$u(t) = u^{eq}, \qquad x(t) = x^{eq}, \qquad y(t) = y^{eq} := g(x^{eq}, u^{eq}), \qquad \forall t \in \mathbb{N}$$

is a solution to (2.6). ☐

Definition 2.4 (Local linearization around an equilibrium point). The discrete-time LTI system

$$\delta x^+ = A\delta x + B\delta u, \qquad\qquad \delta y = C\delta x + D\delta u,$$

defined by the following Jacobian matrices

$$A := \frac{\partial f(x^{eq}, u^{eq})}{\partial x}, \quad B := \frac{\partial f(x^{eq}, u^{eq})}{\partial u}, \quad C := \frac{\partial g(x^{eq}, u^{eq})}{\partial x}, \quad D := \frac{\partial g(x^{eq}, u^{eq})}{\partial u}$$

is called the *local linearization* of (2.6) around the equilibrium point (x^{eq}, u^{eq}). ☐

2.3   LOCAL LINEARIZATION AROUND A TRAJECTORY

Often it is convenient to consider perturbations around an arbitrary solution to (2.1), instead of an equilibrium point. To do this, suppose that

$$u^{\text{sol}} : [0, \infty) \to \mathbb{R}^k, \qquad x^{\text{sol}} : [0, \infty) \to \mathbb{R}^n, \qquad y^{\text{sol}} : [0, \infty) \to \mathbb{R}^m$$

is a solution to (2.1) (not necessarily constant). Assuming that we apply to (2.1) an input

$$u(t) = u^{\text{sol}}(t) + \delta u(t), \quad \forall t \geq 0$$

that is close but not equal to $u^{\text{sol}}(t)$ and that the initial condition

$$x(0) = x^{\text{sol}}(0) + \delta x^{\text{sol}}$$

is close but not quite equal to $x^{\text{sol}}(0)$. Then the corresponding output $y(t)$ will be close but not equal to $y^{\text{sol}}(t)$. To investigate how much $x(t)$ and $y(t)$ are perturbed by this, we now define

$$\delta x(t) := x(t) - x^{\text{sol}}(t), \qquad \delta y(t) := y(t) - y^{\text{sol}}(t), \qquad \forall t \geq 0.$$

Proceeding as before, we conclude that

$$\dot{\delta x} = \frac{\partial f\left(x^{\text{sol}}(t), u^{\text{sol}}(t)\right)}{\partial x} \delta x + \frac{\partial f\left(x^{\text{sol}}(t), u^{\text{sol}}(t)\right)}{\partial u} \delta u + O(\|\delta x\|^2) + O(\|\delta u\|^2),$$

$$\delta y = \frac{\partial g\left(x^{\text{sol}}(t), u^{\text{sol}}(t)\right)}{\partial x} \delta x + \frac{\partial g\left(x^{\text{sol}}(t), u^{\text{sol}}(t)\right)}{\partial u} \delta u + O(\|\delta x\|^2) + O(\|\delta u\|^2),$$

with the main difference with respect to (2.2) and (2.3) being that the derivatives are computed along $x^{\text{sol}}(t)$ and $u^{\text{sol}}(t)$. By dropping all but the first-order terms, we obtain a *local linearization of* (2.1) *around a trajectory*.

Definition 2.5 (Local linearization around a trajectory). The state-space linear system

$$\dot{\delta x} = A(t)\delta x + B(t)\delta u, \qquad\qquad \delta y = C(t)\delta x + D(t)\delta u$$

defined by the following Jacobian matrices

$$A(t) := \frac{\partial f\left(x^{\text{sol}}(t), u^{\text{sol}}(t)\right)}{\partial x}, \qquad\qquad B(t) := \frac{\partial f\left(x^{\text{sol}}(t), u^{\text{sol}}(t)\right)}{\partial u},$$

$$C(t) := \frac{\partial g\left(x^{\text{sol}}(t), u^{\text{sol}}(t)\right)}{\partial x}, \qquad\qquad D(t) := \frac{\partial g\left(x^{\text{sol}}(t), u^{\text{sol}}(t)\right)}{\partial u}$$

is called the *local linearization* of (2.1) around the trajectory $\left(x^{\text{sol}}(\cdot), u^{\text{sol}}(\cdot)\right)$.   □

Attention! In general, local linearizations around trajectories lead to LTV systems because the partial derivatives need to be computed along the trajectory. However, for some nonlinear systems there are trajectories for which local linearizations actually lead to LTI systems. For models of vehicles (cars, airplanes, helicopters, hovercraft, submarines, etc.) trajectories that lead to LTI local linearizations are called *trimming trajectories*. They often correspond to motion along straight lines, circumferences, or helices (see Exercise 2.2).   □

2.4 FEEDBACK LINEARIZATION

In this section we explore another mechanism by which linear systems arise out of nonlinear ones. We start by restricting our attention to mechanical systems.

2.4.1 MECHANICAL SYSTEMS

Note. These include robot arms, mobile robots, airplanes, helicopters, underwater vehicles, hovercraft, etc.

Note. A symmetric matrix M is *positive-definite* if $x'Mx > 0, \forall x \neq 0$ (cf. Section 8.4).

Note 1. A force is *conservative* if the total work of moving an object from one point to another is independent of the path taken. ▶ p. 15

The equations of motion of many mechanical systems can be written in the form

$$M(q)\ddot{q} + B(q,\dot{q})\dot{q} + G(q) = F, \tag{2.7}$$

where $q \in \mathbb{R}^k$ is a k-vector with linear and/or angular positions called the *generalized coordinates vector*, $M(q)$ is a $k \times k$ nonsingular symmetric positive-definite matrix called the *mass matrix*, $F \in \mathbb{R}^k$ is a k-vector with applied forces and/or torques called the *applied forces vector*, $G(q)$ is a k-vector sometimes called the *conservative forces vector*, and $B(q,\dot{q})$ is a $k \times k$ matrix sometimes called the *centrifugal/Coriolis/friction matrix*. For systems with no friction, we generally have

$$\dot{q}'B(q,\dot{q})\dot{q} = 0, \qquad \forall \dot{q} \in \mathbb{R}^k,$$

whereas for systems with friction

$$\dot{q}'B(q,\dot{q})\dot{q} \geq 0, \qquad \forall \dot{q} \in \mathbb{R}^k,$$

with equality only when $\dot{q} = 0$.

Note 1 (Conservative force). A force is *conservative* if the total work $W = \int_A^B G(q) \cdot dq$ of moving an object from point A to point B is independent of the path that the object took. The term "conservative force" comes from the fact that objects moving under such forces maintain their total mechanical energy. Typically G accounts for gravity and spring forces. □

2.4.2 EXAMPLES

Example 2.1 (Inverted pendulum). The dynamics of the inverted pendulum shown in Figure 2.2 are given by

$$m\ell^2\ddot{\theta} = mg\ell \sin\theta - b\dot{\theta} + T,$$

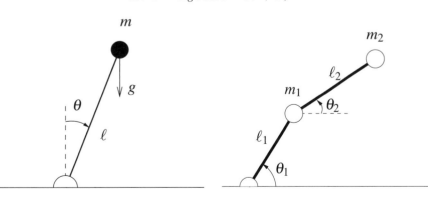

Figure 2.2. Inverted pendulum.   **Figure 2.3.** Two-link robot manipulator.

where T denotes a torque applied at the base and g is the gravitational acceleration. This equation can be identified with (2.7), provided that we define

$$q := \theta, \quad F := T, \quad M(q) := m\ell^2, \quad B(q) := b, \quad G(q) := -mg\ell\sin\theta.$$

Note. When the pendulum is attached to a cart and the control input is the cart's acceleration $\ddot{z}$, we have $T = -m\ell\ddot{z}\cos\theta$. This makes the system more difficult to control. Things become especially difficult around $\theta = \pm\pi/2$. Why?

When the base of the pendulum is connected to a motor, one can regard the torque T as the control input.   □

Example 2.2 (Robot arm). The dynamics of the robot arm with two revolution joints shown in Figure 2.3 can be written as in (2.7), provided that we define

$$q := \begin{bmatrix} \theta_1 \\ \theta_2 \end{bmatrix}, \qquad F := \begin{bmatrix} \tau_1 \\ \tau_2 \end{bmatrix},$$

where τ_1 and τ_2 denote torques applied at the joints. For this system

$$M(q) := \begin{bmatrix} \ell_2^2 m_2^2 + 2\ell_1\ell_2\cos\theta_2 + \ell_1^2(m_1 + m_2) & \ell_2^2 + \ell_1\ell_2 m_2\cos\theta_2 \\ \ell_2^2 + \ell_1\ell_2 m_2\cos\theta_2 & \ell_2^2 m_2 \end{bmatrix}$$

$$B(q, \dot{q}) := \begin{bmatrix} -2m_2\ell_1\ell_2\dot\theta_2\sin\theta_2 & -m_2\ell_1\ell_2\dot\theta_2\sin\theta_2 \\ m_2\ell_1\ell_2\dot\theta_1\sin\theta_2 & 0 \end{bmatrix}$$

$$G(q) := g\begin{bmatrix} m_2\ell_2\cos(\theta_1 - \theta_2) + (m_1 + m_2)\ell_1\cos\theta_1 \\ m_2\ell_2\cos(\theta_1 - \theta_2) \end{bmatrix},$$

where g is the gravitational acceleration [3, p. 202, 205].   □

Example 2.3 (Hovercraft). Figure 2.4 shows a small hovercraft built at Caltech [4]. Its dynamics can be written as in (2.7), provided that we define

$$q := \begin{bmatrix} x \\ y \\ \theta \end{bmatrix}, \qquad F := \begin{bmatrix} (F_s + F_p)\cos\theta - F_\ell\sin\theta \\ (F_s + F_p)\sin\theta + F_\ell\cos\theta \\ \ell(F_s - F_p) \end{bmatrix},$$

where F_s, F_p, and F_ℓ denote the starboard, port, and lateral fan forces. The vehicle in the photograph does not have a lateral fan, which means that $F_\ell = 0$. It is therefore called *underactuated* because the number of controls (F_s and F_p) is smaller than the number of degrees of freedom (x, y, and θ). For this system

$$M(q) := \begin{bmatrix} m & 0 & 0 \\ 0 & m & 0 \\ 0 & 0 & J \end{bmatrix}, \qquad B(q) := \begin{bmatrix} d_v & 0 & 0 \\ 0 & d_v & 0 \\ 0 & 0 & d_\omega \end{bmatrix},$$

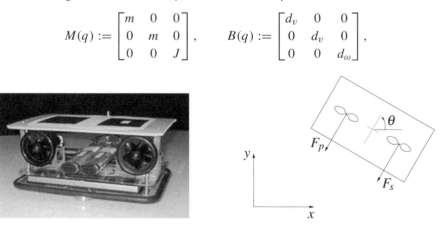

Figure 2.4. Hovercraft.

where $m = 5.5\,\text{kg}$ is the mass of the vehicle, $J = 0.047\,\text{kg}\,\text{m}^2$ is its rotational inertia, $d_v = 4.5$ is the viscous friction coefficient, $d_\omega = .41$ is the coefficient of rotational friction, and $\ell = 0.123\,\text{m}$ is the moment arm of the forces. The geometric center and the center of mass of the vehicle are assumed to coincide [4]. □

2.4.3   FEEDBACK LINEARIZATION CONTROL DESIGN

A mechanical system is called *fully actuated* when one has control over the whole vector or generalized forces. For such systems we regard $u := F$ as the control input and we can use feedback linearization to design nonlinear controllers. In particular, by choosing

$$F = u = u_{\text{nl}}(q, \dot{q}) + M(q)\,v, \qquad u_{\text{nl}}(q, \dot{q}) := B(q, \dot{q})\dot{q} + G(q),$$

we obtain

$$M(q)\ddot{q} + B(q, \dot{q})\dot{q} + G(q) = B(q, \dot{q})\dot{q} + G(q) + M(q)\,v \quad \Leftrightarrow \quad \ddot{q} = v.$$

In practice, we transformed the original nonlinear process into a (linear) double integrator, whose state-space model is given by

$$\dot{x} = \begin{bmatrix} 0 & I \\ 0 & 0 \end{bmatrix} x + \begin{bmatrix} 0 & I \end{bmatrix} v, \qquad y = \begin{bmatrix} I & 0 \end{bmatrix} x, \qquad x := \begin{bmatrix} q \\ \dot{q} \end{bmatrix} \in \mathbb{R}^{2k}. \qquad (2.8)$$

We can now use linear methods to find a controller for v that results in adequate closed-loop performance for the output $y = q$, e.g., a proportional-derivative (PD) controller

$$v = -K_P q - K_D \dot{q}$$

that leads to the following closed-loop dynamics:

$$\dot{x} = \begin{bmatrix} 0 & I \\ -K_P & -K_D \end{bmatrix} x, \qquad\qquad y = \begin{bmatrix} I & 0 \end{bmatrix} x.$$

Figure 2.5 shows a diagram of the overall closed-loop system. From an input-output perspective, the system in the dashed block behaves like the LTI system (2.8).

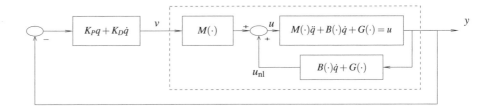

Figure 2.5. Feedback linearization controller.

Attention! Measurement noise can lead to problems in feedback linearization controllers. When the measurements of q and $\dot{q}$ are affected by noise, we have

$$M(q)\ddot{q} + B(q, \dot{q})\dot{q} + G(q) = u_{\text{nl}}(q + n, \dot{q} + w) + M(q + n)\, v,$$

where n is measurement noise in the q sensors and w is the measurement noise in the $\dot{q}$ sensors. In this case

$$\begin{aligned} M(q)\ddot{q} + B(q, \dot{q})\dot{q} + G(q) \\ = B(q + n, \dot{q} + w)(\dot{q} + w) + G(q + n) + M(q + n)\, v, \end{aligned} \qquad (2.9)$$

and (with some work) one can show that

$$\ddot{q} = \left(I + \Delta\right) v + d, \qquad (2.10)$$

where

$$\begin{aligned} \Delta &:= M(q)^{-1}\big(M(q + n) - M(q)\big), \\ d &:= M(q)^{-1}\Big(\big(B(q + n, \dot{q} + w) - B(q, \dot{q})\big)\dot{q} + B(q + n, \dot{q} + w)w \\ &\qquad\qquad + G(q + n) - G(q)\Big). \end{aligned}$$

Since Δ and d can be very large, with feedback linearization controllers it is particularly important to make sure that the controller selected for v is robust with respect to the multiplicative uncertainty Δ and good at rejecting the disturbance d in (2.10). $\qquad\square$

2.4.4 FEEDBACK LINEARIZATION FOR SYSTEMS IN STRICT FEEDBACK FORM

Defining $x_1 = q$ and $x_2 = \dot{q}$, the equations seen before for a mechanical system can be written as

$$\begin{aligned} \dot{x}_1 &= x_2 \\ \dot{x}_2 &= -M^{-1}(x_1)B(x_1, x_2)x_2 - G(x_1) + u \end{aligned}$$

and feedback linearization uses a portion of the control effort u to cancel the nonlinearities in the x_2 equation. This idea can be generalized to larger classes of systems such as those in the so-called *strict feedback form*,

Note. See Exercise 2.5 for systems in strict feedback form of order higher than 2.

$$\begin{aligned} \dot{x}_1 &= f_1(x_1) + x_2 \\ \dot{x}_2 &= f_2(x_1, x_2) + u. \end{aligned}$$

For this system we cannot quite cancel the term $f_1(x_1)$ in the first equation using u, but we can make it linear by introducing a new variable,

$$z_2 := f_1(x_1) + x_2,$$

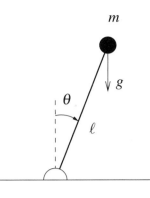

From Newton's law:

$$ m\ell^2\ddot{\theta} = mg\ell\sin\theta - b\dot{\theta} + T, $$

where T denotes a torque applied at the base, and g is the gravitational acceleration.

Figure 2.6. Inverted pendulum.

which we take as the second component of the state, instead of x_2. In this case, we get

$$ \dot{x}_1 = z_2 $$
$$ \dot{z}_2 = \frac{\partial f_1}{\partial x_1}(x_1)\dot{x}_1 + \dot{x}_2 = \frac{\partial f_1}{\partial x_1}(x_1)(f_1(x_1) + x_2) + f_2(x_1, x_2) + u. $$

Now we can use a portion of u to cancel the nonlinear terms that arise in the equation for z_2 and use the rest to control the resulting linear dynamics:

$$ u = u_{\mathrm{nl}}(x_1, x_2) + v, \qquad u_{\mathrm{nl}}(x_1, x_2) = -\frac{\partial f_1}{\partial x_1}(x_1)(f_1(x_1) + x_2) - f_2(x_1, x_2), $$

which leads to

$$ \dot{x}_1 = z_2 $$
$$ \dot{z}_2 = v. $$

The reader is referred, e.g., to [8] for a much more detailed discussion on feedback linearization.

2.5 EXERCISES

2.1 (Local linearization around equilibria). Consider the inverted pendulum in Figure 2.6 and assume the input and output to the system are the signals u and y defined as

$$ T = \mathrm{sat}(u), \qquad\qquad y = \theta, $$

where "sat" denotes the unit-slope saturation function that truncates u at $+1$ and -1.

(a) Linearize this system around the equilibrium point for which $\theta = 0$.

(b) Linearize this system around the equilibrium point for which $\theta = \pi$ (assume that the pendulum is free to rotate all the way to this configuration without hitting the table).

(c) Linearize this system around the equilibrium point for which $\theta = \frac{\pi}{4}$.

Does such an equilibrium point always exist?

(d) Assume that $b = 1/2$ and $mg\ell = 1/4$. Compute the torque $T(t)$ needed for the pendulum to fall from $\theta(0) = 0$ with constant velocity $\dot{\theta}(t) = 1$, $\forall t \geq 0$. Linearize the system around this trajectory.   □

2.2 (Local linearization around a trajectory). A single-wheel cart (unicycle) moving on the plane with linear velocity v and angular velocity ω can be modeled by the nonlinear system

$$\dot{p}_x = v\cos\theta, \qquad\qquad \dot{p}_y = v\sin\theta, \qquad\qquad \dot{\theta} = \omega, \qquad (2.11)$$

where (p_x, p_y) denote the Cartesian coordinates of the wheel and θ its orientation. Regard this as a system with input $u := \begin{bmatrix} v & \omega \end{bmatrix}' \in \mathbb{R}^2$.

(a) Construct a state-space model for this system with state

$$x = \begin{bmatrix} x_1 \\ x_2 \\ x_3 \end{bmatrix} := \begin{bmatrix} p_x\cos\theta + (p_y - 1)\sin\theta \\ -p_x\sin\theta + (p_y - 1)\cos\theta \\ \theta \end{bmatrix}$$

and output $y := \begin{bmatrix} x_1 & x_2 \end{bmatrix}' \in \mathbb{R}^2$.

(b) Compute a local linearization for this system around the equilibrium point $x^{\text{eq}} = 0$, $u^{\text{eq}} = 0$.

(c) Show that $\omega(t) = v(t) = 1$, $p_x(t) = \sin t$, $p_y(t) = 1 - \cos t$, $\theta(t) = t$, $\forall t \geq 0$ is a solution to the system.

(d) Show that a local linearization of the system around this trajectory results in an LTI system.   □

2.3 (Feedback linearization controller). Consider the inverted pendulum in Figure 2.6.

(a) Assume that you can directly control the system in torque, i.e., that the control input is $u = T$.

Design a feedback linearization controller to drive the pendulum to the upright position. Use the following values for the parameters: $\ell = 1$ m, $m = 1$ kg, $b = 0.1$ N m^{-1} s^{-1}, and $g = 9.8$ m s^{-2}. Verify the performance of your system in the presence of measurement noise using Simulink®.

(b) Assume now that the pendulum is mounted on a cart and that you can control the cart's jerk, which is the derivative of its acceleration a. In this case,

$$T = -m\,\ell\,a\cos\theta, \qquad\qquad \dot{a} = u.$$

Design a feedback linearization controller for the new system.

What happens around $\theta = \pm\pi/2$?

Note that, unfortunately, the pendulum needs to pass by one of these points for a swing-up, i.e., the motion from $\theta = \pi$ (pendulum down) to $\theta = 0$ (pendulum upright).   □

Attention! Writing the system in the carefully chosen coordinates x_1, x_2, x_3 is crucial to getting an LTI linearization. If one tried to linearize this system in the original coordinates p_x, p_y, θ with dynamics given by (2.11), one would get an LTV system.

2.4 (Feedback linearization with noise). Verify that (2.10) is indeed equivalent to (2.9), by solving the latter equation with respect to $\ddot{q}$.   □

2.5 (Feedback linearization for systems in strict feedback form). Extend the procedure outlined in Section 2.4.4 for systems in strict feedback form of order 2 to systems in strict feedback of any order $n \geq 2$:

$$\dot{x}_1 = f_1(x_1) + x_2$$
$$\dot{x}_2 = f_2(x_1, x_2) + x_3$$
$$\vdots$$
$$\dot{x}_n = f_n(x_1, x_2, \ldots, x_n) + u.$$ □

LECTURE 3

Causality, Time Invariance, and Linearity

Contents

This lecture introduces a few basic properties of state-space linear systems and some of their direct consequences.

1. Basic Properties of LTV/LTI Systems
2. Characterization of All Outputs to a Given Input
3. Impulse Response
4. Laplace Transform (review)
5. Transfer Function
6. Discrete-Time Case
7. Additional Notes
8. Exercise

Topics 2–4 are direct consequences of causality, time invariance, and linearity. They apply to any system that exhibits these properties, even if such systems are not one of the state-space systems introduced in Lecture 1. For example, they also apply to infinite-dimensional systems.

3.1 BASIC PROPERTIES OF LTV/LTI SYSTEMS

Notation 1. We recall that $u \rightsquigarrow y$ means that "y is one of the outputs that corresponds to u." ▶ p. 5

Note. In words: if the inputs match on $[0, T)$ then the outputs must match on $[0, T)$ for appropriate initial conditions.

In this section we state three basic properties of LTV and LTI systems. All these properties are simple consequences of results in subsequent lectures, so we will defer their proofs for later.

All state-space systems introduced so far (both LTV and LTI) have the property that the output *before* some time t does not depend on the input *after* time t (cf. Figure 3.1). Such systems are called *causal*.

Property P3.1 (Causality). The state-space system (CLTV) is *causal* in the sense that if $u \rightsquigarrow y$ then, for every other input $\bar{u}$ for which

$$\bar{u}(t) = u(t), \quad \forall 0 \le t < T$$

for some time $T > 0$, the system exhibits (at least) one output $\bar{y}$ that satisfies

$$\bar{y}(t) = y(t), \quad \forall 0 \le t < T. \qquad \square$$

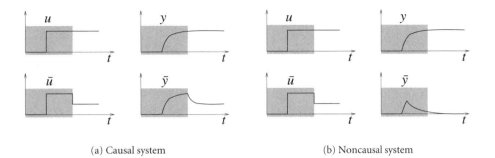

(a) Causal system   (b) Noncausal system

Figure 3.1. Causality.

Attention! The statement of the causality property is *not* "for every input $\bar{u}$ that matches u on $[0, T)$, *every* output $\bar{y}$ matches y on $[0, T)$." In general, only one output $\bar{y}$ (obtained with the same initial condition) will match y. ☐

A key difference between the LTV and LTI systems introduced so far is that the latter have the property that time-shifting of their inputs results in time-shifting of the output (cf. Figure 3.2). This property justifies the terminology *time-varying* versus *time-invariant* for these systems.

Property P3.2 (Time invariance). The state-space system (CLTI) is *time-invariant* in the sense that if $u \rightsquigarrow y$ then, for every scalar $T > 0$, we have $\bar{u} \rightsquigarrow \bar{y}$ for

$$\bar{u}(t) = u(t + T), \quad \forall t \geq 0, \qquad \text{and} \qquad \bar{y}(t) = y(t + T), \quad \forall t \geq 0. \qquad \square$$

Attention! Recall that $\bar{u} \rightsquigarrow \bar{y}$ means that $\bar{y}$ is *one of* the outputs corresponding to the input $\bar{u}$. In general, the input $\bar{u}$ has many other outputs (obtained from different initial conditions) that will not match the time-shifted version of y. Moreover, the initial conditions used to obtain y and $\bar{y}$ will not be the same. ☐

Both the LTV and LTI systems have the property that they can be viewed as linear maps from their inputs to appropriate outputs. This justifies the qualifier *linear* in LTV and LTI.

Note. In words: if the inputs are time shifted, then the outputs will also be time shifted, for appropriate initial conditions.

Notation 2 There is some abuse in the LTV/LTI terminology. ▶ p. 24

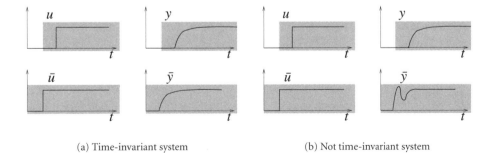

(a) Time-invariant system   (b) Not time-invariant system

Figure 3.2. Time invariance.

Note. In words: If u_1 and u_2 have outputs y_1 and y_2, respectively, then $\alpha y_1 + \beta y_2$ must be (one of the) outputs to $\alpha u_1 + \beta u_2$, for appropriate initial conditions.

Property P3.3 (Linearity). The state-space system (CLTV) is *linear* in the sense that for every $\alpha, \beta \in \mathbb{R}$, if $u_1 \rightsquigarrow y_1$ and $u_2 \rightsquigarrow y_2$, then $\alpha u_1 + \beta u_2 \rightsquigarrow \alpha y_1 + \beta y_2$. □

Attention! Recall that $\alpha u_1 + \beta u_2 \rightsquigarrow \alpha y_1 + \beta y_2$ means that $\alpha y_1 + \beta y_2$ is *one of* the outputs corresponding to the input $\alpha u_1 + \beta u_2$. In general, there may be many other outputs (obtained from different initial conditions) that will not be of this form. □

Notation 2 (LTV and LTI). We use LTV and LTI to denote the continuous-time systems given by the equations (CLTV) and (CLTI), respectively, or the discrete-time systems (DLTV) and (DLTI), respectively. There is some abuse in this terminology because there are many linear time-invariant and time-varying systems that cannot be expressed by these state-space equations. This is the case, e.g., of infinite-dimensional systems. □

3.2   CHARACTERIZATION OF ALL OUTPUTS TO A GIVEN INPUT

Linearity allows one to use a single output y_f corresponding to a given input u to construct all remaining outputs corresponding to u.

1. Let y be another output associated with the given input u. Since we can write the zero input as $0 = u - u$, by linearity we conclude that $y - y_f$ must necessarily be an output corresponding to the zero input :

Notation. An output corresponding to the zero input is called a *homogeneous response*.

$$\begin{cases} u \rightsquigarrow y_f \\ u \rightsquigarrow y \end{cases} \quad \Rightarrow \quad 0 \rightsquigarrow y - y_f.$$

2. Conversely, suppose that one is given an output y_h corresponding to the zero input. Since $u = u + 0$, we conclude that $y = y_f + y_h$ is another output corresponding to u:

$$\begin{cases} u \rightsquigarrow y_f \\ 0 \rightsquigarrow y_h \end{cases} \quad \Rightarrow \quad u \rightsquigarrow y_f + y_h.$$

The following result summarizes the two observations above.

Theorem 3.1. *Let y_f be an output corresponding to a given input u. All outputs corresponding to u can be obtained by*

$$y = y_f + y_h,$$

where y_h is an output corresponding to the zero input. □

This means that to construct all outputs corresponding to u, it is enough to known how to solve the following two problems:

1. Find one particular output corresponding to the input u.

2. Find all outputs corresponding to the zero input.

The remainder of this section addresses the first problem.

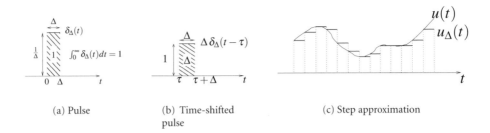

Figure 3.3. Step approximation to a continuous-time signal.

3.3 IMPULSE RESPONSE

Consider a linear SISO system and let δ_Δ denote the unit-area pulse signal δ_Δ in Figure 3.3(a). Using δ_Δ, we can write an approximation to an input signal $u : [0, \infty) \to \mathbb{R}$ as in Figure 3.3(c),

$$u_\Delta(t) := \sum_{k=0}^{\infty} u(k\Delta)\Delta\, \delta_\Delta(t - k\Delta), \quad \forall t \geq 0. \tag{3.1}$$

Note. This assumes we can choose outputs $g_\Delta(t, \tau)$ for which the series (3.2) converges.
Note 2. The last equality in (3.3) is a consequence of the definition of the Riemann integral. It implicitly assumes that the limit in (3.4) and the integral in (3.3) both exist. ▶ p. 29

For each $\tau \geq 0$, let $g_\Delta(t, \tau), t \geq 0$ be an output corresponding to the input $\delta_\Delta(t - \tau)$:

$$\delta_\Delta(t - \tau) \rightsquigarrow g_\Delta(t, \tau).$$

Because of (3.1) and linearity

$$u_\Delta \rightsquigarrow y_\Delta(t) := \sum_{k=0}^{\infty} \Delta\, u(k\Delta)g_\Delta(t, k\Delta), \quad \forall t \geq 0. \tag{3.2}$$

Moreover, since $u_\Delta \to u$ as $\Delta \to 0$, we conclude that

$$u \rightsquigarrow y(t) = \lim_{\Delta \to 0} y_\Delta(t) = \lim_{\Delta \to 0} \sum_{k=0}^{\infty} \Delta\, u(k\Delta)g_\Delta(t, k\Delta)$$

$$= \int_0^{\infty} u(\tau)g(t, \tau)d\tau, \quad \forall t \geq 0 \tag{3.3}$$

is an output corresponding to u, where g is defined by

$$g(t, \tau) = \lim_{\Delta \to 0} g_\Delta(t, \tau). \tag{3.4}$$

Notation. This output has the special property that it is equal to zero when $u = 0$, it is called a *forced* or *zero-state* response. The latter terminology will become clear in Lecture 4.

The function $g(t, \tau)$ can be viewed as the output at time t, corresponding to an input pulse of zero length but unit area (a Dirac pulse) applied at time τ.

For MIMO systems this generalizes to the following result.

Theorem 3.2 (Impulse response). *Consider a continuous-time linear system with k inputs and m outputs. There exists a matrix-valued signal $G(t, \tau) \in \mathbb{R}^{m \times k}$ such that for every input u, a corresponding output is given by*

$$u \rightsquigarrow y(t) = \int_0^{\infty} G(t, \tau)u(\tau)d\tau, \quad \forall t \geq 0. \tag{3.5}$$

Note. Equation (3.5)
is generally taken as
the *definition of
impulse response.*

The matrix-valued signal $G(t, \tau) \in \mathbb{R}^{m \times k}$ is called a *(continuous-time) impulse response.* Its entry $g_{ij}(t, \tau)$ can be viewed as the ith entry of an output at time t, corresponding to a pulse of zero length but unit area (a Dirac pulse) applied at the jth input at time τ. The impulse response $G(t, \tau)$ that appears in (3.5) has several important properties that will be explored below.

Properties (Impulse response).

Note. Soon we will see
that the impulse
response of LTV/LTI
systems is unique.
Therefore one could
replace "one can
choose the impulse
response to satisfy" by
"the impulse response
satisfies."

P3.4 For causal systems , one can choose the impulse response to satisfy

$$G(t, \tau) = 0, \quad \forall \tau > t. \tag{3.6}$$

P3.5 For time-invariant systems, one can choose the impulse response to satisfy

$$G(t + T, \tau + T) = G(t, \tau), \quad \forall t, \tau, T \geq 0. \tag{3.7}$$

Notation. With some
abuse of notation, it is
common to write
simply $G(t_2, t_1) =
G(t_2 - t_1)$,
$\forall t_2 \geq t_1 \geq 0$.

In particular for $\tau = 0, t_1 = T, t_2 = t + T$

$$G(t_2, t_1) = G(t_2 - t_1, 0), \quad \forall t_2 \geq t_1 \geq 0,$$

which shows that $G(t_2, t_1)$ is just a function of $t_2 - t_1$.

P3.6 For causal, time-invariant systems, we can write (3.5) as

$$u \rightsquigarrow y(t) = \int_0^t G(t - \tau)u(\tau)d\tau =: (G \star u)(t), \quad \forall t \geq 0, \tag{3.8}$$

where $\star$ denotes the convolution operator.

Proof. For simplicity we assume a SISO system.

P3.4 By linearity $u = 0 \rightsquigarrow y = 0$. Since the impulse $\delta(t - \tau)$ at time τ is equal to the zero input u on $[0, \tau)$, it must have an output that is zero on $[0, \tau)$; i.e.,

$$\delta(t - \tau) = 0, \quad \forall 0 \leq t < \tau \quad \Rightarrow \quad \exists \bar{y} : \delta(t - \tau) \rightsquigarrow \bar{y} \text{ and}$$
$$\bar{y}(t) = 0, \quad \forall 0 \leq t < \tau.$$

Choosing this input to construct the impulse response, we obtain (3.6).

P3.5 By the definition of impulse response

$$\delta(t - \tau - T) \rightsquigarrow y(t) := g(t, \tau + T),$$

where $\delta(t - \tau - T)$ is an impulse at time $\tau + T$. By time invariance, the impulse $\delta(t - \tau)$ at time τ must have an output $\bar{y}$ such that

$$\delta(t - \tau) \rightsquigarrow \bar{y}(t) = y(t + T) = g(t + T, \tau + T).$$

By using this output to construct the impulse response, we obtain (3.7).

P3.6 Equation (3.8) is obtained by using P3.5 to replace $G(t, \tau)$ by $G(t - \tau)$ in (3.5). Because of P3.4 we can further replace the ∞ in the upper integration limit by t, since $G(t, \tau) = G(t - \tau)$ is equal to zero for $\tau > t$. ∎

For causal, time-invariant systems, P3.6 provides a convenient way to compute the forced response. Due to the closed connection between the (time domain) convolution of signals and the (frequency domain) product of their Laplace transforms, it is especially easy to compute forced responses in the frequency domain. This is explored in the following sections.

3.4 LAPLACE TRANSFORM (REVIEW)

Given a continuous-time signal $x(t), t \geq 0$ its *unilateral Laplace transform* is given by

MATLAB® Hint 10.
laplace(F,t,s)
symbolically
computes the Laplace
transform of
F. ▶ p. 52

$$\mathcal{L}[x(t)] = \hat{x}(s) = \int_0^\infty e^{-st} x(t) dt, \quad s \in \mathbb{C}.$$

The Laplace transform of the derivative $\dot{x}(t)$ of the signal $x(t)$ can be related to $\hat{x}(s)$ by

$$\mathcal{L}[\dot{x}(t)] = \hat{\dot{x}}(s) = s\hat{x}(s) - x(0), \quad s \in \mathbb{C}. \tag{3.9}$$

MATLAB® Hint 11.
ilaplace(F,s,t)
symbolically
computes the inverse
Laplace transform of
F. ▶ p. 53

Given two signals $x(t)$ and $y(t), t \geq 0$, the Laplace transform of their convolution is given by

Note 3. The term
$-x(0)$ in (3.9) does
not appear in bilateral
transforms. ▶ p. 29

$$\mathcal{L}[(x \star y)(t)] = \mathcal{L}\Big[\int_0^t x(\tau)y(t-\tau)d\tau\Big] = \hat{x}(s)\hat{y}(s). \tag{3.10}$$

The Laplace Transform is covered extensively, e.g., in [12].

Note 4. Why does
(3.10) hold? ▶ p. 29

3.5 TRANSFER FUNCTION

Because of Theorem 3.2, the continuous-time linear system has an output

$$y(t) = \int_0^\infty G(t-\tau)u(\tau)d\tau, \quad \forall t \geq 0.$$

Taking its Laplace transform, one obtains

$$\hat{y}(s) = \int_0^\infty \int_0^\infty e^{-st} G(t-\tau)u(\tau)d\tau \, dt.$$

Changing the order of integration, one gets

$$\hat{y}(s) = \int_0^\infty \int_0^\infty e^{-st} G(t-\tau)u(\tau)dt \, d\tau$$
$$= \int_0^\infty \Big(\int_0^\infty e^{-s(t-\tau)} G(t-\tau)dt\Big)e^{-s\tau}u(\tau)d\tau. \tag{3.11}$$

But because of causality,

$$\int_0^\infty e^{-s(t-\tau)} G(t-\tau)dt = \int_{-\tau}^\infty e^{-s\bar{t}} G(\bar{t})d\bar{t} = \int_0^\infty e^{-s\bar{t}} G(\bar{t})d\bar{t} =: \hat{G}(s). \tag{3.12}$$

Substituting (3.12) into (3.11) and removing $\hat{G}(s)$ from the integral, we conclude that

$$\hat{y}(s) = \int_0^\infty \hat{G}(s)e^{-s\tau}u(\tau)d\tau = \hat{G}(s)\int_0^\infty e^{-s\tau}u(\tau)d\tau = \hat{G}(s)\hat{u}(s).$$

Theorem 3.3. *For every input u, the Laplace transform of a corresponding output y is given by* $\hat{y}(s) = \hat{G}(s)\hat{u}(s)$. □

This result motivates the following definition.

Notation. For MIMO systems the transfer function is often called the *transfer matrix*.

Definition 3.1 (Transfer function). The *transfer function* of a continuous-time causal linear time-invariant system is the Laplace transform

$$\hat{G}(s) = \mathcal{L}[G(t)] := \int_0^\infty e^{-st} G(t)dt, \quad s \in \mathbb{C}.$$

of an impulse response $G(t_2, t_1) = G(t_2 - t_1), \forall t_2 \geq t_1 \geq 0$. □

3.6 DISCRETE-TIME CASE

A result absolutely analogous to Theorem 3.2 can be derived for discrete-time systems, except that now the step approximation is actually exact, so there is no need to take limits and the result appears as a summation.

Theorem 3.4 (Impulse response). *Consider a discrete-time linear system with k inputs and m outputs. There exists a matrix-valued signal $G(t, \tau) \in \mathbb{R}^{m \times k}$ such that for every input u, a corresponding output is given by*

$$u \rightsquigarrow y(t) = \sum_{\tau=0}^\infty G(t, \tau)u(\tau), \quad \forall t \geq 0. \tag{3.13}$$

The matrix-valued signal $G(t, \tau) \in \mathbb{R}^{m \times k}$ is called a *discrete-time impulse response*. Its entry $g_{ij}(t, \tau)$ can be viewed as the *i*th entry of an output at time *t*, corresponding to a unit discrete-time pulse applied at the *j*th input at time τ. The discrete-time impulse response also satisfies Properties P3.4–P3.6.

MATLAB® Hint 12. `ztrans(F,t,z)` computes the $\mathcal{Z}$-transform of F.   ▶ p. 53

For discrete-time systems the $\mathcal{Z}$-transform plays a role analogous to the Laplace transform in defining a transfer function, essentially by replacing all the integrals by summations. Given a discrete-time signal $y(\cdot)$, its *(unilateral) $\mathcal{Z}$-transform* is given by

$$\hat{y}(z) = \mathcal{Z}[y(t)] := \sum_{t=0}^\infty z^{-t} y(t), \quad s \in \mathbb{C}.$$

MATLAB® Hint 13. `iztrans(F,z,t)` computes the inverse $\mathcal{Z}$-transform of F.   ▶ p. 53

The $\mathcal{Z}$-transform is covered extensively, e.g., in [12].

Definition 3.2 (Transfer function). The *transfer function* of a discrete-time causal linear time-invariant system is the $\mathcal{Z}$-transform

$$\hat{G}(z) = \mathcal{Z}[G(t)] := \sum_{t=0}^\infty z^{-t} G(t), \quad z \in \mathbb{C}$$

of an impulse response $G(t_2, t_1) = G(t_2 - t_1), \forall t_2 \geq t_1 \geq 0$. □

Theorem 3.5. *For every input u, the $\mathcal{Z}$-transform of a corresponding output y is given by* $\hat{y}(z) = \hat{G}(z)\hat{u}(z)$. □

3.7 ADDITIONAL NOTES

Note 2 (Impulse response). To prove the last equality in (3.3), we use the fact that given a function of two variables $f(x, y)$,

$$\lim_{z \to 0} f(z, z) = \lim_{x \to 0} \lim_{y \to 0} f(x, y),$$

as long as the two limits on the right-hand side exist. Using this in (3.3), we conclude that

$$y(t) = \lim_{\Delta \to 0} \lim_{\epsilon \to 0} \sum_{k=0}^{\infty} \Delta u(k\Delta) g_\epsilon(t, k\Delta) = \lim_{\Delta \to 0} \sum_{k=0}^{\infty} \Delta u(k\Delta) \Big(\lim_{\epsilon \to 0} g_\epsilon(t, k\Delta) \Big),$$

for every $t \geq 0$. But $\lim_{\epsilon \to 0} g_\epsilon(t, k\Delta)$ is precisely $g(t, k\Delta)$. Therefore

$$y(t) = \lim_{\Delta \to 0} \sum_{k=0}^{\infty} \Delta u(k\Delta) g(t, k\Delta), \qquad\qquad t \geq 0. \qquad (3.14)$$

We recall now, that by the definition of the Riemann integral,

$$\int_0^{\infty} f(\tau) d\tau = \lim_{\Delta \to 0} \sum_{k=0}^{\infty} \Delta f(k\Delta).$$

Comparing this with (3.14), we conclude that indeed

$$y(t) = \int_0^{\infty} u(\tau) g(t, \tau) d\tau, \qquad\qquad t \geq 0. \qquad (3.15)$$

In this derivation, we assumed that the limit $\lim_{\epsilon \to 0} g_\epsilon(t, k\Delta)$ and the integral in (3.15) both exist. ∎

Note 3 (Laplace transform of the derivative). By the product rule,

$$\frac{d}{dt}\big(x(t) e^{-st}\big) = \dot{x} e^{-st} - s x(t) e^{-st}.$$

Integrating the above equation from 0 to ∞, we conclude that

$$\Big[x(t) e^{-st} \Big]_0^{\infty} = \hat{\dot{x}}(s) - s \hat{x}(s).$$

Since $\lim_{t \to \infty} x(t) e^{-st} = 0$, whenever $\hat{x}(s)$ exists, we conclude that

$$\hat{\dot{x}}(s) = s \hat{x}(s) - x(0). \qquad\qquad \square$$

Note 4 (Laplace transform of the convolution). Given two signals $x(t)$ and $y(t)$, $t \geq 0$,

$$\mathcal{L}[(x \star y)(t)] = \mathcal{L}\Big[\int_0^t x(\tau) y(t - \tau) d\tau \Big] = \int_0^{\infty} \int_0^t e^{-st} x(\tau) y(t - \tau) d\tau dt.$$

Exchanging the order of integration, we obtain

$$\mathcal{L}[(x \star y)(t)] = \int_0^\infty e^{-s\tau} x(\tau) \Big(\int_\tau^\infty e^{-s(t-\tau)} y(t - \tau) dt \Big) d\tau.$$

If we then make the change of integration variable $\bar{t} = t - \tau$ in the inner integral, we obtain

$$\mathcal{L}[(x \star y)(t)] = \int_0^\infty e^{-s\tau} x(\tau) \Big(\int_0^\infty e^{-s\bar{t}} y(\bar{t}) dt \Big) d\tau = \hat{x}(s) \hat{y}(s). \qquad \square$$

3.8   EXERCISE

3.1 (Impulse response). Prove Theorem 3.4. $\qquad \qquad \square$

LECTURE 4

Impulse Response and Transfer Function of State-Space Systems

CONTENTS

This lecture applies the concepts of impulse response and transfer function introduced in Lecture 3 to state-space linear systems.

1. Impulse Response and Transfer Function for LTI Systems
2. Discrete-Time Case
3. Elementary Realization Theory
4. Equivalent State-Space Systems
5. LTI Systems in MATLAB®
6. Exercises

4.1 IMPULSE RESPONSE AND TRANSFER FUNCTION FOR LTI SYSTEMS

MATLAB® Hint 1.
ss(A,B,C,D) creates the continuous-time LTI state-space system (CLTI). ▶ p. 6

Consider the continuous-time LTI system

$$\dot{x} = Ax + Bu, \qquad\qquad y = Cx + Du. \qquad \text{(CLTI)}$$

Taking the Laplace transform of both sides of the two equations in (CLTI), we obtain

$$s\hat{x}(s) - x(0) = A\hat{x}(s) + B\hat{u}(s), \qquad \hat{y}(s) = C\hat{x}(s) + D\hat{u}(s).$$

Note 3. Why? ▶ p. 29

Solving for $\hat{x}(s)$, we obtain

$$(sI - A)\hat{x}(s) = x(0) + B\hat{u}(s) \quad\Rightarrow\quad \hat{x}(s) = (sI - A)^{-1}B\hat{u}(s) + (sI - A)^{-1}x(0),$$

from which we conclude that

Note. Equation (4.1) confirms the decomposition seen in Lecture 3 of any output as the sum of a particular output with a homogeneous response.

$$\hat{y}(s) = \hat{\Psi}(s)x(0) + \hat{G}(s)\hat{u}(s), \qquad \hat{\Psi}(s) := C(sI - A)^{-1},$$
$$\hat{G}(s) := C(sI - A)^{-1}B + D.$$

Coming back to the time domain by applying inverse Laplace transforms, we obtain

$$y(t) = \Psi(t)x(0) + (G \star u)(t) = \Psi(t)x(0) + \int_0^t G(t - \tau)u(\tau)dt, \qquad (4.1)$$

where

$$G(t) := \mathcal{L}^{-1}[\hat{G}(s)], \qquad\qquad \Psi(t) := \mathcal{L}^{-1}[\hat{\Psi}(s)].$$

MATLAB® Hint 14.
tf(sys_ss) and
zpk(sys_ss)
compute the transfer
function of the
state-space system
sys_ss.   ▶ p. 39

Comparing (4.1) with the equation (3.8) that is used to define the impulse response of a causal, linear, time-invariant system, we conclude the following.

Theorem 4.1. *The impulse response and transfer function of the system* (CLTI) *are given by*

Notation. This output
is called the *forced* or
zero-state response.

$$G(t) = \mathcal{L}^{-1}\big[C(sI - A)^{-1}B + D\big] \quad and \quad \hat{G}(s) = C(sI - A)^{-1}B + D,$$

respectively. Moreover, the output given by (3.8) *corresponds to the zero initial condition* $x(0) = 0$. □

4.2   DISCRETE-TIME CASE

Consider the discrete-time LTI system

$$x^{+} = Ax + Bu, \qquad\qquad y = Cx + Du. \qquad\qquad \text{(DLTI)}$$

Theorem 4.2. *The impulse response and transfer function of the system* (DLTI) *are given by*

$$G(t) = \mathcal{Z}^{-1}\big[C(zI - A)^{-1}B + D\big] \quad and \quad \hat{G}(z) = C(zI - A)^{-1}B + D,$$

respectively. Moreover, the output given by (3.13) *corresponds to the zero initial condition* $x(0) = 0$. □

4.3   ELEMENTARY REALIZATION THEORY

Definition 4.1 (Realization). Given a transfer function $\hat{G}(s)$, we say that a continuous-time or discrete-time LTI state-space system

$$\begin{cases} \dot{x} = Ax + Bu, \\ y = Cx + Du \end{cases} \quad or \quad \begin{cases} x^{+} = Ax + Bu, \\ y = Cx + Du, \end{cases} \qquad \text{(LTI)}$$

respectively, is a *realization of* $\hat{G}(s)$ if

Notation. For short,
one often says that
(A, B, C, D) is a
realization of $\hat{G}(s)$.

$$\hat{G}(s) = C(sI - A)^{-1}B + D. \qquad\qquad (4.2)$$

For discrete-time systems, one would replace s by z in (4.2). □

In general, many systems may realize the same transfer function, which motivates the following definition.

Note. Why? Because if $\hat{G}(s)$ and $\hat{u}(s)$ are the same, then $\hat{y}_f(s) = \hat{G}(s)\hat{u}(s)$ will be the same. However, as we shall see later, the homogeneous responses may differ *even for the same initial conditions.*

Definition 4.2 (Zero-state equivalence). Two state-space systems are said to be *zero-state equivalent* if they realize the same transfer function, which means that they exhibit the same forced response to every input. □

4.3.1 FROM REALIZATION TO TRANSFER FUNCTION

A first question one can ask is, What types of transfer functions can be realized by LTI state-space systems? To answer this question, we attempt to compute the transfer function realized by (LTI). To do this, we recall that

$$M^{-1} = \frac{1}{\det M}(\operatorname{adj} M)', \qquad \operatorname{adj} M := [\operatorname{cof}_{ij} M],$$

where $\operatorname{adj} M$ denotes the *adjoint matrix* of M, whose entry $\operatorname{cof}_{ij} M$ is the *ijth cofactor* of M, i.e, the determinants of the M submatrix obtained by removing row i and column j multiplied by $(-1)^{i+j}$. Therefore

Notation. The *characteristic polynomial* of an $n \times n$ matrix A is the degree n monic polynomial given by $\Delta(s) := \det(sI - A)$. Its roots are the eigenvalues of A.

MATLAB® Hint 15. `poly(A)` computes the characteristic polynomial of A, and `eig(A)` computes its eigenvalues. ▶ p. 77

$$C(sI - A)^{-1}B + D = \frac{1}{\det(sI - A)}C[\operatorname{adj}(sI - A)]'B + D.$$

The denominator $\det(sI - A)$ is an n-degree polynomial called the *characteristic polynomial* of A. Its roots are the eigenvalues of A. The adjoint $\operatorname{adj}(sI - A)$ will contain determinants of $(n - 1) \times (n - 1)$ matrices and therefore its entries will be polynomials of degree $n - 1$ or smaller. The entries of $C[\operatorname{adj}(sI - A)]'B$ will therefore be linear combinations of polynomials of degree smaller than or equal to $n - 1$. Therefore all the entries of

$$\frac{1}{\det(sI - A)}C[\operatorname{adj}(sI - A)]'B$$

Note. The degree of the denominator will never be larger than n, but it may be smaller due to cancellations with common factors in the numerator.

will be ratios of polynomials with the degrees of the denominators strictly larger than the degrees of the numerators. A function of this form is called a *strictly proper rational function.*

When $D \neq 0$, some of the polynomials in the numerators of

$$\frac{1}{\det(sI - A)}C[\operatorname{adj}(sI - A)]'B + D$$

will have the same degree as the denominator (but not higher). This is called a *proper rational function.* We thus conclude that an LTI state-space system can realize only proper rational functions. It turns out that an LTI state-space system can actually realize *every* proper rational function.

Note. To prove that that two statements P and Q are equivalent, one generally starts by showing that $P \Rightarrow Q$ and then the converse statement that $Q \Rightarrow P$. In *direct proofs*, to prove that $P \Rightarrow Q$ one assumes that P is true and then, using a sequence of logical steps, arrives at the conclusion that Q is also true.

Theorem 4.3 (MIMO realization). *A transfer function $\hat{G}(s)$ can be realized by an LTI state-space system* if and only if *$\hat{G}(s)$ is a proper rational function.* □

4.3.2   FROM TRANSFER FUNCTION TO REALIZATION

We have seen only that being a proper rational function is necessary for $\hat{G}(s)$ to be realized by an LTI system. To prove the converse, we need to show how to construct an LTI system that realizes an arbitrary given proper rational function $\hat{G}(s)$. This is can be done by the following steps:

Example. Step 1.

$$\hat{G}(s) = \begin{bmatrix} \frac{4s-10}{2s+1} & \frac{3}{s+2} \\ \frac{1}{(2s+1)(s+2)} & \frac{s+1}{(s+2)^2} \end{bmatrix}$$

$$\hat{G}_{\text{sp}}(s) = \begin{bmatrix} \frac{-12}{2s+1} & \frac{3}{s+2} \\ \frac{1}{(2s+1)(s+2)} & \frac{s+1}{(s+2)^2} \end{bmatrix}$$

$$D = \begin{bmatrix} 2 & 0 \\ 0 & 0 \end{bmatrix}$$

1. Decompose the $m \times k$ matrix $\hat{G}(s)$ as

$$\hat{G}(s) = \hat{G}_{\text{sp}}(s) + D, \qquad (4.3)$$

where $\hat{G}_{\text{sp}}(s)$ is strictly proper and

$$D := \lim_{s \to \infty} \hat{G}(s) \qquad (4.4)$$

is a constant matrix.

The matrix D is part of the state-space realization, and we choose A, B, C so that

$$\hat{G}_{\text{sp}}(s) = C(sI - A)^{-1} B.$$

Notation. The *monic least common denominator (lcd)* of a family of polynomials is the monic polynomial of smallest order that can be divided by all the given ones.

Example. Step 2.

$$d(s) = (s + \frac{1}{2})(s + 2)^2$$

$$= s^3 + \frac{9}{2}s^2 + 6s + 2$$

2. Find the monic least common denominator of all entries of $\hat{G}_{\text{sp}}(s)$:

$$d(s) = s^n + \alpha_1 s^{n-1} + \alpha_2 s^{n-2} + \cdots + \alpha_{n-1} s + \alpha_n.$$

Example. Step 3.

$$\hat{G}_{\text{sp}}(s) = \frac{\begin{bmatrix} -6(s+2)^2 & 3(s+\frac{1}{2})(s+2) \\ \frac{s+2}{2} & (s+1)(s+\frac{1}{2}) \end{bmatrix}}{d(s)}$$

$$N_1 = \begin{bmatrix} -6 & 3 \\ 0 & 1 \end{bmatrix}, \ N_2 = \begin{bmatrix} -24 & \frac{15}{2} \\ \frac{1}{2} & \frac{3}{2} \end{bmatrix},$$

$$N_3 = \begin{bmatrix} -24 & 9 \\ 1 & \frac{1}{2} \end{bmatrix}$$

3. Expand $\hat{G}_{\text{sp}}(s)$ as

$$\hat{G}_{\text{sp}}(s) = \frac{1}{d(s)} \Big[N_1 s^{n-1} + N_2 s^{n-2} + \cdots + N_{n-1} s + N_n \Big], \qquad (4.5)$$

where the N_i are constant $m \times k$ matrices.

Example. Step 4.

$$A = \begin{bmatrix} -\frac{9}{2} & 0 & -6 & 0 & 2 & 0 \\ 0 & -\frac{9}{2} & 0 & -6 & 0 & 2 \\ 1 & 0 & 0 & 0 & 0 & 0 \\ 0 & 1 & 0 & 0 & 0 & 0 \\ 0 & 0 & 1 & 0 & 0 & 0 \\ 0 & 0 & 0 & 1 & 0 & 0 \end{bmatrix}$$

$$B = \begin{bmatrix} 1 & 0 \\ 0 & 1 \\ 0 & 0 \\ 0 & 0 \\ 0 & 0 \\ 0 & 0 \end{bmatrix}$$

$$C = \begin{bmatrix} -6 & 3 & -24 & \frac{15}{2} & -24 & 3 \\ 0 & 1 & \frac{1}{2} & \frac{3}{2} & 1 & \frac{1}{2} \end{bmatrix}$$

4. Select

$$A = \begin{bmatrix} -\alpha_1 I_{k \times k} & -\alpha_2 I_{k \times k} & \cdots & -\alpha_{n-1} I_{k \times k} & -\alpha_n I_{k \times k} \\ I_{k \times k} & 0_{k \times k} & \cdots & 0_{k \times k} & 0_{k \times k} \\ 0_{k \times k} & I_{k \times k} & \cdots & 0_{k \times k} & 0_{k \times k} \\ \vdots & \vdots & \ddots & \vdots & \vdots \\ 0_{k \times k} & 0_{k \times k} & \cdots & I_{k \times k} & 0_{k \times k} \end{bmatrix}_{nk \times nk}, \quad (4.6a)$$

$$B = \begin{bmatrix} I_{k \times k} \\ 0_{k \times k} \\ \vdots \\ 0_{k \times k} \\ 0_{k \times k} \end{bmatrix}_{nk \times k}, \quad C = \begin{bmatrix} N_1 & N_2 & \cdots & N_{n-1} & N_n \end{bmatrix}_{m \times nk}. \quad (4.6b)$$

This is called a realization in *controllable canonical form* for reasons that will become clear later.

MATLAB® Hint 16.
tf(num,den) creates a rational transfer function with numerator and denominator specified by num, den. ▶ p. 38

MATLAB® Hint 17.
zpk(z,p,k) creates a rational transfer function with zeros, poles, and gain specified by z, p, k. ▶ p. 38

MATLAB® Hint 18.
ss(sys_tf) computes a realization for the transfer function sys_tf. ▶ p. 39

Proposition 4.1. *The matrices (A, B, C, D) defined by (4.4) and (4.6) are a realization for $\hat{G}(s)$.* □

Proof of Proposition 4.1. We start by computing the vector

$$Z(s) = \begin{bmatrix} Z_1' & Z_2' & \cdots & Z_n' \end{bmatrix}' := (sI - A)^{-1} B,$$

which is a solution to

$$(sI - A)Z(s) = B \quad \Leftrightarrow \quad \begin{cases} (s + \alpha_1)Z_1 + \alpha_2 Z_2 + \cdots + \alpha_{n-1} Z_{n-1} + \alpha_n Z_n = I_{k \times k} \\ sZ_2 - Z_1 = 0, \ sZ_3 - Z_2 = 0, \ \ldots, \ sZ_n = Z_{n-1}. \end{cases}$$

From the bottom equations, we conclude that

$$Z_n = \frac{1}{s} Z_{n-1}, \ Z_{n-1} = \frac{1}{s} Z_{n-1}, \ \ldots, \ Z_2 = \frac{1}{s} Z_1 \quad \Rightarrow \quad Z_k = \frac{1}{s^{k-1}} Z_1,$$

and, by substituting this in the top equation, we obtain

$$\left(s + \alpha_1 + \frac{\alpha_2}{s} + \cdots + \frac{\alpha_{n-1}}{s^{n-2}} + \frac{\alpha_n}{s^{n-1}} \right) Z_1 = I_{k \times k}.$$

Since the polynomial in the left-hand side is given by $\frac{d(s)}{s^{n-1}}$, we conclude that

$$Z(s) = \begin{bmatrix} Z_1 \\ Z_2 \\ \vdots \\ Z_n \end{bmatrix} = \frac{1}{d(s)} \begin{bmatrix} s^{n-1} I_{k \times k} \\ s^{n-2} I_{k \times k} \\ \vdots \\ I_{k \times k} \end{bmatrix}.$$

From this, we conclude that

$$C(sI - A)^{-1}B = CZ(s) = \frac{1}{d(s)} \begin{bmatrix} N_1 & N_2 & \cdots & N_{n-1} & N_n \end{bmatrix} \begin{bmatrix} s^{n-1}I_{k \times k} \\ s^{n-2}I_{k \times k} \\ \vdots \\ I_{k \times k} \end{bmatrix}$$

$$= \hat{G}_{\mathrm{sp}}(s)$$

because of (4.5). Finally, using (4.3), one concludes that $\hat{G}(s) = C(sI - A)^{-1}B + D$.

$\square$

4.3.3 SISO CASE

From SISO strictly proper systems, the construction outlined in Section 4.3.2 becomes extremely simple, and determining a realization can be done by inspection. The following theorem (to be proved in Exercise 4.4) summarizes this observation.

Theorem 4.4 (SISO realization). *The SISO transfer function*

$$\hat{g}(s) = \frac{\beta_1 s^{n-1} + \beta_2 s^{n-2} + \cdots + \beta_{n-1}s + \beta_n}{s^n + \alpha_1 s^{n-1} + \alpha_2 s^{n-2} + \cdots + \alpha_{n-1}s + \alpha_n}$$

admits either of the realizations

Notation. This realization is said to be in *controllable canonical form* for reasons that will become clear later.

$$A = \begin{bmatrix} -\alpha_1 & -\alpha_2 & \cdots & -\alpha_{n-1} & -\alpha_n \\ 1 & 0 & \cdots & 0 & 0 \\ 0 & 1 & \cdots & 0 & 0 \\ \vdots & \vdots & \ddots & \vdots & \vdots \\ 0 & 0 & \cdots & 1 & 0 \end{bmatrix}_{n \times n}, \qquad B = \begin{bmatrix} 1 \\ 0 \\ \vdots \\ 0 \\ 0 \end{bmatrix}_{n \times 1},$$

$$C = \begin{bmatrix} \beta_1 & \beta_2 & \cdots & \beta_{n-1} & \beta_n \end{bmatrix}_{1 \times n}$$

or

Notation. This realization is said to be in *observable canonical form* for reasons that will become clear later.

$$\bar{A} = \begin{bmatrix} -\alpha_1 & 1 & 0 & \cdots & 0 \\ -\alpha_2 & 0 & 1 & \cdots & 0 \\ -\alpha_3 & 0 & 0 & \cdots & 0 \\ \vdots & \vdots & \vdots & \ddots & \vdots \\ -\alpha_n & 0 & 0 & \cdots & 0 \end{bmatrix}_{n \times n}, \qquad \bar{B} = \begin{bmatrix} \beta_1 \\ \beta_2 \\ \beta_3 \\ \vdots \\ \beta_n \end{bmatrix}_{n \times 1},$$

$$\bar{C} = \begin{bmatrix} 1 & 0 & \cdots & 0 & 0 \end{bmatrix}_{1 \times n}.$$

$\square$

4.4 EQUIVALENT STATE-SPACE SYSTEMS

Consider the continuous-time LTI system

$$\dot{x} = Ax + Bu, \qquad\qquad y = Cx + Du.$$

Note. This transformation can be viewed as a change of basis for the state.

Given a nonsingular matrix T, suppose that we define

$$\bar{x} := Tx.$$

The same system can be defined using $\bar{x}$ as the state, by noting that

$$\dot{\bar{x}} = T\dot{x} = TAx + TBu = TAT^{-1}\bar{x} + TBu,$$

$$y = Cx + Du = CT^{-1}\bar{x} + Du,$$

which can be written as

$$\dot{\bar{x}} = \bar{A}\bar{x} + \bar{B}u, \qquad\qquad y = \bar{C}\bar{x} + \bar{D}u$$

for

$$\bar{A} := TAT^{-1}, \qquad \bar{B} := TB, \qquad \bar{C} := CT^{-1}, \qquad \bar{D} := D. \qquad (4.7)$$

Definition 4.3 (Algebraically equivalent). Two continuous-time or discrete-time LTI systems

Notation. Often one also says that the system on the right can be obtained from the system on the left using the similarity transformation $\bar{x} = Tx$.

$$\begin{cases} \dot{x}/x^+ = Ax + Bu \\ y = Cx + Du \end{cases} \quad \text{or} \quad \begin{cases} \dot{\bar{x}}/\bar{x}^+ = \bar{A}\bar{x} + \bar{B}u \\ y = \bar{C}\bar{x} + \bar{D}u, \end{cases}$$

respectively, are called *algebraically equivalent* if there exists a nonsingular matrix T such that (4.7) holds. The corresponding map $\bar{x} = Tx$ is called a *similarity transformation* or an *equivalence transformation*. □

Properties. Suppose that two state-space LTI systems are algebraically equivalent.

P4.1 With every input signal u, both systems associate the same set of outputs y.

However, the output is generally not the same for the same initial conditions, except for the forced or zero-state response, which is always the same.

Note. To obtain the same output, the initial conditions must be related by the similarity transformation: $\bar{x}(0) = Tx(0)$.

P4.2 The systems are zero-state equivalent; i.e., both systems have the same transfer function and impulse response.

This is a consequence of P4.1, but can also be proved directly as follows:

$$\begin{aligned} \bar{C}(sI - \bar{A})^{-1}\bar{B} + \bar{D} &= CT^{-1}(sI - TAT^{-1})^{-1}TB + D \\ &= C\big(T^{-1}(sI - TAT^{-1})T\big)^{-1}B + D \\ &= C(sI - A)^{-1}B + D. \end{aligned} \qquad \square$$

Attention! In general the converse of P4.2 does *not* hold, i.e., zero-state equivalence *does not imply* algebraic equivalence. □

4.5   LTI SYSTEMS IN MATLAB®

Attention!
MATLAB® simulations of a transfer function model always produce a forced response (zero initial conditions).

MATLAB® represents LTI systems using either state-space or transfer function models. State-space models are created using the MATLAB® function `ss()` introduced in Lecture 1, whereas transfer function models are created using the MATLAB® functions `tf()` and `zpk()` described below.

The functions `ss()`, `tf()`, and `zpk()` can also be used to convert between state-space and transfer function models. However, most MATLAB® functions that manipulate LTI models accept both state-space and transfer function models.

CREATION OF TRANSFER FUNCTION MODELS

MATLAB® Hint 16 (`tf`).   The command `sys_tf=tf(num,den)` assigns to `sys_tf` a MATLAB® rational transfer function. The argument `num` is a vector with the coefficients of the numerator of the system's transfer function, and `den` is a vector with the coefficients of the denominator. The last coefficient must always be the zeroth one; e.g., to get $\frac{2s}{s^2+3}$, one should use

```
num=[2 0];den=[1 0 3];
```

For transfer matrices, `num` and `den` are cell arrays. Type `help tf` for examples.

Optionally, one can specify the names of the inputs, outputs, and state to be used in subsequent plots as follows:

```
sys_tf=tf(num,den,...
          'InputName', {'input1', 'input2',...},...
          'OutputName',{'output1','output2',...},...
          'StateName', {'input1', 'input2',...});
```

The number of elements in the bracketed lists must match the number of inputs, outputs, and state variables.

For discrete-time systems, one should instead use the command `sys_tf=ss(num, den,Ts)`, where `Ts` is the sampling time, or `-1` if one does not want to specify it.
□

MATLAB® Hint 17 (`zpk`).   The command `sys_tf=zpk(z,p,k)` assigns to `sys_tf` a MATLAB® rational transfer function. The argument `z` is a vector with the zeros of the system, `p` is a vector with its poles, and `k` is the gain; e.g., to get $\frac{2s}{(s+1)(s+3)}$, one should use

```
z=0;p=[1,3];k=2;
```

For transfer matrices, `z` and `p` are cell arrays and `k` is a regular array. Type `help zpk` for examples.

Optionally, one can specify the names of the inputs, outputs, and state to be used in subsequent plots as follows:

```
sys_zpk=zpk(z,p,k,...
            'InputName', {'input1', 'input2',...},...
            'OutputName',{'output1','output2',...},...
            'StateName', {'input1', 'input2',...});
```

The number of elements in the bracketed lists must match the number of inputs, outputs, and state variables.

For discrete-time systems, one should instead use the command `sys_tf=zpk(z,p, k,Ts)`, where `Ts` is the sampling time, or `-1` if one does not want to specify it. $\quad\square$

MODEL CONVERSION

Note. These functions essentially compute $C(sI - A)^{-1}B + D$.

MATLAB® Hint 14 (tf). The functions `tf(sys_ss)` and `zpk(sys_ss)` compute the transfer function of the state-space model `sys_ss` specified as in MATLAB® Hint 1 (p. 6).

The function `tf(sys_ss)` returns the transfer function as a ratio of polynomials on s.

The function `zpk(sys_ss)` returns the polynomials factored as the product of monomials (for the real roots) and binomials (for the complex roots). This form highlights the zeros and poles of the system. $\quad\square$

Note. This function uses an algorithm similar to the one described in Section 4.3.

MATLAB® Hint 18 (ss). The function `ss(sys_tf)` computes a state-space realization for the transfer function `sys` specified as in MATLAB® Hints 16 (p. 38) or 17 (p. 38). $\quad\square$

4.6 EXERCISES

4.1 (Causality, linearity, and time invariance). Use equation (4.1) to show that the system

$$\dot{x} = Ax + Bu, \qquad\qquad y = Cx + Du \qquad\qquad \text{(CLTI)}$$

is causal, linear, and time-invariant. $\quad\square$

4.2 ($\mathcal{Z}$-transform of a LTI system' output). Show that the $\mathcal{Z}$-transform of any output to

$$x^+ = Ax + Bu, \qquad\qquad y = Cx + Du \qquad\qquad \text{(DLTI)}$$

is given by

$$\hat{y}(z) = \hat{\Psi}(z)x(0) + \hat{G}(z)\hat{u}(z), \qquad \hat{\Psi}(z) := C(zI - A)^{-1}z,$$
$$\hat{G}(z) := C(zI - A)^{-1}B + D. \qquad\square$$

4.3 (Observable canonical form). Given a transfer function $\hat{G}(s)$, let $(\bar{A}, \bar{B}, \bar{C}, \bar{D})$ be a realization for its transpose $\bar{G}(s) := \hat{G}(s)'$. Show that (A, B, C, D), where $A := \bar{A}'$, $B := \bar{C}'$, $C := \bar{B}'$, and $D = \bar{D}'$ is a realization for $\hat{G}(s)$.

Note that if the realization $(\bar{A}, \bar{B}, \bar{C}\ \bar{D})$ for $\bar{G}(s)$ is in controllable canonical form, *then the realization (A, B, C, D) for $\hat{G}(s)$ so obtained is in* observable canonical form. $\quad\square$

4.4 (SISO realizations). This exercise aims at proving Theorem 4.4. Use the construction outlined in Section 4.3.2 to arrive at results consistent with those in Theorem 4.4.

(a) Compute the controllable canonical form realization for the transfer function

$$\hat{g}(s) = \frac{k}{s^n + \alpha_1 s^{n-1} + \alpha_2 s^{n-2} + \cdots + \alpha_{n-1} s + \alpha_n}.$$

(b) For the realization in (a), compute the transfer function from the input u to the new output $y = x_i$, where x_i is the ith element of the state x.

Hint: You can compute $(sI - A)^{-1} b$ using the technique used in class for MIMO systems, or you may simply invert $(sI - A)^{-1}$ using the adjoint formula for matrix inversion,

$$M^{-1} = \frac{1}{\det M}(\text{adj } M)', \qquad\qquad \text{adj } M := [\text{cof}_{ij}\, M],$$

where $\text{cof}_{ij}\, M$ denotes the ijth cofactor of M. In this problem you actually need only to compute a single entry of $(sI - A)^{-1}$.

(c) Compute the controllable canonical form realization for the transfer function

$$\hat{g}(s) = \frac{\beta_1 s^{n-1} + \beta_2 s^{n-2} + \cdots + \beta_{n-1} s + \beta_n}{s^n + \alpha_1 s^{n-1} + \alpha_2 s^{n-2} + \cdots + \alpha_{n-1} s + \alpha_n}. \tag{4.8}$$

(d) Compute the observable canonical form realization for the transfer function in equation (4.8). □

4.5 (Zero-state equivalence). Show that the following pairs of systems are zero-state equivalent, but not algebraically equivalent.

(a)

$$\begin{cases} \dot{x} = \begin{bmatrix} 1 & 0 \\ 0 & 1 \end{bmatrix} x + \begin{bmatrix} 1 \\ 0 \end{bmatrix} u \\ y = \begin{bmatrix} 1 & 0 \end{bmatrix} x \end{cases} \qquad \begin{cases} \dot{\bar{x}} = \begin{bmatrix} 1 & 0 \\ 0 & 2 \end{bmatrix} \bar{x} + \begin{bmatrix} 1 \\ 0 \end{bmatrix} u \\ y = \begin{bmatrix} 1 & 0 \end{bmatrix} \bar{x} \end{cases}$$

(b)

$$\begin{cases} \dot{x} = \begin{bmatrix} 1 & 0 \\ 0 & 1 \end{bmatrix} x + \begin{bmatrix} 1 \\ 0 \end{bmatrix} u \\ y = \begin{bmatrix} 1 & 0 \end{bmatrix} x \end{cases} \qquad \begin{cases} \dot{\bar{x}} = \bar{x} + u \\ y = \bar{x} \end{cases}$$

Hint: To prove that the systems are not algebraically equivalent, you must show that there exists no similarity transformation that transforms one system into the other. □

4.6 (Equivalent realizations). Consider the following two systems:

$$\dot{x} = \begin{bmatrix} 2 & 1 & 2 \\ 0 & 2 & 2 \\ 0 & 0 & 1 \end{bmatrix} x + \begin{bmatrix} 1 \\ 1 \\ 0 \end{bmatrix} u, \qquad\qquad y = \begin{bmatrix} 1 & -1 & 0 \end{bmatrix} x,$$

$$\dot{x} = \begin{bmatrix} 2 & 1 & 1 \\ 0 & 2 & 1 \\ 0 & 0 & -1 \end{bmatrix} x + \begin{bmatrix} 1 \\ 1 \\ 0 \end{bmatrix} u, \qquad\qquad y = \begin{bmatrix} 1 & -1 & 0 \end{bmatrix} x.$$

(a) Are these systems zero-state equivalent?

(b) Are they algebraically equivalent? □

LECTURE 5

Solutions to LTV Systems

CONTENTS

This lecture studies the properties of solutions to state-space linear time-varying systems.

1. Solution to Homogeneous Linear Systems
2. Solution to Nonhomogeneous Linear Systems
3. Discrete-Time Case
4. Exercises

5.1 SOLUTION TO HOMOGENEOUS LINEAR SYSTEMS

Note. A state-space linear system without inputs is called *homogeneous.*

We start by considering the solution to a continuous-time linear time-varying system with a given initial condition but zero input,

$$\dot{x} = A(t)x, \qquad x(t_0) = x_0 \in \mathbb{R}^n, \qquad t \geq 0. \qquad (5.1)$$

A key property of homogeneous linear systems is that the map from the initial condition $x(t_0) = x_0 \in \mathbb{R}^n$ to the solution $x(t) \in \mathbb{R}^n$ at a given time $t \geq 0$ is always *linear* and can therefore be expressed by a matrix multiplication.

Attention! Even when $t_0 > 0$, the solution (5.2) holds before t_0.

Theorem 5.1 (Peano-Baker series). *The unique solution to (5.1) is given by*

$$x(t) = \Phi(t, t_0)x_0, \quad x_0 \in \mathbb{R}^n, \ t \geq 0, \qquad (5.2)$$

Note. This theorem is a consequence of Property P5.1 below.

where

$$\Phi(t, t_0) := I + \int_{t_0}^{t} A(s_1)ds_1 + \int_{t_0}^{t} A(s_1) \int_{t_0}^{s_1} A(s_2)ds_2 ds_1$$
$$+ \int_{t_0}^{t} A(s_1) \int_{t_0}^{s_1} A(s_2) \int_{t_0}^{s_2} A(s_3)ds_3 ds_2 ds_1 + \cdots . \qquad (5.3)$$

The $n \times n$ matrix $\Phi(t, t_0)$ is called the *state transition matrix*, and the series in (5.3) is called the *Peano-Baker series*. The state transition matrix defined by (5.3) has several important properties that will be explored below.

Properties (State transition matrix).

P5.1 For every $t_0 \geq 0$, $\Phi(t, t_0)$ is the unique solution to

$$\frac{d}{dt}\Phi(t, t_0) = A(t)\Phi(t, t_0), \qquad \Phi(t_0, t_0) = I, \qquad t \geq 0. \qquad (5.4)$$

Theorem 5.1 is a direct consequence of this property because (5.1) follows from (5.4) and (5.2).

Proof. For $t = t_0$, $\Phi(t_0, t_0) = I$, because all the integrals in (5.3) are equal to zero. Taking the derivative of each side of (5.3) with respect to time, we obtain

$$\frac{d}{dt}\Phi(t, t_0) = A(t) + A(t)\int_{t_0}^{t} A(s_2)ds_2$$
$$+ A(t)\int_{t_0}^{t} A(s_2)\int_{t_0}^{s_2} A(s_3)ds_3 ds_2 + \cdots$$
$$= A(t)\Phi(t, t_0).$$

This proves that $\Phi(t, t_0)$ satisfies (5.4).

Proving that the series actually converges for all $t, t_0 \geq 0$ and that the solution is unique is beyond the scope of this course. Both results follow from general properties of solutions to ordinary differential equations and are a consequence of the fact that $\Phi \mapsto A(t)\Phi$ is a globally Lipschitz map for every fixed t [1, Chapter 1].

P5.2 For every fixed $t_0 \geq 0$, the ith column of $\Phi(t, t_0)$ is the unique solution to

$$\dot{x}(t) = A(t)x(t), \qquad x(t_0) = e_i, \qquad t \geq 0,$$

where e_i is the ith vector of the canonical basis of $\mathbb{R}^n$.

This is just a restatement of Property P5.1 above.

P5.3 For every $t, s, \tau \geq 0$,

$$\Phi(t, s)\Phi(s, \tau) = \Phi(t, \tau). \qquad (5.5)$$

This is called the *semigroup property.*

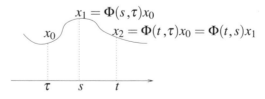

Figure 5.1. Semigroup property.

Proof. Take an arbitrary $x_0 \in \mathbb{R}^n$ and consider the solution to

$$\dot{x} = A(t)x, \qquad\qquad x(\tau) = x_0.$$

Its value at times s and t is given by

$$x_1 := \Phi(s, \tau)x_0, \qquad\qquad x_2 := \Phi(t, \tau)x_0,$$

respectively. However, we can regard the same $x(\cdot)$ as the solution to

$$\dot{x} = A(t)x, \qquad\qquad x(s) = x_1.$$

(Cf. Figure 5.1.) Therefore its value at time t is given by

$$x_2 = \Phi(t, s)x_1 = \Phi(t, s)\Phi(s, \tau)x_0.$$

By unicity of solution, the two vales for the solution at time t must coincide, so we have

$$\Phi(t, s)\Phi(s, \tau)x_0 = \Phi(t, \tau)x_0, \qquad \forall x_0.$$

Since this must be true for every $x_0 \in \mathbb{R}^n$, we conclude that (5.5) holds. ■

P5.4   For every $t, \tau \geq 0$, $\Phi(t, \tau)$ is nonsingular and

$$\Phi(t, \tau)^{-1} = \Phi(\tau, t).$$

Proof. From Property P5.3, we have

$$\Phi(\tau, t)\Phi(t, \tau) = \Phi(t, \tau)\Phi(\tau, t) = I,$$

which means that $\Phi(t, \tau)$ is the inverse of $\Phi(\tau, t)$ and vice versa, by definition of the inverse of a matrix. ■

5.2   SOLUTION TO NONHOMOGENEOUS LINEAR SYSTEMS

We now go back to the original nonhomogeneous LTV system

$$\dot{x} = A(t)x + B(t)u, \quad y = C(t)x(t) + D(t)u, \quad x(t_0) = x_0 \in \mathbb{R}^n, \quad t \geq 0 \quad (5.6)$$

to determine its solution.

Theorem 5.2 (Variation of constants). *The unique solution to (5.6) is given by*

$$x(t) = \Phi(t, t_0)x_0 + \int_{t_0}^{t} \Phi(t, \tau)B(\tau)u(\tau)d\tau \qquad\qquad (5.7)$$

$$y(t) = C(t)\Phi(t, t_0)x_0 + \int_{t_0}^{t} C(t)\Phi(t, \tau)B(\tau)u(\tau)d\tau + D(t)u(t), \qquad (5.8)$$

where $\Phi(t, t_0)$ is the state transition matrix. □

Note. This term
corresponds to the
system's output for a
zero input.

Equation (5.7) is known as the *variation of constants formula.* The term

$$y_h(t) := C(t)\Phi(t, t_0)x_0$$

in (5.8) is called the *homogeneous response,* whereas the term

Note. This term
corresponds to the
system's output for
zero initial conditions.

$$y_f(t) := \int_{t_0}^t C(t)\Phi(t, \tau)B(\tau)u(\tau)d\tau + D(t)u(t)$$

is called the *forced response.*

Proof of Theorem 5.2. To verify that (5.7) is a solution to (5.6), note that at $t = t_0$, the integral in (5.7) disappears, and we get $x(t_0) = x_0$. Taking the derivative of each side of (5.7) with respect to time, we obtain

Note. We recall that

$$\frac{d}{dt}\int_a^t f(t, s)ds$$

$$= f(t, t) + \int_a^t \frac{\partial f(t, s)}{\partial t}ds.$$

$$\dot{x} = \frac{d\Phi(t, t_0)}{dt}x_0 + \Phi(t, t)B(t)u(t) + \int_{t_0}^t \frac{d\Phi(t, \tau)}{dt}B(\tau)u(\tau)d\tau$$

$$= A(t)\Phi(t, t_0)x_0 + B(t)u(t) + A(t)\int_{t_0}^t \Phi(t, \tau)B(\tau)u(\tau)d\tau$$

$$= A(t)x(t) + B(t)u(t),$$

Note. Unicity for the
nonhomogeneous
case can also be
concluded from the
unicity for the
homogeneous case
using a proof by
contradiction.

which shows that (5.7) is indeed a solution to (5.6). Unicity of solution results from the fact that $x \mapsto A(t)x + B(t)u(t)$ is a globally Lipschitz map for every fixed t [1, Chapter 1].

The expression for $y(t)$ in (5.8) is obtained by direct substitution of $x(t)$ in $y(t) = C(t)x(t) + D(t)u$. ∎

5.3   DISCRETE-TIME CASE

Attention! As
opposed to the
continuous-time case,
in discrete time (5.11)
is valid only for
$t \geq t_0$. Therefore the
state transition matrix
cannot be used to go
back in time.

The (unique) solution to the homogeneous discrete-time linear time-varying system

$$x(t + 1) = A(t)x(t), \qquad x(t_0) = x_0 \in \mathbb{R}^n, \qquad t \in \mathbb{N} \qquad (5.9)$$

is given by

$$x(t) = \Phi(t, t_0)x_0, \quad x_0 \in \mathbb{R}^n, \, t \geq t_0, \qquad (5.10)$$

where

Notation. Sometimes
(5.11) is written as

$$\Phi(t, t_0) = \prod_{\tau=t_0}^{t-1} A(\tau),$$

but this notation can
be dangerous because
it does not accurately
describe the case
$t = t_0$ and it hides the
fact that the order of
the matrices in the
product (5.11) is
generally crucial.

$$\Phi(t, t_0) := \begin{cases} I & t = t_0 \\ A(t - 1)A(t - 2)\cdots A(t_0 + 1)A(t_0) & t > t_0 \end{cases} \qquad (5.11)$$

is called the *(discrete-time) state transition matrix.*

Properties (State transition matrix).

P5.5   For every $t_0 \geq 0$, $\Phi(t, t_0)$ is the unique solution to

$$\Phi(t + 1, t_0) = A(t)\Phi(t, t_0), \qquad \Phi(t_0, t_0) = I, \qquad t \geq t_0.$$

The fact that (5.10) is the unique solution to (5.9) is a direct consequence of this property, which can be proved by induction on t, starting at $t = t_0$.

P5.6 For every fixed $t_0 \geq 0$, the ith column of $\Phi(t, t_0)$ is the unique solution to

$$x(t + 1) = A(t)x(t), \qquad x(t_0) = e_i, \qquad t \geq t_0,$$

where e_i is the ith vector of the canonical basis of $\mathbb{R}^n$.

This is just a restatement of Property P5.5 above.

P5.7 For every $t \geq s \geq \tau \geq 0$,

$$\Phi(t, s)\Phi(s, \tau) = \Phi(t, \tau).$$

Attention! The discrete-time state transition matrix $\Phi(t, t_0)$ may be singular. In fact, this will always be the case whenever one of $A(t-1), A(t-2), \ldots, A(t_0)$ is singular. $\square$

Theorem 5.3 (Variation of constants). *The unique solution to*

$$x(t + 1) = A(t)x(t) + B(t)u(t), \qquad y(t) = C(t)x(t) + D(t)u(t),$$

with $x(t_0) = x_0 \in \mathbb{R}^n$, $t \in \mathbb{N}$, is given by

$$x(t) = \Phi(t, t_0)x_0 + \sum_{\tau = t_0}^{t-1} \Phi(t, \tau + 1)B(\tau)u(\tau), \qquad\qquad \forall t \geq t_0$$

$$y(t) = C(t)\Phi(t, t_0)x_0 + \sum_{\tau = t_0}^{t-1} C(t)\Phi(t, \tau + 1)B(\tau)u(\tau) + D(t)u(t), \quad \forall t \geq t_0$$

where $\Phi(t, t_0)$ is the discrete-time state transition matrix. $\square$

5.4 EXERCISES

5.1 (Causality and linearity). Use equation (5.7) to show that the system

$$\dot{x} = A(t)x + B(t)u, \qquad\qquad y = C(t)x + D(t)u \qquad\qquad \text{(CLTV)}$$

is causal and linear. $\square$

5.2 (State transition matrix). Consider the system

$$\dot{x} = \begin{bmatrix} 0 & t \\ 0 & 2 \end{bmatrix} x + \begin{bmatrix} 0 \\ t \end{bmatrix} u, \qquad y = \begin{bmatrix} 1 & 0 \end{bmatrix} x, \qquad x \in \mathbb{R}^2, \ u, y \in \mathbb{R}.$$

(a) Compute its state transition matrix

(b) Compute the system output to the constant input $u(t) = 1$, $\forall t \geq 0$ for an arbitrary initial condition $x(0) = \begin{bmatrix} x_1(0) & x_2(0) \end{bmatrix}'$. $\square$

LECTURE 6

Solutions to LTI Systems

CONTENTS

This lecture studies the properties of solutions to state-space linear time-invariant systems.

1. Matrix Exponential
2. Properties of the Matrix Exponential
3. Computation of Matrix Exponentials Using Laplace Transforms
4. The Importance of the Characteristic Polynomial
5. Discrete-Time Case
6. Symbolic Computations in MATLAB®
7. Exercises

6.1 MATRIX EXPONENTIAL

By applying the results in Lecture 5 to the homogeneous time-invariant system

$$\dot{x} = Ax, \qquad x(t_0) = x_0 \in \mathbb{R}^n, \qquad t \geq 0,$$

we conclude that its unique solution is given by

$$x(t) = \Phi(t, t_0)x_0, \quad x_0 \in \mathbb{R}^n, \ t \geq 0,$$

where now the state transition matrix is given by the Peano-Baker series,

$$\Phi(t, t_0) := I + \int_{t_0}^{t} A \, ds_1 + \int_{t_0}^{t}\int_{t_0}^{s_1} A^2 ds_2 ds_1 + \int_{t_0}^{t}\int_{t_0}^{s_1}\int_{t_0}^{s_2} A^3 ds_3 ds_2 ds_1 + \cdots.$$

Since

$$\int_{t_0}^{t}\int_{t_0}^{s_1}\cdots\int_{t_0}^{s_{k-2}}\int_{t_0}^{s_{k-1}} A^k ds_k ds_{k-1}\cdots ds_2 ds_1 = \frac{(t - t_0)^k}{k!} A^k,$$

we conclude that

$$\Phi(t, t_0) = \sum_{k=0}^{\infty} \frac{(t - t_0)^k}{k!} A^k. \tag{6.1}$$

MATLAB® Hint 19.
expm(A) computes
the matrix
exponential of
M.   ▶ p. 52

Motivated by the power series of the scalar exponential, we define the *matrix exponential* of a given $n \times n$ matrix M by

$$e^M := \sum_{k=0}^{\infty} \frac{1}{k!} M^k,$$

which allows us to rewrite (6.1) simply as

Attention! Equation (6.2) *does not generalize to the time-varying case* in any simple way. In particular, the state transition matrix of a time-varying system is *not* generally equal to $e^{\int_{t_0}^t A(\tau)d\tau}$.

$$\Phi(t, t_0) = e^{A(t-t_0)}. \tag{6.2}$$

Attention! Do not fail to notice that e^M is defined by (6.1). It is *not* true that its *ij*th entry is given by $e^{m_{ij}}$, where m_{ij} is the *ij*th entry of M. □

Going back to the nonhomogeneous case, we conclude from the variation of constants formula that the solution to

$$\dot{x} = Ax + Bu, \qquad y = Cx + Du, \qquad x(t_0) = x_0 \in \mathbb{R}^n, \qquad t \geq 0 \tag{6.3}$$

is given by

$$x(t) = e^{A(t-t_0)} x_0 + \int_{t_0}^t e^{A(t-\tau)} B(\tau) u(\tau) d\tau,$$

$$y(t) = C e^{A(t-t_0)} x_0 + \int_{t_0}^t C e^{A(t-\tau)} B u(\tau) d\tau + Du(t). \tag{6.4}$$

6.2 PROPERTIES OF THE MATRIX EXPONENTIAL

The following properties are direct consequences of the properties seen before for the state transition matrix of general time-varying system.

Properties (Matrix exponential).

P6.1 The function e^{At} is the unique solution to

$$\frac{d}{dt} e^{At} = A e^{At}, \qquad\qquad e^{A \cdot 0} = I, \qquad\qquad t \geq 0.$$

P6.2 The *i*th column of e^{At} is the unique solution to

$$\dot{x}(t) = Ax(t), \qquad\qquad x(0) = e_i, \qquad\qquad t \geq 0,$$

where e_i is the *i*th vector of the canonical basis of $\mathbb{R}^n$.

Note. This is a consequence of the semigroup property: $\Phi(t, 0)\Phi(0, -\tau) = \Phi(t, -\tau)$.

P6.3 For every $t, \tau \in \mathbb{R}$,

$$e^{At} e^{A\tau} = e^{A(t+\tau)}.$$

Attention! In general, $e^{At} e^{Bt} \neq e^{(A+B)t}$.

P6.4 For every $t \in \mathbb{R}$, e^{At} is nonsingular and

$$\left(e^{At} \right)^{-1} = e^{-At}.$$

□

Attention! For
$A = \begin{bmatrix} 1 & 2 \\ 3 & 4 \end{bmatrix}$ and

$p(s) = s^2 + 2s + 5$,

$p(A) = \begin{bmatrix} 1 & 2 \\ 3 & 4 \end{bmatrix} \begin{bmatrix} 1 & 2 \\ 3 & 4 \end{bmatrix}$

$+ 2\begin{bmatrix} 1 & 2 \\ 3 & 4 \end{bmatrix} + \begin{bmatrix} 5 & 0 \\ 0 & 5 \end{bmatrix}$

$\neq \begin{bmatrix} 1^2+2+5 & 2^2+2\times2+5 \\ 3^2+2\times3+5 & 4^2+2\times4+5 \end{bmatrix}$.

For LTI systems, the state transition matrix has further important properties that derive from the Cayley-Hamilton theorem, reviewed next. Given a polynomial

$$p(s) = a_0 s^n + a_1 s^{n-1} + a_2 s^{n-2} + \cdots + a_{n-1} s + a_n$$

and an $n \times n$ matrix A, we define

$$p(A) := a_0 A^n + a_1 A^{n-1} + a_2 A^{n-2} + \cdots + a_{n-1} A + a_n I_{n \times n},$$

which is also an $n \times n$ matrix.

Notation. One often says that $\Delta(s)$ *annihilates* A.

Theorem 6.1 (Cayley-Hamilton). *For every $n \times n$ matrix A,*

$$\Delta(A) = A^n + a_1 A^{n-1} + a_2 A^{n-2} + \cdots + a_{n-1} A + a_n I_{n \times n} = 0_{n \times n},$$

where

$$\Delta(s) = s^n + a_1 s^{n-1} + a_2 s^{n-2} + \cdots + a_{n-1} s + a_n$$

is the characteristic polynomial of A. □

Notation. The *characteristic polynomial* of an $n \times n$ matrix A is the degree n monic polynomial given by

$\Delta(s) := \det(sI - A)$.

Its roots are the eigenvalues of A.

The reader is referred, e.g., to [1] for a proof of the Cayley-Hamilton theorem.

The following properties of the matrix exponential are a consequence of the Cayley-Hamilton theorem and are specific to the time-invariant case.

Properties (Matrix exponential, continued).

P6.5 For every $n \times n$ matrix A, there exist n scalar functions $\alpha_0(t), \alpha_1(t), \ldots, \alpha_{n-1}(t)$ for which

$$e^{At} = \sum_{i=0}^{n-1} \alpha_i(t) A^i, \qquad \forall t \in \mathbb{R}. \tag{6.5}$$

Proof. By the Cayley-Hamilton theorem,

$$A^n + a_1 A^{n-1} + a_2 A^{n-2} + \cdots + a_{n-1} A + a_n I = 0,$$

where the a_i are the coefficients of the characteristic polynomial of A. Therefore

$$A^n = -a_1 A^{n-1} - a_2 A^{n-2} - \cdots - a_{n-1} A - a_n I.$$

Using this, we conclude that

$$\begin{aligned}
A^{n+1} &= -a_1 A^n - a_2 A^{n-1} - \cdots - a_{n-1} A^2 - a_n A \\
&= a_1(a_1 A^{n-1} + a_2 A^{n-2} + \cdots + a_{n-1} A + a_n I) - a_2 A^{n-1} - \cdots \\
&\quad - a_{n-1} A^2 - a_n A \\
&= (a_1^2 - a_2) A^{n-1} + (a_1 a_2 - a_3) A^{n-2} + \cdots + (a_1 a_{n-1} - a_n) A \\
&\quad + a_1 a_n I - \cdots - a_{n-1} A^2.
\end{aligned}$$

Therefore A^{n+1} can also be written as a linear combination of A^{n-1}, A^{n-2}, ..., A, I. Applying the same procedure for increasing powers of A, we conclude that for every $k \geq 0$, A^k can be written as

$$A^k = \bar{a}_{n-1}(k) A^{n-1} + \bar{a}_{n-2}(k) A^{n-2} + \cdots + \bar{a}_1(k) A + a_0(k) I, \qquad (6.6)$$

for appropriate coefficients $a_i(k)$. Replacing this in the definition of e^{At}, we conclude that

$$e^{At} = \sum_{k=0}^{\infty} \frac{t^k}{k!} A^k = \sum_{k=0}^{\infty} \frac{t^k}{k!} \sum_{i=0}^{n-1} \bar{a}_i(k) A^i.$$

Exchanging the order of summation, we obtain

$$e^{At} = \sum_{i=0}^{n-1} \left(\sum_{k=0}^{\infty} \frac{t^k \bar{a}_i(k)}{k!} \right) A^i.$$

Equation (6.5) follows if one defines $\alpha_i(t) := \sum_{k=0}^{\infty} \frac{t^k \bar{a}_i(k)}{k!}$. ∎

P6.6 For every $n \times n$ matrix A,

$$A e^{At} = e^{At} A, \qquad \forall t \in \mathbb{R}.$$

This is a direct consequence of P6.5.

6.3 COMPUTATION OF MATRIX EXPONENTIALS USING LAPLACE TRANSFORMS

We saw in Property P6.1 that e^{At} is uniquely defined by

$$\frac{d}{dt} e^{At} = A e^{At}, \qquad e^{A \cdot 0} = I, \qquad t \geq 0.$$

Taking the Laplace transform of each side of the differential equation, we conclude that

$$\mathcal{L}\left[\frac{d}{dt} e^{At}\right] = \mathcal{L}\left[A e^{At}\right] \quad \Leftrightarrow \quad s\widehat{e^{At}} - e^{At}\big|_{t=0} = A\widehat{e^{At}}$$

$$\Leftrightarrow \quad (sI - A)\widehat{e^{At}} = I \quad \Leftrightarrow \quad \widehat{e^{At}} = (sI - A)^{-1}.$$

Note. Since we are working with unilateral Laplace transforms, this method gives values for $t \geq 0$.

Therefore we can use inverse Laplace transform tables to compute e^{At}:

$$e^{At} = \mathcal{L}^{-1}\left[(sI - A)^{-1}\right].$$

6.4   THE IMPORTANCE OF THE CHARACTERISTIC POLYNOMIAL

We have seen in Lecture 4 that

$$(sI - A)^{-1} = \frac{1}{\det(sI - A)} [\text{adj}(sI - A)]',$$

where

$$\det(sI - A) = (s - \lambda_1)^{m_1}(s - \lambda_2)^{m_2} \cdots (s - \lambda_k)^{m_k},$$

is the characteristic polynomial of A, whose roots λ_i are the eigenvalues of A, and $\text{adj}(sI - A)$ is the adjoint matrix of $sI - A$ whose entries are polynomials in s of degree $n - 1$ or lower.

To compute the inverse Laplace transform of $(sI - A)^{-1}$, we need to perform a partial fraction expansion of each entry of $(sI - A)^{-1}$. These are of the form

$$\frac{\alpha_1 s^{n-1} + \alpha_2 s^{n-2} + \cdots + \alpha_{n-1} s + \alpha_n}{(s - \lambda_1)^{m_1}(s - \lambda_2)^{m_2} \cdots (s - \lambda_k)^{m_k}}$$

$$= \frac{a_{11}}{s - \lambda_1} + \frac{a_{12}}{(s - \lambda_1)^2} + \cdots + \frac{a_{1m_1}}{(s - \lambda_1)^{m_1}} + \cdots + \frac{a_{k1}}{s - \lambda_k} + \frac{a_{k2}}{(s - \lambda_k)^2} + \cdots$$

$$+ \frac{a_{km_k}}{(s - \lambda_k)^{m_k}}.$$

The inverse Laplace transform is then given by

$$\mathcal{L}^{-1}\left[\frac{\alpha_1 s^{n-1} + \alpha_2 s^{n-2} + \cdots + \alpha_{n-1} s + \alpha_n}{(s - \lambda_1)^{m_1}(s - \lambda_2)^{m_2} \cdots (s - \lambda_k)^{m_k}}\right]$$

$$= a_{11} e^{\lambda_1 t} + a_{12} t \, e^{\lambda_1 t} + \cdots + a_{1m_1} t^{m_1-1} e^{\lambda_1 t} + \cdots$$

$$+ a_{k1} e^{\lambda_k t} + a_{k2} t \, e^{\lambda_k t} + \cdots + a_{km_k} t^{m_k-1} e^{\lambda_k t}.$$

Notation. A matrix is called *Hurwitz* or a *stability matrix* if all its eigenvalues have strictly negative real parts.

Thus, when all the eigenvalues λ_i of A have strictly negative real parts, all entries of e^{At} converge to zero as $t \to \infty$, which means that the output

$$y(t) = C e^{A(t-t_0)} x_0 + \int_{t_0}^{t} C e^{A(t-\tau)} B u(\tau) d\tau + D u(t)$$

converges to the forced response

Note. Here we have shown only that if A is a stability matrix, then $\lim_{t \to \infty} e^{At} = 0$, but we show in Lecture 7 that only stability matrices have this property.   ▶ p. 59

$$y_f(t) = \int_{t_0}^{t} C e^{A(t-\tau)} B u(\tau) d\tau + D u(t).$$

6.5   DISCRETE-TIME CASE

Applying the results of Lecture 5 to the discrete-time homogeneous time-invariant system

$$x^+ = Ax, \qquad\qquad x(t_0) = x_0 \in \mathbb{R}^n, \qquad\qquad t \in \mathbb{N},$$

we conclude that its unique solution is given by

$$x(t) = \Phi(t, t_0)x_0, \quad x_0 \in \mathbb{R}^n, \ t \geq 0,$$

where now the state transition matrix is simply given by

$$\Phi(t, t_0) := A^{t-t_0}, \quad \forall t \geq t_0.$$

Going back to the nonhomogeneous case, we conclude from the discrete-time variation of constants formula that the solution to

$$x^+ = Ax + Bu, \qquad y = Cx + Du, \qquad x(t_0) = x_0 \in \mathbb{R}^n, \qquad t \geq 0$$

is given by

$$x(t) = A^{t-t_0}x_0 + \sum_{\tau=t_0}^{t-1} A^{t-1-\tau}Bu(\tau)$$

$$y(t) = C A^{t-t_0}x_0 + \sum_{\tau=t_0}^{t-1} C A^{t-1-\tau}Bu(\tau) + Du(t).$$

The matrix power can be computed using $\mathcal{Z}$-transforms as follows. From the definition of the $\mathcal{Z}$-transform, we conclude that

$$\mathcal{Z}[A^{t+1}] := \sum_{t=0}^{\infty} z^{-t}A^{t+1} = z \sum_{t=0}^{\infty} z^{-(t+1)}A^{t+1} = z\left(\sum_{t=0}^{\infty} z^{-t}A^t - I\right)$$

$$= z\left(\mathcal{Z}[A^t] - I\right).$$

On the other hand, $\mathcal{Z}[A^{t+1}] = A\mathcal{Z}[A^t]$. Therefore we conclude that

$$A\widehat{A^t} = z(\widehat{A^t} - I) \quad \Leftrightarrow \quad (zI - A)\widehat{A^t} = zI \quad \Leftrightarrow \quad \widehat{A^t} = z(zI - A)^{-1}.$$

Taking inverse $\mathcal{Z}$-transforms, we obtain

$$A^t = \mathcal{Z}^{-1}\left[z(zI - A)^{-1}\right].$$

Notation. A matrix is called *Schur stable* if all its eigenvalues have magnitude strictly smaller than 1.

Now, when all eigenvalues of A have magnitude smaller than 1, all entries of A^t will converge to zero as $t \to \infty$, which means that the output will converge to the forced response.

6.6 SYMBOLIC COMPUTATIONS IN MATLAB®

MATLAB® is capable of performing symbolic computation using Maple's engine. This is especially useful to compute matrix exponentials and Laplace transforms.

MATLAB® Hint 9 (syms). The command syms x1 x2 defines x1 and x2 as symbolic variables. From this point forward, any computations involving these variables are performed symbolically and result in symbolic expressions. One can include in the syms command information about the types of the variables.

1. The command `syms x1 x2 real` defines `x1` and `x2` as symbolic variables in $\mathbb{R}$.

2. The command `syms x1 x2 positive` defines `x1` and `x2` as symbolic variables in $(0, \infty)$.

By itself, the command `syms` lists all symbolic variables.

For example,

```
>> A=[1,1;0,1];syms s;Q=inv(s*eye(2)-A)
Q =
[   1/(s-1), 1/(s-1)^2]
[        0,   1/(s-1)]
```

defines a new symbolic variable `Q` that is equal to $(sI - A)^{-1}$ for $A := \left[\begin{smallmatrix} 1 & 1 \\ 0 & 1 \end{smallmatrix}\right]$. $\quad\square$

MATLAB® Hint 8 (`jacobian`). The function `jacobian(f,x)` computes the Jacobian of the vector `f` of symbolic expressions with respect to the vector `x` of symbolic variables.

For example,

```
>> syms px py theta v omega
>> x=[px;py;theta];u=[v;omega];       % state variable and
                                       % control input
>> f=[v*cos(theta);v*sin(theta);omega];  % system dynamics
>> A=jacobian(f,x),B=jacobian(f,u)
A =
[              0,              0, -v*sin(theta)]
[              0,              0,  v*cos(theta)]
[              0,              0,             0]
B =
[ cos(theta),           0]
[ sin(theta),           0]
[          0,           1]
```

computes the local linearization of the unicycle considered in Exercise 2.2 in its original coordinates, as in equation (2.11). $\quad\square$

MATLAB® Hint 19 (`expm`). The function `expm(M)` computes the matrix exponential of `M`. When `M` is a symbolic variable, the computation is carried out symbolically.

For example,

```
>> A=[1,1;0,1];syms t;Q=expm(A*t)
Q =
[   exp(t), t*exp(t)]
[        0,   exp(t)]
```

defines a new symbolic variable `Q` that is equal to e^{At} for $A := \left[\begin{smallmatrix} 1 & 1 \\ 0 & 1 \end{smallmatrix}\right]$. $\quad\square$

MATLAB® Hint 10 (`laplace`). The function `laplace(F,t,s)` computes the unilateral Laplace transform of the symbolic expression `F` on the symbolic time variable `t` and returns it as a function of the complex variable `s`.

For example:

```
>> A=[1,1;0,1];syms t s;Q=laplace(expm(A*t),t,s)
Q =
[   1/(s-1),  1/(s-1)^2]
[         0,    1/(s-1)]
```

MATLAB® Hint 11 (`ilaplace`). The function `ilaplace(F,s,t)` computes the unilateral inverse Laplace transform of the symbolic expression F on the symbolic complex variable s and returns it as a function of the time variable t.

For example:

```
>> A=[1,1;0,1];syms t s;Q=ilaplace(inv(s*eye(2)-A),s,t)
Q =
[   exp(t), t*exp(t)]
[        0,   exp(t)]
```

MATLAB® Hint 12 (`ztrans`). The function `ztrans(F,t,z)` computes the unilateral $\mathcal{Z}$-transform of the symbolic expression F on the symbolic time variable t and returns it as a function of the complex variable z.

For example:

```
>> At=[1,t;0,1];syms t z;Q=ztrans(At,t,z)
Q =
[   z/(z-1),  z/(z-1)^2]
[         0,    z/(z-1)]
```

MATLAB® Hint 13 (`iztrans`). The function `iztrans(F,z,t)` computes the unilateral inverse $\mathcal{Z}$-transform of the symbolic expression F on the symbolic complex variable z and returns it as a function of the time variable t.

Note. `iztrans` is especially useful to compute matrix powers, because `A^t` does not work for a symbolic t.

For example:

```
>> A=[1,1;0,1];syms t z;Q=iztrans(z*inv(z*eye(2)-A),z,t)
Q =
[ 1, t]
[ 0, 1]
```

6.7   EXERCISES

6.1. We saw in Section 4.1 that the solution to the time-invariant system (6.3) with $t_0 = 0$ was given by

$$y(t) = \Psi(t)x_0 + (G \star u)(t) = \Psi(t)x_0 + \int_0^t G(t-\tau)u(\tau)dt,$$

where

$$\Psi(t) := \mathcal{L}^{-1}[C(sI-A)^{-1}], \qquad G(t) := \mathcal{L}^{-1}[C(sI-A)^{-1}B+D].$$

In view of (6.4), what do you conclude about the relationship between $G(t)$, $\Psi(t)$, and e^{At}?

Hint: Recall that a Dirac pulse has the property that

$$\int_{t_1}^{t_2} \delta(t-\tau)f(\tau)d\tau = f(t), \qquad\qquad \forall t \in [t_1, t_2]. \qquad\qquad \square$$

6.2 (Matrix powers and exponential). Compute A^t and e^{At} for the following matrices

$$A_1 = \begin{bmatrix} 1 & 1 & 0 \\ 0 & 1 & 0 \\ 0 & 0 & 1 \end{bmatrix}, \quad A_2 = \begin{bmatrix} 1 & 1 & 0 \\ 0 & 0 & 1 \\ 0 & 0 & 1 \end{bmatrix}, \quad A_3 = \begin{bmatrix} 2 & 0 & 0 & 0 \\ 2 & 2 & 0 & 0 \\ 0 & 0 & 3 & 3 \\ 0 & 0 & 0 & 3 \end{bmatrix}. \qquad (6.7)$$

LECTURE 7

Solutions to LTI Systems: The Jordan Normal Form

CONTENTS

This lecture studies how the Jordan normal form of A affects the solution to state-space linear time-invariant systems.

1. Jordan Normal Form
2. Computation of Matrix Powers Using the Jordan Normal Form
3. Computation of Matrix Exponentials Using the Jordan Normal Form
4. Eigenvalues with Multiplicity Larger than 1
5. Exercises

7.1 JORDAN NORMAL FORM

We start by reviewing the key relevant linear algebra concepts related to the Jordan normal form.

Note. The Jordan normal form is covered, e.g., in [10, 14].

Theorem 7.1 (Jordan normal form). *For every matrix $A \in \mathbb{C}^{n \times n}$, there exists a nonsingular change of basis matrix $P \in \mathbb{C}^{n \times n}$ that transforms A into*

$$J = PAP^{-1} = \begin{bmatrix} J_1 & 0 & 0 & \cdots & 0 \\ 0 & J_2 & 0 & \cdots & 0 \\ 0 & 0 & J_3 & \cdots & 0 \\ \vdots & \vdots & \vdots & \ddots & \vdots \\ 0 & 0 & 0 & \cdots & J_\ell \end{bmatrix},$$

where each J_i is a Jordan block of the form

Attention! There can be several Jordan blocks for the same eigenvalue, but in that case there must be more than one independent eigenvector for that eigenvalue.

$$J_i = \begin{bmatrix} \lambda_i & 1 & 0 & \cdots & 0 \\ 0 & \lambda_i & 1 & \cdots & 0 \\ 0 & 0 & \lambda_i & \cdots & 0 \\ \vdots & \vdots & \vdots & \ddots & \vdots \\ 0 & 0 & 0 & \cdots & \lambda_i \end{bmatrix}_{n_i \times n_i},$$

where each λ_i is an eigenvalue of A, and the number ℓ of Jordan blocks is equal to the total number of independent eigenvectors of A. The matrix J is unique up to a reordering of the Jordan blocks and is called the Jordan normal form *of A.*   □

MATLAB® Hint 20.
Jordan(A)
computes the Jordan
normal form of
A.   ▶ p. 56

Definition 7.1 (Semisimple). A matrix is called *semisimple* or *diagonalizable* if its Jordan normal form is diagonal.   □

Note 5. How to find
the Jordan normal
form of a matrix by
hand?   ▶ p. 56

Theorem 7.2. *For an $n \times n$ matrix A, the following three conditions are equivalent:*

1. *A is semisimple.*

2. *A has n linearly independent eigenvectors.*

3. *There is no nonzero polynomial of degree less than n that annihilates A, i.e., for every nonzero polynomial $p(s)$ of degree less than n, $p(A) \neq 0$.*   □

MATLAB® Hint 20 (jordan). The command [P,J]=jordan(A) computes the Jordan normal form of the matrix A and returns it into the matrix J. The corresponding change of basis matrix is returned in P so that $J = P^{-1}AP$.   □

Attention! The computation of the Jordan normal form is very sensitive to numerical errors. To see this, find the Jordan normal form of the following two matrices (which are very similar):

$$A_1 = \begin{bmatrix} 0 & 1 \\ 0 & 0 \end{bmatrix}, \qquad\qquad A_2 = \begin{bmatrix} -10^{-6} & 1 \\ 0 & 0 \end{bmatrix}. \qquad □$$

Note 5 (Determining the Jordan normal form). Some guessing is generally involved in finding the Jordan normal form. The following procedure can be used to compute the Jordan normal form of a given matrix A by hand.

1. Compute the eigenvalues of A.

2. List all possible Jordan normal forms that are compatible with the eigenvalues of A. To do this, keep in mind that

Note. Other rules can
help you narrow
down the possibilities.
E.g., the number of
Jordan blocks
associated with an
eigenvalue λ must be
equal to the number
of independent
eigenvectors of A
associated with the
eigenvalue λ.

   • eigenvalues with multiplicity equal to 1 must always correspond to 1×1 Jordan blocks,

   • eigenvalues with multiplicity equal to 2 can correspond to one 2×2 block or two 1×1 blocks, and

   • eigenvalues with multiplicity equal to 3 can correspond to one 3×3 block, one 2×2 block and two 1×1 blocks, or three 1×1 blocks, etc.

3. For each candidate Jordan normal form, check whether there exists a nonsingular matrix P for which $J = PAP^{-1}$. To find out whether this is so, you may solve the (equivalent, but simpler) linear equation

$$JP = PA$$

for the unknown matrix P and check whether it has a nonsingular solution.

Since the Jordan normal form is unique (up to a permutation of the blocks), once you find a matrix J with a Jordan structure and a matrix P for which $J = PAP^{-1}$, you have found *the* Jordan normal form of A.   □

7.2 COMPUTATION OF MATRIX POWERS USING THE JORDAN NORMAL FORM

Given an $n \times n$ matrix A, let $J = PAP^{-1}$ be the Jordan normal form of A. Since

$$J = PAP^{-1} \quad \Leftrightarrow \quad A = P^{-1}JP,$$

we conclude that

$$A^t = \underbrace{P^{-1}JP\ P^{-1}JP\ \cdots\ P^{-1}JP}_{k\ \text{times}} = P^{-1}J^t P = P^{-1} \begin{bmatrix} J_1^t & 0 & \cdots & 0 \\ 0 & J_2^t & \cdots & 0 \\ \vdots & \vdots & \ddots & \vdots \\ 0 & 0 & \cdots & J_\ell^t \end{bmatrix} P,$$

$$\tag{7.1}$$

where the J_i are the Jordan blocks of A. It turns out that it is easy to compute J_i^t for a Jordan block J_i:

$$J_i = \begin{bmatrix} \lambda_i & 1 & 0 & \cdots & 0 \\ 0 & \lambda_i & 1 & \cdots & 0 \\ 0 & 0 & \lambda_i & \cdots & 0 \\ \vdots & \vdots & \vdots & \ddots & \vdots \\ 0 & 0 & 0 & \cdots & \lambda_i \end{bmatrix}_{n_i \times n_i}$$

$$\Rightarrow \quad J_i^t = \begin{bmatrix} \lambda_i^t & t\lambda_i^{t-1} & \frac{t!\lambda_i^{t-2}}{(t-2)!2!} & \frac{t!\lambda_i^{t-3}}{(t-3)!3!} & \cdots & \frac{t!\lambda_i^{t-n_i+1}}{(t-n_i+1)!(n_i-1)!} \\ 0 & \lambda_i^t & t\lambda_i^{t-1} & \frac{t!\lambda_i^{t-2}}{(t-2)!2!} & \cdots & \frac{t!\lambda_i^{t-n_i+2}}{(t-n_i+2)!(n_i-2)!} \\ 0 & 0 & \lambda_i^t & t\lambda_i^{t-1} & \cdots & \frac{t!\lambda_i^{t-n_i+3}}{(t-n_i+3)!(n_i-3)!} \\ \vdots & \vdots & \vdots & \ddots & \ddots & \vdots \\ 0 & 0 & 0 & 0 & \ddots & t\lambda_i^{t-1} \\ 0 & 0 & 0 & 0 & \cdots & \lambda_i^t \end{bmatrix},$$

which can be verified by induction on t.

This expression confirms what we had seen before (and provides additional insight) about the connection between the eigenvalues of A and what happens to A^t as $t \to \infty$.

1. When all the eigenvalues of A have magnitude strictly smaller than 1 , then all the $J_i^t \to 0$ as $t \to \infty$, and therefore $A^t \to 0$ as $t \to \infty$.

Notation. A matrix is called *Schur stable* if all its eigenvalues have magnitude strictly smaller than 1.

2. When all the eigenvalues of A have magnitude smaller or equal to 1 and all the Jordan blocks corresponding to eigenvalues with magnitude equal to 1 are 1×1, then all the J_i^t remain bounded as $t \to \infty$, and consequently, A^t remains bounded as $t \to \infty$.

3. When at least one eigenvalue of A has magnitude larger than 1 or magnitude equal to 1, but the corresponding Jordan block is larger than 1×1, then A^t is unbounded as $t \to \infty$.

7.3 COMPUTATION OF MATRIX EXPONENTIALS USING THE JORDAN NORMAL FORM

Given an $n \times n$ matrix A, we saw that

$$e^{At} := \sum_{k=1}^{\infty} \frac{t^k}{k!} A^k.$$

Denoting by $J = PAP^{-1}$ the Jordan normal form of A, we conclude from (7.1) that

$$e^{At} := P^{-1} \left(\sum_{k=1}^{\infty} \frac{t^k}{k!} \begin{bmatrix} J_1^k & 0 & \cdots & 0 \\ 0 & J_2^k & \cdots & 0 \\ \vdots & \vdots & \ddots & \vdots \\ 0 & 0 & \cdots & J_\ell^k \end{bmatrix} \right) P$$

$$= P^{-1} \left(\begin{bmatrix} e^{J_1 t} & 0 & \cdots & 0 \\ 0 & e^{J_2 t} & \cdots & 0 \\ \vdots & \vdots & \ddots & \vdots \\ 0 & 0 & \cdots & e^{J_\ell t} \end{bmatrix} \right) P, \tag{7.2}$$

where the J_i are the Jordan blocks of A. It turns out that it is also easy to compute $e^{J_i t}$ for a Jordan block J_i, leading to

$$J_i = \begin{bmatrix} \lambda_i & 1 & 0 & \cdots & 0 \\ 0 & \lambda_i & 1 & \cdots & 0 \\ 0 & 0 & \lambda_i & \cdots & 0 \\ \vdots & \vdots & \vdots & \ddots & \vdots \\ 0 & 0 & 0 & \cdots & \lambda_i \end{bmatrix}_{n_i \times n_i} \Rightarrow e^{J_i t} = e^{\lambda_i t} \begin{bmatrix} 1 & t & \frac{t^2}{2!} & \frac{t^3}{3!} & \cdots & \frac{t^{n_i-1}}{(n_i-1)!} \\ 0 & 1 & t & \frac{t^2}{2!} & \cdots & \frac{t^{n_i-2}}{(n_i-2)!} \\ 0 & 0 & 1 & t & \cdots & \frac{t^{n_i-3}}{(n_i-3)!} \\ \vdots & \vdots & \vdots & \ddots & \ddots & \vdots \\ 0 & 0 & 0 & 0 & \ddots & t \\ 0 & 0 & 0 & 0 & \cdots & 1 \end{bmatrix}. \tag{7.3}$$

This can be verified by checking that the expression given for $e^{J_i t}$ satisfies $e^{J_i \cdot 0} = I$ and

$$\frac{d}{dt} e^{\lambda_i t} \begin{bmatrix} 1 & t & \frac{t^2}{2!} & \cdots & \frac{t^{n_i-1}}{(n_i-1)!} \\ 0 & 1 & t & \cdots & \frac{t^{n_i-2}}{(n_i-2)!} \\ 0 & 0 & 1 & \cdots & \frac{t^{n_i-3}}{(n_i-3)!} \\ \vdots & \vdots & \ddots & \ddots & \vdots \\ 0 & 0 & 0 & \cdots & 1 \end{bmatrix} = \lambda_i e^{J_i t} + e^{\lambda_i t} \begin{bmatrix} 0 & 1 & t & \cdots & \frac{t^{n_i-2}}{(n_i-2)!} \\ 0 & 0 & 1 & \cdots & \frac{t^{n_i-3}}{(n_i-3)!} \\ 0 & 0 & 0 & \cdots & \frac{t^{n_i-4}}{(n_i-4)!} \\ \vdots & \vdots & \vdots & \ddots & \vdots \\ 0 & 0 & 0 & \cdots & 0 \end{bmatrix}$$

$$= \lambda_i e^{J_i t} + \begin{bmatrix} 0 & 1 & 0 & \cdots & 0 \\ 0 & 0 & 1 & \cdots & 0 \\ 0 & 0 & 0 & \cdots & 0 \\ \vdots & \vdots & \vdots & \ddots & \vdots \\ 0 & 0 & 0 & \cdots & 0 \end{bmatrix} e^{J_i t} = J_i e^{J_i t}.$$

Equations (7.2)–(7.3) confirm what we had seen before (and provide additional insight) about the connection between the eigenvalues of A and what happens to e^{At} as $t \to \infty$.

Notation. A matrix is called *Hurwitz* or a *stability matrix* if all its eigenvalues have strictly negative real parts.

1. When all the eigenvalues of A have strictly negative real parts , then all the $e^{J_i t} \to 0$ as $t \to \infty$, and therefore $e^{At} \to 0$ as $t \to \infty$.

2. When all the eigenvalues of A have negative or zero real parts and all the Jordan blocks corresponding to eigenvalues with zero real parts are 1×1, then all the $e^{J_i t}$ remain bounded as $t \to \infty$, and consequently, e^{At} remains bounded as $t \to \infty$.

3. When at least one eigenvalue of A has a positive real part or a zero real part, but the corresponding Jordan block is larger than 1×1, then e^{At} is unbounded as $t \to \infty$.

7.4   EIGENVALUES WITH MULTIPLICITY LARGER THAN 1

Diagonalizability is a *generic property for real matrices*. This means that if one draws entries at random, the probability of obtaining a matrix that is not diagonalizable is zero. However, in spite of being so unlikely, nondiagonalizable matrices arise frequently in state-space linear systems. The explanation for this paradox lies in the fact that certain system interconnections always produce nondiagonalizable blocks.

Consider the parallel connection in Figure 7.1(a) of two integrators. This system corresponds to the state-space model

$$\begin{cases} \dot{y}_1 = u, \\ \dot{y}_2 = u, \\ y = y_1 + y_2 \end{cases} \qquad \Leftrightarrow \qquad \begin{cases} \dot{x} = \begin{bmatrix} 0 & 0 \\ 0 & 0 \end{bmatrix} x + \begin{bmatrix} 1 \\ 1 \end{bmatrix} u, \\ y = \begin{bmatrix} 1 & 1 \end{bmatrix} x, \end{cases}$$

Note. What are the eigenvectors of A corresponding to the zero eigenvalue?

where we chose for state $x := \begin{bmatrix} y_1 & y_2 \end{bmatrix}'$. The A matrix for this system is diagonalizable with two zero eigenvalues with independent eigenvectors.

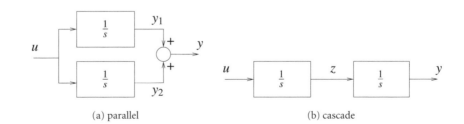

(a) parallel   (b) cascade

Figure 7.1. Block interconnections.

Consider now the cascade interconnection in Figure 7.1(b) of the same two integrators. This system corresponds to the following state-space model

$$\begin{cases} \dot{y}_1 = u, \\ \dot{y}_2 = y_1, \\ y = y_2 \end{cases} \qquad \Leftrightarrow \qquad \begin{cases} \dot{x} = \begin{bmatrix} 0 & 1 \\ 0 & 0 \end{bmatrix} x + \begin{bmatrix} 0 \\ 1 \end{bmatrix} u, \\ y = \begin{bmatrix} 1 & 0 \end{bmatrix} x, \end{cases}$$

where we chose for state $x := \begin{bmatrix} y_2 & y_1 \end{bmatrix}'$. In this case, the A matrix is not diagonalizable and has a single 2×2 Jordan block.

The following general conclusions can be extrapolated from this example.

1. *Cascade interconnections of k identical subsystem systems always lead to A matrices for the cascade with $k \times k$ Jordan blocks, one for each (simple) eigenvalue of the individual subsystems.* The cascade will have larger Jordan blocks if the individual subsystems already have Jordan blocks larger than 1×1.

   In view of what we saw in Sections 7.2 and 7.3, cascade interconnections can thus have a significant impact on the properties of continuous-time and discrete-time systems when the subsystems have poles with a zero real part or magnitude equal to 1, respectively.

2. In contrast, *parallel interconnections of identical subsystems do not increase the size of the Jordan blocks.*

   Thus, parallel interconnections generally do not significantly change the system's properties as far as the boundedness of solutions is concerned.

7.5 EXERCISE

7.1 (Jordan normal forms). Compute the Jordan normal form of the A matrix for the system represented by the following block diagram:

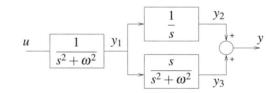

Figure 7.2. Block interconnection for Exercise 7.1.

PART II

STABILITY

LECTURE 8

Internal or Lyapunov Stability

Contents

This lecture introduces a notion of stability that expresses how the (internal) state of the system evolves with time.

1. Matrix Norms (review)
2. Lyapunov Stability
3. Eigenvalue Conditions for Lyapunov Stability
4. Positive-Definite Matrices (review)
5. Lyapunov Stability Theorem
6. Discrete-Time Case
7. Stability of Locally Linearized Systems
8. Stability Tests with MATLAB®
9. Exercises

8.1 MATRIX NORMS (REVIEW)

Several matrix norms are available. The following are the most common matrix norms for an $m \times n$ matrix $A = [a_{ij}]$.

1. The *one-norm*,

MATLAB® Hint 21.
`norm(A,1)`
computes the
one-norm of A.

$$\|A\|_1 := \max_{1 \leq j \leq n} \sum_{i=1}^{m} |a_{ij}|.$$

For a (column) vector $v = [v_i] \in \mathbb{R}^{\ell}$, $\|v\|_1 := \sum_{i=1}^{\ell} |v_i|$.

2. The ∞-*norm*,

MATLAB® Hint 22.
`norm(A,inf)`
computes the
∞-norm of A.

$$\|A\|_{\infty} := \max_{1 \leq i \leq m} \sum_{j=1}^{n} |a_{ij}|.$$

For a (column) vector $v = [v_i] \in \mathbb{R}^{\ell}$, $\|v\|_{\infty} := \max_{1 \leq i \leq \ell} |v_i|$.

Notation. In the absence of a subscript, $\| \cdot \|$ generally refers to the two-norm.

MATLAB® Hint 23. norm(A,2), or simply norm(A), computes the two-norm of A.

MATLAB® Hint 24. svd(A) computes the *singular values* of A, which are the square roots of the eigenvalues of $A'A$.

MATLAB® Hint 25. norm(A,'fro') computes the Frobenius norm of A.

3. The *two-norm*,

$$\|A\|_2 := \sigma_{\max}[A],$$

where $\sigma_{\max}[A]$ denotes the largest singular value of A. For a (column) vector $v = [v_i] \in \mathbb{R}^\ell$, this norm corresponds to the usual Euclidean norm $v := \sqrt{\sum_{i=1}^\ell v_i^2}$.

4. The *Frobenius norm*,

$$\|A\|_F := \sqrt{\sum_{i=1}^m \sum_{j=1}^n a_{ij}^2} = \sqrt{\sum_{i=1}^n \sigma_i[A]^2},$$

where the $\sigma_i[A]$ are the singular values of A. For (column) vectors, the Frobenius norm coincides with the two-norm (and also with the Euclidean norm), but in general this is not true for matrices.

All matrix norms are *equivalent* in the sense that each one of them can be upper and lower bounded by any other times a multiplicative constant:

$$\frac{\|A\|_1}{\sqrt{n}} \le \|A\|_2 \le \sqrt{n}\|A\|_1,$$

$$\frac{\|A\|_\infty}{\sqrt{n}} \le \|A\|_2 \le \sqrt{m}\|A\|_\infty,$$

$$\frac{\|A\|_F}{\sqrt{n}} \le \|A\|_2 \le \|A\|_F.$$

The four matrix norms above are *submultiplicative*; i.e., given two matrices A and B

$$\|AB\|_p \le \|A\|_p \, \|B\|_p, \qquad p \in \{1, 2, \infty, F\}.$$

For any submultiplicative norm $\| \cdot \|_p$, we have

$$\|Ax\|_p \le \|A\|_p \, \|x\|_p, \qquad \forall x$$

and therefore

$$\|A\|_p \ge \max_{x \ne 0} \frac{\|Ax\|_p}{\|x\|_p}.$$

Note. For subordinate norms, we can view the value of $\|A\|_p$ as the maximum vector norm amplification that can result from multiplying a vector by A.

The one-, two-, and ∞-norms are also *subordinate* to the corresponding vector norms; i.e., we actually have

$$\|A\|_p = \max_{x \ne 0} \frac{\|Ax\|_p}{\|x\|_p}, \qquad p \in \{1, 2, \infty\}. \tag{8.1}$$

The equality in (8.1) arises from the fact that subordinate norms have the property that for every matrix A there exists a vector $x^* \in \mathbb{R}^n$ for which

$$\|A\|_p = \frac{\|Ax^*\|_p}{\|x^*\|_p}, \qquad p \in \{1, 2, \infty\}. \tag{8.2}$$

Attention! The Frobenius norm is submultiplicative but not subordinate, which means that

$$\|A\|_F > \max_{x \neq 0} \frac{\|Ax\|_F}{\|x\|_F} = \max_{x \neq 0} \frac{\|Ax\|_2}{\|x\|_2} = \|A\|_2.$$

One can check this, e.g., for the matrix $\left[\begin{smallmatrix} 2 & 0 \\ 0 & 1 \end{smallmatrix}\right]$, for which $\|A\|_F = \sqrt{5} \approx 2.24$, and yet

$$\max_{x \neq 0} \frac{\|Ax\|_F}{\|x\|_F} = \|A\|_2 = 2.$$

This example shows that the Frobenius norm typically overestimates how much amplification can result from multiplying by A. □

8.2 LYAPUNOV STABILITY

Consider the following continuous-time LTV system

$$\dot{x} = A(t)x + B(t)u, \quad y = C(t)x + D(t)u, \quad x \in \mathbb{R}^n, \ u \in \mathbb{R}^k, \ y \in \mathbb{R}^m. \quad \text{(CLTV)}$$

Definition 8.1 (Lyapunov stability). The system (CLTV) is said to be

1. *(marginally) stable in the sense of Lyapunov* or *internally stable* if, for every initial condition $x(t_0) = x_0 \in \mathbb{R}^n$, the homogeneous state response

$$x(t) = \Phi(t, t_0)x_0, \qquad \forall t \geq 0$$

   is uniformly bounded,

2. *asymptotically stable (in the sense of Lyapunov)* if, in addition, for every initial condition $x(t_0) = x_0 \in \mathbb{R}^n$, we have that $x(t) \to 0$ as $t \to \infty$,

3. *exponentially stable* if, in addition, there exist constants $c, \lambda > 0$ such that, for every initial condition $x(t_0) = x_0 \in \mathbb{R}^n$, we have

$$\|x(t)\| \leq c e^{\lambda(t - t_0)} \|x(t_0)\|, \qquad \forall t \geq 0, \qquad \text{or}$$

4. *unstable* if it is not marginally stable in the Lyapunov sense. □

The matrices $B(\cdot)$, $C(\cdot)$, and $D(\cdot)$ play no role in this definition; only $A(\cdot)$ matters because this matrix completely defines the state transition matrix Φ. Therefore one often simply talks about the Lyapunov stability of the homogeneous system

$$\dot{x} = A(t)x, \qquad x \in \mathbb{R}^n.$$

Attention!

1. For marginally stable systems, the effect of initial conditions does not grow unbounded with time (but it may grow temporarily during a transient phase).

2. For asymptotically stable systems, the effect of initial conditions eventually disappears with time.

3. For unstable systems, the effect of initial conditions (may) grow over time (depending on the specific initial conditions and the value of the matrix C). □

8.3 EIGENVALUE CONDITIONS FOR LYAPUNOV STABILITY

The results in Lecture 7 about matrix exponentials provides us with simple conditions to classify the continuous-time homogeneous LTI system

$$\dot{x} = Ax, \qquad x \in \mathbb{R}^n \tag{H-CLTI}$$

in terms of its Lyapunov stability, *without explicitly computing the solution to the system.*

Theorem 8.1 (Eigenvalue conditions). *The system* (H-CLTI) *is*

1. marginally stable *if and only if all the eigenvalues of A have negative or zero real parts and all the Jordan blocks corresponding to eigenvalues with zero real parts are* 1×1,

2. asymptotically stable *if and only if all the eigenvalues of A have strictly negative real parts,*

3. exponentially stable *if and only if all the eigenvalues of A have strictly negative real parts, or*

4. unstable *if and only if at least one eigenvalue of A has a positive real part or zero real part, but the corresponding Jordan block is larger than* 1×1. □

Notation. A matrix is called *Hurwitz* or a *stability matrix* if all its eigenvalues have strictly negative real parts.

Attention! When all the eigenvalues of A have strictly negative real parts, all entries of e^{At} converge to zero exponentially fast, and therefore $\|e^{At}\|$ converges to zero exponentially fast (for every matrix norm); i.e., there exist constants $c, \lambda > 0$ such that

Note. When all Jordan blocks have multiplicity equal to 1, we can choose λ to be the largest (least negative) real part of the eigenvalues. Otherwise, λ has to be strictly smaller than that. See Exercise 8.3.

$$\|e^{At}\| \le c\, e^{-\lambda t}, \qquad \forall t \in \mathbb{R}.$$

In this case, for a submultiplicative norm, we have

$$\|x(t)\| = \|e^{t-t_0} x_0\| \le \|e^{A(t-t_0)}\| \, \|x_0\| \le c\, e^{-\lambda(t-t_0)} \|x_0\|, \qquad \forall t \in \mathbb{R}.$$

This means that asymptotic stability and exponential stability are equivalent concepts for LTI systems. □

Note. See Exercise 8.4.

Attention! These conditions do not generalize to time-varying systems, even if the eigenvalues of $A(t)$ do not depend on t. One can find matrix-valued signals $A(t)$ that are stability matrices for every fixed $t \ge 0$, but the time-varying system $\dot{x} = A(t)x$ is not even stable. □

8.4 POSITIVE-DEFINITE MATRICES (REVIEW)

Notation. When one talks about positive-definite, negative-definite or semidefinite matrices, it is generally implicit that the matrix is symmetric.

A symmetric $n \times n$ matrix Q is *positive-definite* if

$$x'Qx > 0, \qquad \forall x \in \mathbb{R}^n \setminus \{0\}. \tag{8.3}$$

When $>$ is replaced by $<$, we obtain the definition of a *negative-definite* matrix. Positive-definite matrices are always nonsingular, and their inverses are also positive-definite. Negative-definite matrices are also always nonsingular, and their inverses are negative-definite.

When (8.3) holds only for $\leq$ or $\geq$, the matrix is said to be *positive-semidefinite* or *negative-semidefinite*, respectively.

The following statements are equivalent for a symmetric $n \times n$ matrix Q.

1. Q is positive-definite.

MATLAB® Hint 15.
eig(A) computes the eigenvalues of the matrix A. ▶ p. 77

2. All eigenvalues of Q are strictly positive .

3. The determinants of all upper left submatrices of Q are positive.

4. There exists an $n \times n$ nonsingular real matrix H such that

$$Q = H'H.$$

Note. Every $n \times n$ symmetric matrix has real eigenvalues and n orthogonal (independent) eigenvectors.

For a positive-definite matrix Q we have

Note. In (8.4) we are using the two-norm for x.

$$0 < \lambda_{\min}[Q]\|x\|^2 \leq x'Qx \leq \lambda_{\max}[Q]\|x\|^2, \qquad \forall x \neq 0, \tag{8.4}$$

where $\lambda_{\min}[Q]$ and $\lambda_{\max}[Q]$ denote the smallest and largest eigenvalues of Q, respectively. The properties of positive-definite matrices are covered extensively, e.g., in [10, 14].

8.5 LYAPUNOV STABILITY THEOREM

The Lyapunov stability theorem provides an alternative condition to check whether or not the continuous-time homogeneous LTI system

$$\dot{x} = Ax, \qquad x \in \mathbb{R}^n \tag{H-CLTI}$$

is asymptotically stable.

Notation. A matrix is called *Hurwitz* or a *stability matrix* if all its eigenvalues have strictly negative real parts.

Theorem 8.2 (Lyapunov stability). *The following five conditions are equivalent:*

MATLAB® Hint 26.
P=lyap(A,Q) solves the Lyapunov equation AP + PA' = −Q. ▶ p. 77

1. *The system* (H-CLTI) *is asymptotically stable.*

2. *The system* (H-CLTI) *is exponentially stable.*

3. *All the eigenvalues of A have strictly negative real parts.*

Note 6. We will later add a sixth equivalent condition that will allow Q in (8.5) to be only positive-semidefinite.
▶ p. 114

4. *For every symmetric positive-definite matrix Q, there exists a unique solution P to the following* Lyapunov equation

$$A'P + PA = -Q. \tag{8.5}$$

Moreover, P is symmetric and positive-definite.

Note. The equation
(8.6) is called a *linear
matrix inequality
(LMI).* The term
"linear" comes from
the linearity of the
left-hand side in P,
and $<$ refers to the
fact that the left-hand
side must be *negative-
definite.*

Note. To prove that
multiple statements
$P_1, P_2, \ldots, P_\ell$ are
equivalent, one simply
needs to prove a cycle
of implications:
$P_1 \Rightarrow P_2$,
$P_2 \Rightarrow P_3, \ldots$,
$P_{\ell-1} \Rightarrow P_\ell$, and
$P_\ell \Rightarrow P_1$.

5. *There exists a symmetric positive-definite matrix P for which the following Lyapunov matrix inequality holds:*

$$A'P + PA < 0. \tag{8.6}$$

Proof of Theorem 8.2. The equivalence between conditions 1, 2, and 3 has already
been proved.

We prove that condition 2 $\Rightarrow$ condition 4 by showing that the unique solution to
(8.5) is given by

$$P := \int_0^\infty e^{A't} Q e^{At} dt. \tag{8.7}$$

To verify that this is so, four steps are needed.

1. *The (improper) integral in (8.7) is well defined* (i.e., it is finite). This is a consequence of the fact that the system (H-CLTI) is exponentially stable, and there-
   fore $\|e^{A't} Q e^{At}\|$ converges to zero exponentially fast as $t \to \infty$. Because of
   this, the integral is absolutely convergent.

2. *The matrix P in (8.7) solves the equation (8.5).* To verify this, we compute

   $$A'P + PA = \int_0^\infty A' e^{A't} Q e^{At} + e^{A't} Q e^{At} A \, dt.$$

   But

   $$\frac{d}{dt}\left(e^{A't} Q e^{At}\right) = A' e^{A't} Q e^{At} + e^{A't} Q e^{At} A,$$

   therefore

   $$A'P + PA = \int_0^\infty \frac{d}{dt}\left(e^{A't} Q e^{At}\right) dt = \left[e^{A't} Q e^{At}\right]_0^\infty$$
   $$= \left(\lim_{t \to \infty} e^{A't} Q e^{At}\right) - e^{A'0} Q e^{A0}.$$

   Equation (8.5) follows from this and the facts that $\lim_{t \to \infty} e^{At} = 0$ because of
   asymptotic stability and that $e^{A0} = I$.

3. *The matrix P in (8.7) is symmetric and positive-definite.* Symmetry comes from
   the fact that

Note. Check that
$\left(e^{At}\right)' = e^{A't}$.
(Cf. Exercise 8.5.)

   $$P' = \int_0^\infty \left(e^{A't} Q e^{At}\right)' dt = \int_0^\infty \left(e^{At}\right)' Q' \left(e^{A't}\right)' dt = \int_0^\infty e^{A't} Q e^{At} dt = P.$$

   To check that P is positive-definite, we pick an arbitrary (constant) vector $z \in
   \mathbb{R}^n$ and compute

   $$z'Pz = \int_0^\infty z' e^{A't} Q e^{At} z \, dt = \int_0^\infty w(t)' Q w(t) dt,$$

where $w(t) := e^{At}z$, $\forall t \geq 0$. Since Q is positive-definite, we conclude that $z'Pz \geq 0$. Moreover,

$$z'Pz = 0 \quad \Rightarrow \quad \int_0^\infty w(t)'Qw(t)dt = 0,$$

which can only happen if $w(t) = e^{At}z = 0$, $\forall t \geq 0$, from which one concludes that $z = 0$, because e^{At} is nonsingular. Therefore P is positive-definite.

4. *No other matrix solves this equation.* To prove this by contradiction, assume that there exists another solution $\bar{P}$ to (8.5); i.e.,

$$A'P + PA = -Q, \qquad \text{and} \qquad A'\bar{P} + \bar{P}A = -Q.$$

Note 7. To prove a statement P by *contradiction*, one starts by assuming that P is *not* true and then one searches for some logical inconsistency.

Then

$$A'(P - \bar{P}) + (P - \bar{P})A = 0.$$

Multiplying the above equation on the left and right by $e^{A't}$ and e^{At}, respectively, we conclude that

$$e^{A't}A'(P - \bar{P})e^{At} + e^{A't}(P - \bar{P})Ae^{At} = 0, \qquad \forall t \geq 0.$$

On the other hand,

$$\frac{d}{dt}\big(e^{A't}(P - \bar{P})e^{At}\big) = e^{A't}A'(P - \bar{P})e^{At} + e^{A't}(P - \bar{P})Ae^{At} = 0,$$

and therefore $e^{A't}(P - \bar{P})e^{At}$ must remain constant for all times. But, because of stability, this quantity must converge to zero as $t \to \infty$, so it must be always zero. Since e^{At} is nonsingular, this is possible only if $P = \bar{P}$.

The implication that condition 4 $\Rightarrow$ condition 5 follows immediately, because if we select $Q = -I$ in condition 4, then the matrix P that solves (8.5) also satisfies (8.6).

To prove that condition 5 $\Rightarrow$ condition 2, let P be a symmetric positive-definite matrix for which (8.6) holds and let

$$Q := -(A'P + PA) > 0.$$

Consider an arbitrary solution to equation (H-CLTI), and define the scalar signal

$$v(t) := x'(t)Px(t) \geq 0, \qquad \forall t \geq 0.$$

Taking derivatives, we obtain

$$\dot{v} = \dot{x}'Px + x'P\dot{x} = x'(A'P + PA)x = -x'Qx \leq 0, \qquad \forall t \geq 0. \qquad (8.8)$$

Therefore $v(t)$ is a nonincreasing signal, and we conclude that

$$v(t) = x'(t)Px(t) \leq v(0) = x'(0)Px(0), \qquad \forall t \geq 0.$$

Note. Here we are using the two-norm for x.

But since $v = x'Px \geq \lambda_{\min}[P]\|x\|^2$, we conclude that

$$\|x(t)\|^2 \leq \frac{x'(t)Px(t)}{\lambda_{\min}[P]} = \frac{v(t)}{\lambda_{\min}[P]} \leq \frac{v(0)}{\lambda_{\min}[P]}, \quad \forall t \geq 0, \tag{8.9}$$

which means that the system (H-CLTI) is stable. To verify that it is actually exponentially stable, we go back to (8.8) and, using the facts that $x'Qx \geq \lambda_{\min}[Q]\|x\|^2$ and that $v = x'Px \leq \lambda_{\max}[P]\|x\|^2$, we conclude that

$$\dot{v} = -x'Qx \leq -\lambda_{\min}[Q]\|x\|^2 \leq -\frac{\lambda_{\min}[Q]}{\lambda_{\max}[P]}v, \quad \forall t \geq 0. \tag{8.10}$$

To proceed, we need the Comparison lemma.

Lemma 8.1 (Comparison). *Let $v(t)$ be a differentiable scalar signal for which*

$$\dot{v}(t) \leq \mu\, v(t), \quad \forall t \geq t_0$$

for some constant $\mu \in \mathbb{R}$. Then

$$v(t) \leq e^{\mu(t-t_0)}v(t_0), \quad \forall t \geq t_0. \tag{8.11}$$

Applying the Comparison lemma (Lemma 8.1) to (8.10), we conclude that

$$v(t) \leq e^{-\lambda(t-t_0)}v(t_0), \quad \forall t \geq 0, \qquad \lambda := -\frac{\lambda_{\min}[Q]}{\lambda_{\max}[P]},$$

which shows that $v(t)$ converges to zero exponentially fast and so does $\|x(t)\|$ [see (8.9)]. ∎

Proof of Lemma 8.1. Define a new signal $u(t)$ as follows:

$$u(t) := e^{-\mu(t-t_0)}v(t), \quad \forall t \geq t_0.$$

Taking derivatives, we conclude that

$$\dot{u} = -\mu e^{-\mu(t-t_0)}v(t) + e^{-\mu(t-t_0)}\dot{v}(t) \leq -\mu e^{-\mu(t-t_0)}v(t) + \mu e^{-\mu(t-t_0)}v(t) = 0.$$

Therefore u is nonincreasing, and we conclude that

$$u(t) = e^{-\mu(t-t_0)}v(t) \leq u(t_0) = v(t_0), \quad \forall t \geq t_0,$$

which is precisely equivalent to (8.11). ∎

8.6 DISCRETE-TIME CASE

Consider now the following discrete-time LTV system

$$x(t+1) = A(t)x(t) + B(t)u(t), \qquad y(t) = C(t)x(t) + D(t)u(t). \tag{DLTV}$$

Definition 8.2 (Lyapunov stability). The system (DLTV) is said to be

1. *(marginally) stable in the Lyapunov sense* or *internally stable* if, for every initial condition $x(t_0) = x_0 \in \mathbb{R}^n$, the homogeneous state response

$$x(t) = \Phi(t, t_0)x_0, \qquad \forall t_0 \geq 0$$

is uniformly bounded,

2. *asymptotically stable (in the Lyapunov sense)* if, in addition, for every initial condition $x(t_0) = x_0 \in \mathbb{R}^n$, we have $x(t) \to 0$ as $t \to \infty$,

3. *exponentially stable* if, in addition, there exist constants $c > 0$, $\lambda < 1$ such that, for every initial condition $x(t_0) = x_0 \in \mathbb{R}^n$,

$$\|x(t)\| \leq c\lambda^{t-t_0}\|x(t_0)\|, \qquad \forall t \geq t_0, \qquad \text{or}$$

4. *unstable* if it is not marginally stable in the Lyapunov sense. □

The matrices $B(\cdot)$, $C(\cdot)$, and $D(\cdot)$ play no role in this definition; therefore, one often simply talks about the Lyapunov stability of the homogeneous system

$$x(t + 1) = A(t)x, \qquad x \in \mathbb{R}^n. \tag{H-DLTV}$$

Theorem 8.3 (Eigenvalue conditions). *The discrete-time homogeneous LTI system*

$$x^+ = Ax, \qquad x \in \mathbb{R}^n \tag{H-DLTI}$$

is

1. marginally stable *if and only if all the eigenvalues of A have magnitude smaller than or equal to 1 and all the Jordan blocks corresponding to eigenvalues with magnitude equal to 1 are* 1×1,

2. asymptotically *and* exponentially *stable if and only if all the eigenvalues of A have magnitude strictly smaller than 1, or*

3. unstable *if and only if at least 1 eigenvalue of A has magnitude larger than 1 or magnitude equal to 1, but the corresponding Jordan block is larger than* 1×1. □

Notation. A matrix is called *Schur stable* if all its eigenvalues have magnitude strictly smaller than 1.

Theorem 8.4 (Lyapunov stability in discrete time). *The following five conditions are equivalent:*

1. *The system* (H-DLTI) *is asymptotically stable.*

2. *The system* (H-DLTI) *is exponentially stable.*

3. *All the eigenvalues of A have magnitude strictly smaller than 1.*

4. *For every symmetric positive-definite matrix Q, there exists a unique solution P to the following* Stein equation *(more commonly known as the* discrete-time Lyapunov equation*)*

MATLAB® Hint 27.
P=dlyap(A,Q)
solves the Stein
equation APA$'$ − P =
−Q.   ▶ p. 77

$$A'PA - P = -Q. \tag{8.12}$$

Moreover, P is symmetric and positive-definite.

5. *There exists a symmetric positive-definite matrix P for which the following Lyapunov matrix inequality holds:*

$$A'PA - P < 0. \qquad \square$$

Attention! In discrete time, in the proof of the Lyapunov stability theorem (Theorem 8.4) one studies the evolution of the signal

$$v(t) = x'(t)Px(t), \quad \forall t \geq t_0.$$

In this case, along solutions to the system (H-DLTI), we have

$$v(t + 1) = x'(t + 1)Px(t + 1) = x(t)A'PAx(t),$$

and the discrete-time Lyapunov equation (8.12) guarantees that

$$v(t + 1) = x(t)(P - Q)x(t) = v(t) - x(t)Qx(t), \qquad \forall t \geq 0.$$

From this we conclude that $v(t)$ is nonincreasing and, with a little more effort, that it actually decreases to zero exponentially fast. $\qquad \square$

Table 8.1 summarizes the results in this section and contrasts them with the continuous-time conditions for Lyapunov stability.

8.7 STABILITY OF LOCALLY LINEARIZED SYSTEMS

Consider a continuous-time homogeneous nonlinear system

$$\dot{x} = f(x), \qquad x \in \mathbb{R}^n, \tag{8.13}$$

with an equilibrium point at $x^{\text{eq}} \in \mathbb{R}^n$; i.e., $f(x^{\text{eq}}) = 0$. We saw in Lecture 2 that the local linearization of (8.13) around x^{eq} is given by

$$\dot{\delta x} = A\,\delta x, \tag{8.14}$$

with $\delta x := x - x^{\text{eq}}$ and

$$A := \frac{\partial f(x^{\text{eq}})}{\partial x}.$$

It turns out that the original nonlinear system (8.13) inherits some of the desirable stability properties of the linearized system.

Theorem 8.5 (Stability of linearization). *Assume that the function f in (8.13) is twice differentiable. If the linearized system (8.14) is exponentially stable, then there exists a ball $B \subset \mathbb{R}^n$ around x^{eq} and constants c, $\lambda > 0$ such that for every solution $x(t)$ to the nonlinear system (8.13) that starts at $x(t_0) \in B$, we have*

$$\|x(t) - x^{\text{eq}}\| \leq ce^{\lambda(t - t_0)}\|x(t_0) - x^{\text{eq}}\|, \qquad \forall t \geq t_0. \tag{8.15}$$

$\square$

Notation. When this happens, we say that x^{eq} is a *locally exponentially stable* equilibrium point of the nonlinear system (8.13). The qualifier "locally" refers to the fact that the exponentially decaying bound (8.15) needs to hold only for initial conditions in a ball B around x^{eq} [8].

Table 8.1. Lyapunov Stability Tests for LTI Systems

	Continuous time		Discrete time	
Definition	Eigenvalue test	Lyapunov test	Eigenvalue test	Lyapunov test
Unstable — For some $t_0, x(t_0)$, $x(t)$ can be unbounded.	For some $\lambda_i[A]$, $\Re\lambda_i[A] > 0$ or $\Re\lambda_i[A] = 0$ with Jordan block larger than 1×1.		For some $\lambda_i[A]$, $\|\lambda_i[A]\| > 1$ or $\|\lambda_i[A]\| = 1$ with Jordan block larger than 1×1.	
Marginally stable — For every $t_0, x(t_0)$, $x(t)$ is uniformly bounded.	For every $\lambda_i[A]$, $\Re\lambda_i[A] < 0$ or $\Re\lambda_i[A] = 0$ with 1×1 Jordan block.		For every $\lambda_i[A]$, $\|\lambda_i[A]\| < 1$ or $\|\lambda_i[A]\| = 1$ with 1×1 Jordan block.	
Asymptotically stable — For every $t_0, x(t_0)$, $\lim_{t\to\infty} x(t) = 0$.	For every $\lambda_i[A]$, $\Re\lambda_i[A] < 0$.	For every $Q > 0$, $\exists P = P' > 0$: $A'P + PA = -Q$ or $\exists P = P' > 0$: $A'P + PA < 0$.	For every $\lambda_i[A]$, $\|\lambda_i[A]\| < 1$.	For every $Q > 0$, $\exists P = P' > 0$: $A'PA - P = -Q$ or $\exists P = P' > 0$: $A'PA - P < 0$.
Exponentially stable — $\exists c, \lambda > 0$: for every $t_0, x(t_0)$, $\|x(t)\| \le ce^{-\lambda t}\|x(t_0)\|$, $\forall t \ge t_0$.				

Proof of Theorem 8.5. Since f is twice differentiable, we know from Taylor's theorem that

$$r(x) := f(x) - \left(f(x^{\mathrm{eq}}) + A\,\delta x \right) = f(x) - A\,\delta x = O(\|\delta x\|^2),$$

which means that there exist a constant c and a ball $\bar{B}$ around x^{eq} for which

$$\|r(x)\| \le c\|\delta x\|^2, \qquad \forall x \in \bar{B}. \tag{8.16}$$

Since the linearized system is exponentially stable, there exists a positive-definite matrix P for which

$$A'P + PA = -I.$$

Inspired by the proof of the Lyapunov stability theorem (Theorem 8.2), we define the scalar signal

$$v(t) := \delta x' P\,\delta x, \qquad \forall t \ge 0$$

and compute its derivative along trajectories to the nonlinear system in equation (8.13):

Note. In (8.17) we
used the
submultiplicative
property of the
two-norm.

$$\begin{aligned}
\dot{v} &= f(x)'P\,\delta x + \delta x'P\,f(x) \\
&= (A\,\delta x + r(x))'P\,\delta x + \delta x'P(A\,\delta x + r(x)) \\
&= \delta x'(A'P + PA)\delta x + 2\,\delta x'P\,r(x) \\
&= -\|\delta x\|^2 + 2\,\delta x'P\,r(x) \\
&\le -\|\delta x\|^2 + 2\,\|P\|\,\|\delta x\|\,\|r(x)\|.
\end{aligned} \tag{8.17}$$

To make the proof work, we would like to make sure that the right-hand side is negative; e.g.,

$$-\|\delta x\|^2 + 2\,\|P\|\,\|\delta x\|\,\|r(x)\| \le -\frac{1}{2}\|\delta x\|^2.$$

Notation. The set $\mathcal{E}$ was constructed so that $x(t) \in \mathcal{E} \;\Leftrightarrow\; v(t) \le \epsilon.$

To achieve this, let ϵ be a positive constant sufficiently small so that the ellipsoid

$$\mathcal{E} := \{ x \in \mathbb{R}^n : (x - x^{\mathrm{eq}})'P(x - x^{\mathrm{eq}}) \le \epsilon \}$$

centered at x^{eq} satisfies the following two properties.

1. The ellipsoid $\mathcal{E}$ is fully contained inside the ball $\bar{B}$ arising from Taylor's theorem (cf. Figure 8.1). When x is inside this ellipsoid, equation (8.16) holds, and therefore

$$x(t) \in \mathcal{E} \;\Rightarrow\; \dot{v} \le -\|\delta x\|^2 + 2\,c\,\|P\|\,\|\delta x\|^3 = -\left(1 - 2\,c\,\|P\|\,\|\delta x\|\right)\|\delta x\|^2.$$

2. We further shrink ϵ so that inside the ellipsoid $\mathcal{E}$ we have

$$1 - 2\,c\,\|P\|\,\|\delta x\| \ge \frac{1}{2} \quad\Leftrightarrow\quad \|\delta x\| \le \frac{1}{4\,c\,\|P\|}.$$

For this choice of ϵ, we actually have

$$x(t) \in \mathcal{E} \;\Rightarrow\; \dot{v} \le -\frac{1}{2}\|\delta x\|^2. \tag{8.18}$$

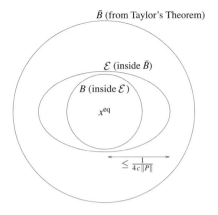

Figure 8.1. Construction of ball B for the proof of Theorem 8.5.

Notation. A set such as $\mathcal{E}$, with the property that if the state starts inside the set it remains there forever, is called *forward invariant*.

We therefore conclude that

$$x(t) \in \mathcal{E} \quad \Rightarrow \quad \begin{cases} v(t) \leq \epsilon \\ \dot{v}(t) \leq 0 \end{cases} \quad \Rightarrow \quad v \text{ cannot increase above } \epsilon \quad \Rightarrow \quad x \text{ cannot exit } \mathcal{E}.$$

Therefore if $x(0)$ starts inside $\mathcal{E}$, it cannot exit this set. Moreover, from (8.18) and the fact that $\delta x' \, P \, \delta x \leq \|P\| \|\delta x\|^2$, we further conclude that if $x(0)$ starts inside $\mathcal{E}$,

$$\dot{v} \leq -\frac{v}{2\|P\|}$$

and therefore, by the Comparison lemma (Lemma 8.1), v and consequently $\delta x :=$ $x - x^{\text{eq}}$ decrease to zero exponentially fast. The ball B around x^{eq} in the statement of the theorem can be any ball inside $\mathcal{E}$. ∎

When the linearized system is unstable, the nonlinear system also has undesirable properties (proof in [1, Chapter 6]):

Theorem 8.6 (Instability of linearization). *Assume that the function f in (8.13) is twice differentiable. If the linearized system (8.14) is unstable, then there are solutions that start arbitrarily close to x^{eq}, but do not converge to this point as $t \to \infty$.* □

Attention! When the linearized system is only marginally stable, not much can be said about the nonlinear system merely from analyzing the linearized system. For example, the two systems

$$\dot{x} = -x^3 \qquad\qquad \text{and} \qquad\qquad \dot{x} = +x^3 \qquad (8.19)$$

have the same local linearization,

$$\dot{\delta x} = 0,$$

around $x^{\text{eq}} = 0$, which is only marginally stable. Yet for the left-hand side system in (8.19), x always converges to zero, while for the right-hand side system, x always diverges away from the equilibrium point. □

Example 8.1 (Inverted pendulum). Consider the inverted pendulum in Figure 8.2 and assume that $u = T$ and $y = \theta$ are its input and output, respectively.

The local linearization of this system around the equilibrium point for which $\theta = \pi$ is given by

$$\dot{\delta x} = A\,\delta x + Bu, \qquad\qquad \delta y = C\,\delta x,$$

where

Note. This equilibrium point is $x^{\mathrm{eq}} = \pi$, $u^{\mathrm{eq}} = 0$, $y^{\mathrm{eq}} = \pi$, and therefore $\delta x := x - x^{\mathrm{eq}} = x - \pi$, $\delta u := u - u^{\mathrm{eq}} = u$, $\delta y := y - y^{\mathrm{eq}} = y - \pi$.

$$A := \begin{bmatrix} 0 & 1 \\ -\frac{g}{\ell} & -\frac{b}{m\ell^2} \end{bmatrix}, \qquad B := \begin{bmatrix} 0 \\ 1 \end{bmatrix}, \qquad C := \begin{bmatrix} 1 & 0 \end{bmatrix}.$$

The eigenvalues of A are given by

$$\det(\lambda I - A) = \lambda\left(\lambda + \frac{b}{m\ell^2}\right) + \frac{g}{\ell} = 0 \quad\Leftrightarrow\quad \lambda = -\frac{b}{2m\ell^2} \pm \sqrt{\left(\frac{b}{2m\ell^2}\right) - \frac{g}{\ell}},$$

and therefore the linearized system is exponentially stable, because

$$-\frac{b}{2m\ell^2} \pm \sqrt{\left(\frac{b}{2m\ell^2}\right) - \frac{g}{\ell}}$$

Note. We now know that this convergence is actually exponential.

has a negative real part. This is consistent with the obvious fact that in the absence of u the (nonlinear) pendulum converges to this equilibrium.

The local linearization of this system around the equilibrium point for which $\theta = 0$ is given by

Note. This equilibrium point is $x^{\mathrm{eq}} = 0$, $u^{\mathrm{eq}} = 0$, $y^{\mathrm{eq}} = 0$, and therefore $\delta x := x - x^{\mathrm{eq}} = x$, $\delta u := u - u^{\mathrm{eq}} = u$, $\delta y := y - y^{\mathrm{eq}} = y$.

$$\dot{\delta x} = A\,\delta x + Bu, \qquad\qquad \delta y = C\,\delta x,$$

where

$$A := \begin{bmatrix} 0 & 1 \\ \frac{g}{\ell} & -\frac{b}{m\ell^2} \end{bmatrix}, \qquad B := \begin{bmatrix} 0 \\ 1 \end{bmatrix}, \qquad C := \begin{bmatrix} 1 & 0 \end{bmatrix}.$$

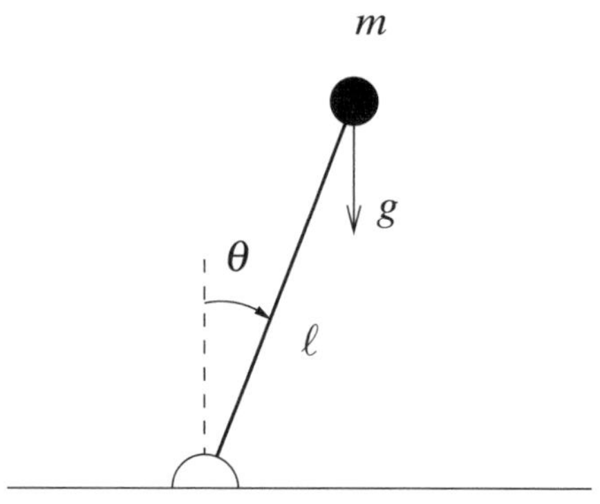

From Newton's law,

$$m\ell^2\ddot{\theta} = mg\ell\sin\theta - b\dot{\theta} + T,$$

where T denotes a torque applied at the base and g is the gravitational acceleration.

Figure 8.2. Inverted pendulum.

The eigenvalues of A are given by

$$\det(\lambda I - A) = \lambda\left(\lambda + \frac{b}{m\ell^2}\right) - \frac{g}{\ell} = 0 \quad \Leftrightarrow \quad \lambda = -\frac{b}{2m\ell^2} \pm \sqrt{\left(\frac{b}{2m\ell^2}\right) + \frac{g}{\ell}},$$

and therefore the linearized system is unstable, because

$$-\frac{b}{2m\ell^2} + \sqrt{\left(\frac{b}{2m\ell^2}\right) + \frac{g}{\ell}} > 0.$$

This is consistent with the obvious fact that in the absence of u the (nonlinear) pendulum does not naturally move up to the upright position if it starts away from it. However, one can certainly make it move up by applying some torque u.

DISCRETE-TIME CASE

Consider a discrete-time homogeneous nonlinear system

$$x^+ = f(x), \qquad x \in \mathbb{R}^n,$$

with an equilibrium point at $x^{\text{eq}} \in \mathbb{R}^n$; i.e., $f(x^{\text{eq}}) = x^{\text{eq}}$. The local linearization of (8.13) around x^{eq} is given by

$$\delta x^+ = A\,\delta x, \tag{8.20}$$

with $\delta x := x - x^{\text{eq}}$ and

$$A := \frac{\partial f(x^{\text{eq}})}{\partial x}.$$

Theorem 8.7. *Assume that the function f in (8.13) is twice differentiable.*

1. *If the linearized system (8.20) is exponentially stable, then there exists a ball B around x^{eq} such that every solution $x(t)$ to the nonlinear system (8.13) that starts at $x(0) \in B$ converges to x^{eq} exponentially fast as $t \to \infty$.*

2. *If the linearized system (8.20) is unstable, then there are solutions that start arbitrarily close to x^{eq}, but do not converge to this point as $t \to \infty$.* $\square$

8.8 STABILITY TESTS WITH MATLAB®

MATLAB® Hint 15 (`eig`). The function `eig(A)` computes the eigenvalues of the matrix `A`. Alternatively, `eig(sys)` computes the eigenvalues of the A matrix for a state-space system `sys` specified by `sys=ss(A,B,C,D)`, where `A,B,C,D` are a realization of the system. $\square$

MATLAB® Hint 26 (`lyap`). The command `P=lyap(A,Q)` solves the Lyapunov equation

$$AP + PA' = -Q. \qquad \square$$

MATLAB® Hint 27 (`dlyap`). The command `P=dlyap(A,Q)` solves the Lyapunov equation

$$APA' - P = -Q. \qquad \square$$

Attention! To solve

$$A'P + PA = -Q,$$

one needs to use `P=lyap(A',Q)`.

Attention! To solve

$$A'PA - P = -Q,$$

one needs to use `P=dlyap(A',Q)`.

8.9 EXERCISES

8.1 (Submultiplicative matrix norms). Not all matrix norms are submultiplicative. Verify that this property does not hold for the norm

$$\|A\|_\Delta := \max_{1\le i\le m}\ \max_{1\le j\le n}|a_{ij}|,$$

which explains why this norm is not commonly used.

Hint: Consider the matrices $A = B = \begin{bmatrix} 1 & 1 \\ 0 & 1 \end{bmatrix}$. □

8.2. For a given matrix A, construct vectors for which (8.2) holds for each of the three norms $\|\cdot\|_1$, $\|\cdot\|_2$, and $\|\cdot\|_\infty$. □

8.3 (Exponential of a stability matrix). Prove that when all the eigenvalues of A have strictly negative real parts, there exist constants $c, \lambda > 0$ such that

$$\|e^{At}\| \le c\, e^{-\lambda t}, \qquad \forall t \in \mathbb{R}.$$

Hint: Use the Jordan normal form. □

8.4 (Stability of LTV systems). Consider a linear system with a state-transition $\Phi(t, \tau)$ matrix for which

$$\Phi(t, 0) = \begin{bmatrix} e^t \cos 2t & e^{-2t} \sin 2t \\ -e^t \sin 2t & e^{-2t} \cos 2t \end{bmatrix}.$$

(a) Compute the state transition matrix $\Phi(t, t_0)$.

(b) Compute a matrix $A(t)$ that corresponds to the given state transition matrix.

(c) Compute the eigenvalues of $A(t)$.

(d) Classify this system in terms of Lyapunov stability.

    Hint: In answering part (d), do not be misled by your answer to part (c).

8.5 (Exponential matrix transpose). Verify that $(e^{At})' = e^{A't}$.

Hint: Use the definition of matrix exponential. □

8.6 (Stability margin). Consider the continuous-time LTI system

$$\dot{x} = Ax, \qquad x \in \mathbb{R}^n$$

and suppose that there exists a positive constant μ and positive-definite matrices $P, Q \in \mathbb{R}^n$ for the Lyapunov equation

$$A'P + PA + 2\mu P = -Q. \tag{8.21}$$

Show that all eigenvalues of A have real parts less than $-\mu$. A matrix A with this property is said to be *asymptotically stable with stability margin* μ.

Hint: Start by showing that all eigenvalues of A have real parts less than $-\mu$ if and only if all eigenvalues of $A + \mu I$ have real parts less than 0 (i.e., $A + \mu I$ is a stability matrix). □

8.7 (Stability of nonlinear systems). Investigate whether or not the solutions to the following nonlinear systems converge to the given equilibrium point when they start close enough to it.

(a) The state-space system

$$\dot{x}_1 = -x_1 + x_1(x_1^2 + x_2^2)$$
$$\dot{x}_2 = -x_2 + x_2(x_1^2 + x_2^2),$$

with equilibrium point $x_1 = x_2 = 0$.

(b) The second-order system

$$\ddot{w} + g(w)\dot{w} + w = 0,$$

with equilibrium point $w = \dot{w} = 0$. Determine for which values of $g(0)$ we can guarantee convergence to the origin based on the local linearization.

This equation is called the *Lienard equation* and can be used to model several mechanical systems, depending on the choice of the function $g(\cdot)$. $\square$

LECTURE 9

Input-Output Stability

CONTENTS

This lecture introduces a notion of stability that expresses how the magnitude of the output relates to the magnitude of the input in the absence of initial conditions.

1. Bounded-Input, Bounded-Output Stability
2. Time Domain Conditions for BIBO Stability
3. Frequency Domain Conditions for BIBO Stability
4. BIBO versus Lyapunov Stability
5. Discrete-Time Case
6. Exercises

9.1 BOUNDED-INPUT, BOUNDED-OUTPUT STABILITY

In Lecture 8 we discussed internal or Lyapunov stability, which is concerned only with the effect of the initial conditions on the response of the system. We now consider a distinct notion of stability that ignores initial conditions and is concerned only with the effect of the input on the forced response. We see below that for LTI systems these two notions of stability are closely related.

Consider the continuous-time LTV system

$$\dot{x} = A(t)x + B(t)u, \quad y = C(t)x + D(t)u, \quad x \in \mathbb{R}^n, \ u \in \mathbb{R}^k, \ y \in \mathbb{R}^m. \quad \text{(CLTV)}$$

We saw in Lecture 5 that the forced response of this system (i.e., the output for zero initial conditions) is given by

Attention! BIBO stability addresses only the solutions with zero initial conditions.

$$y_f(t) = \int_0^t C(t)\Phi(t, \tau)B(\tau)u(\tau)d\tau + D(t)u(t),$$

where $\Phi(t, \tau)$ denotes the system's state transition matrix.

Note. The factor g in (9.1) can be viewed as a system "gain." Any norm can be used in (9.1), but different norms lead to different gains g.

Definition 9.1 (BIBO stability). The system (CLTV) is said to be *(uniformly)* BIBO *stable* if there exists a finite constant g such that, for every input $u(\cdot)$, its forced response $y_f(\cdot)$ satisfies

$$\sup_{t \in [0,\infty)} \|y_f(t)\| \le g \sup_{t \in [0,\infty)} \|u(t)\|. \quad (9.1)$$

9.2 TIME DOMAIN CONDITIONS FOR BIBO STABILITY

One can analyze the (time dependent) impulse response of an LTV system to determine whether or not the system is BIBO stable.

Theorem 9.1 (Time domain BIBO stability condition). *The following two statements are equivalent.*

1. *The system (CLTV) is uniformly BIBO stable.*

2. *Every entry of $D(\cdot)$ is uniformly bounded and*

$$\sup_{t \geq 0} \int_0^t |g_{ij}(t, \tau)| d\tau < \infty, \tag{9.2}$$

for every entry $g_{ij}(t, \tau)$ of $C(t)\Phi(t, \tau)B(\tau)$. □

Notation. A signal $x(t)$ is uniformly bounded if there exists a finite constant c such that $\|x(t)\| \leq c, \forall t \geq 0$.

Proof of Theorem 9.1. We start by proving that statement 2 $\Rightarrow$ statement 1. To prove that boundedness of $D(\cdot)$ and (9.2) constitute a sufficient condition for uniform BIBO stability, we use the fact that

$$\|y_f(t)\| \leq \int_0^t \|C(t)\Phi(t, \tau)B(\tau)\| \, \|u(\tau)\| d\tau + \|D(t)\| \, \|u(t)\|, \qquad \forall t \geq 0.$$

Defining

$$\mu := \sup_{t \in [0, \infty)} \|u(t)\|, \qquad\qquad \delta := \sup_{t \in [0, \infty)} \|D(t)\|,$$

we conclude that

$$\|y_f(t)\| \leq \left(\int_0^t \|C(t)\Phi(t, \tau)B(\tau)\| d\tau + \delta \right) \mu. \qquad \forall t \geq 0.$$

Therefore (9.1) holds with

$$g := \sup_{t \geq 0} \int_0^t \|C(t)\Phi(t, \tau)B(\tau)\| d\tau + \delta.$$

Note. This is a consequence of the triangle inequality.

It remains to show that this g is finite. To do this, we note that

$$\|C(t)\Phi(t, \tau)B(\tau)\| \leq \sum_{i,j} |g_{ij}(t, \tau)|,$$

and therefore

$$\int_0^t \|C(t)\Phi(t, \tau)B(\tau)\| d\tau \leq \sum_{i,j} \int_0^t |g_{ij}(t, \tau)| d\tau, \qquad \forall t \geq 0.$$

Using (9.2), we conclude that indeed

$$g = \sup_{t \geq 0} \int_0^t \|C(t)\Phi(t, \tau)B(\tau)\| d\tau + \delta \leq \sup_{t \geq 0} \sum_{ij} \int_0^t |g_{ij}(t, \tau)| d\tau + \delta < \infty.$$

Note 6. To prove an implication $P \Rightarrow Q$ by *contraposition*, one proves instead the equivalent statement that $\neg Q \Rightarrow \neg P$. In words: if Q is false, then P must also be false. Contraposition can be viewed as a variation of a proof by contradiction (see Note 7, p. 69), since if one were to assume that the implication is false, it should be possible to have Q false and P true, which is inconsistent with the fact that $\neg Q \Rightarrow \neg P$.

It remains to prove that statement 1 $\Rightarrow$ statement 2. We prove this implication by showing that if 2 is false, then 1 must necessarily also be false, i.e., that (CLTV) cannot be BIBO stable. Suppose first that 2 is false because the entry $d_{ij}(\cdot)$ of $D(\cdot)$ is unbounded. We show next that in this case (9.1) can be violated no matter what we choose for the finite gain g. To do this, pick an arbitrary time T and consider the following step input:

$$u_T(\tau) := \begin{cases} 0 & 0 \le \tau < T \\ e_j & \tau \ge Y \end{cases} \qquad \forall \tau \ge 0,$$

where $e_j \in \mathbb{R}^k$ is the jth vector in the canonical basis of $\mathbb{R}^k$. For this input, the forced response at time T is exactly

$$y_f(T) = D(T)e_j.$$

We thus have found an input for which

$$\sup_{t \in [0,\infty)} \|u_T(t)\| = 1$$

and

$$\sup_{t \in [0,\infty)} \|y_f(t)\| \ge \|y_f(T)\| = \|D(T)e_j\| \ge |d_{ij}(T)|,$$

where the last inequality results from the fact that the norm of the vector $D(T)e_j$ must be larger than the absolute value of its ith entry, which is precisely $d_{ij}(T)$. Since $d_{ij}(\cdot)$ is unbounded, we conclude that we can make $\sup_{t \in [0,\infty)} \|y_f(t)\|$ arbitrarily large by using inputs $u_T(\cdot)$ for which $\sup_{t \in [0,\infty)} \|u_T(t)\| = 1$, which is not compatible with the existence of a finite gain g that satisfies (9.1). This means that $D(\cdot)$ must be uniformly bounded for a system to be BIBO stable.

Suppose now that 2 is false because

$$\int_0^t |g_{ij}(t, \tau)| d\tau \tag{9.3}$$

is unbounded for some i and j. Also in this case we can show that (9.1) can be violated no matter what we choose for the finite gain g. To do this, pick an arbitrary time T and consider the following "switching" input:

$$u_T(\tau) = \begin{cases} +e_j & g_{ij}(t, \tau) \ge 0 \\ -e_j & g_{ij}(t, \tau) < 0 \end{cases}, \qquad \forall \tau \ge 0.$$

For this input, the forced response at time T is given by

$$y_f(t) = \int_0^t C(t)\Phi(t, \tau)B(\tau)u(\tau)d\tau + D(t)u(t),$$

and its ith entry is equal to

$$\int_0^t |g_{ij}(t, \tau)| d\tau \pm d_{ij}(t).$$

We thus have found an input for which

$$\sup_{t \in [0,\infty)} \|u_T(t)\| = 1$$

and

$$\sup_{t \in [0,\infty)} \|y_f(t)\| \geq \|y_f(T)\| \geq \left| \int_0^t |g_{ij}(t,\tau)| d\tau \pm d_{ij}(t) \right|.$$

Since (9.3) is unbounded, also now we conclude that we can make $\sup_{t \in [0,\infty)} \|y_f(t)\|$ arbitrarily large using inputs $u_T(\cdot)$ for which $\sup_{t \in [0,\infty)} \|u_T(t)\| = 1$, which is not compatible with the existence of a finite gain g that satisfies (9.1). This means that (9.2) must also hold for a system to be BIBO stable. ∎

TIME-INVARIANT CASE

For the time-invariant system

$$\dot{x} = Ax + Bu, \qquad\qquad y = Cx + Du, \qquad\qquad \text{(CLTI)}$$

we have

$$C\Phi(t,\tau)B = Ce^{A(t-\tau)}B.$$

We can therefore rewrite (9.2) as

$$\sup_{t \geq 0} \int_0^t |\bar{g}_{ij}(t-\tau)| d\tau < \infty,$$

with the understanding that now $\bar{g}_{ij}(t-\tau)$ denotes the ijth entry of

$$Ce^{A(t-\tau)}B.$$

Making the change of variable $\rho := t - \tau$, we conclude that

$$\sup_{t \geq 0} \int_0^t |\bar{g}_{ij}(t-\tau)| = \sup_{t \geq 0} \int_0^t |\bar{g}_{ij}(\rho)| d\rho = \int_0^\infty |\bar{g}_{ij}(\rho)| d\rho.$$

Therefore Theorem 9.1 can be restated as follows.

Theorem 9.2 (Time domain BIBO LTI condition). *The following two statements are equivalent.*

1. *The system* (CLTI) *is uniformly BIBO stable.*

2. *For every entry $\bar{g}_{ij}(\rho)$ of $Ce^{A\rho}B$, we have*

$$\int_0^\infty |\bar{g}_{ij}(\rho)| d\rho < \infty.$$

□

9.3 FREQUENCY DOMAIN CONDITIONS FOR BIBO STABILITY

The Laplace transform provides a very convenient tool for studying BIBO stability. To determine whether a time-invariant system

$$\dot{x} = Ax + Bu, \qquad\qquad y = Cx + Du \qquad\qquad \text{(CLTI)}$$

is BIBO stable, we need to compute the entries $g_{ij}(t)$ of $Ce^{At}B$. To do this, we compute its Laplace transform,

$$\mathcal{L}[Ce^{At}B] = C(sI - A)^{-1}B.$$

As we saw in Lecture 4, the ijth entry of this matrix will be a strictly proper rational function of the general form

$$\hat{g}_{ij}(s) = \frac{\alpha_0 s^q + \alpha_1 s^{q-1} + \cdots + \alpha_{q-1}s + \alpha_q}{(s - \lambda_1)^{m_1}(s - \lambda_2)^{m_2} \cdots (s - \lambda_k)^{m_k}},$$

where the λ_ℓ are the (distinct) poles of $\hat{g}_{ij}(s)$ and the m_ℓ are the corresponding multiplicities. To compute the inverse Laplace transform, we need to perform a partial fraction expansion of $\hat{g}_{ij}(s)$, which is of the form

$$\begin{aligned}
\hat{g}_{ij}(s) = {} & \frac{a_{11}}{s - \lambda_1} + \frac{a_{12}}{(s - \lambda_1)^2} + \cdots + \frac{a_{1m_1}}{(s - \lambda_1)^{m_1}} + \cdots \\
& + \frac{a_{k1}}{s - \lambda_k} + \frac{a_{k2}}{(s - \lambda_k)^2} + \cdots + \frac{a_{km_k}}{(s - \lambda_k)^{m_k}}.
\end{aligned}$$

The inverse Laplace transform is then given by

$$\begin{aligned}
g_{ij}(t) = {} & \mathcal{L}^{-1}\big[\hat{g}_{ij}(s)\big] \\
= {} & a_{11}e^{\lambda_1 t} + a_{12}t\,e^{\lambda_1 t} + \cdots + a_{1m_1}t^{m_1-1}e^{\lambda_1 t} + \cdots \\
& + a_{k1}e^{\lambda_k t} + a_{k2}t\,e^{\lambda_k t} + \cdots + a_{km_k}t^{m_k-1}e^{\lambda_k t}.
\end{aligned}$$

We therefore conclude the following.

1. If for all $\hat{g}_{ij}(s)$, all the poles λ_ℓ have strictly negative real parts, then $g_{ij}(t)$ converges to zero exponentially fast and the system (CLTI) is BIBO stable.

2. If at least one of the $\hat{g}_{ij}(s)$ has a pole λ_ℓ with a zero or positive real part, then $|g_{ij}(t)|$ does not converge to zero and the system (CLTI) is not BIBO stable.

Although $\hat{g}_{ij}$ is not an entry of the transfer function of (CLTI) (because the D term is missing from its definition), adding a constant D will not change its poles. Therefore the conclusions above can be restated as follows.

Theorem 9.3 (Frequency domain BIBO condition). *The following two statements are equivalent:*

1. *The system* (CLTI) *is uniformly BIBO stable.*

2. *Every pole of every entry of the transfer function of the system* (CLTI) *has a strictly negative real part.*   □

9.4 BIBO VERSUS LYAPUNOV STABILITY

We saw in Theorem 9.2 that the LTI system

$$\dot{x} = Ax + Bu, \qquad\qquad y = Cx + Du \qquad\qquad \text{(CLTI)}$$

is uniformly BIBO stable if and only if every entry $\bar{g}_{ij}(t)$ of $Ce^{At}B$ satisfies

$$\int_0^\infty |\bar{g}_{ij}(t)|dt < \infty. \qquad\qquad (9.4)$$

However, if the system (CLTI) is exponentially stable, then every entry of e^{At} converges to zero exponentially fast and therefore (9.4) must hold.

Theorem 9.4. *When the system (CLTI) is exponentially stable, then it must also be BIBO stable.* $\qquad\square$

Attention! In general, the converse of Theorem 9.4 is not true, because there are systems that are BIBO stable but not exponentially stable. This can happen when the premultiplication of e^{At} by C and/or the postmultiplication by B cancel terms in e^{At} that are not converging to zero exponentially fast. This occurs, e.g., for the system

$$\dot{x} = \begin{bmatrix} 1 & 0 \\ 0 & -2 \end{bmatrix} x + \begin{bmatrix} 0 \\ 1 \end{bmatrix} u, \qquad\qquad y = \begin{bmatrix} 1 & 1 \end{bmatrix} x,$$

Note. We see below in Lectures 17 (SISO) and 19 (MIMO) that this discrepancy between Lyapunov and BIBO stability is always associated with lack of controllability or observability, two concepts that will be introduced shortly. In this example, the system is not controllable.

for which

$$e^{At} = \begin{bmatrix} e^t & 0 \\ 0 & e^{-2t} \end{bmatrix}$$

is unbounded and therefore Lyapunov unstable, but

$$Ce^{At}B = \begin{bmatrix} 1 & 1 \end{bmatrix}\begin{bmatrix} e^t & 0 \\ 0 & e^{-2t} \end{bmatrix}\begin{bmatrix} 0 \\ 1 \end{bmatrix} = e^{-2t},$$

and therefore the system is BIBO stable. $\qquad\square$

9.5 DISCRETE-TIME CASE

Consider now the following discrete-time LTV system

$$x(t+1) = A(t)x(t) + B(t)u(t), \qquad y(t) = C(t)x(t) + D(t)u(t). \qquad \text{(DLTV)}$$

We saw in Lecture 5 that the forced response of this system is given by

$$y_f(t) = \sum_{\tau=0}^{t-1} C(t)\Phi(t, \tau+1)B(\tau)u(\tau)d\tau + D(t)u(t), \qquad \forall t \geq 0,$$

where $\Phi(t, \tau)$ denotes the system's discrete-time state transition matrix. The discrete-time definition of BIBO stability is essentially identical to the continuous-time one.

Attention! BIBO
stability addresses
only the solutions
with zero initial
conditions.

Note. The factor g
can be viewed as the
"gain" of the system.

Definition 9.2 (BIBO stability). The system (DLTV) is said to be *(uniformly) BIBO stable* if there exists a finite constant g such that, for every input $u(\cdot)$, its forced response $y_f(\cdot)$ satisfies

$$\sup_{t \in \mathbb{N}} \|y_f(t)\| \leq g \sup_{t \in \mathbb{N}} \|u(t)\|. \qquad \square$$

Theorem 9.5 (Time domain BIBO condition). *The following two statements are equivalent.*

1. *The system* (DLTV) *is uniformly BIBO stable.*

2. *Every entry of $D(\cdot)$ is uniformly bounded and*

$$\sup_{t \geq 0} \sum_{\tau=0}^{t-1} |g_{ij}(t, \tau)| < \infty$$

   for every entry $g_{ij}(t, \tau)$ of $C(t)\Phi(t, \tau)B(\tau)$. $\qquad \square$

For the following time-invariant discrete-time system

$$x^+ = Ax + Bu, \qquad\qquad y = Cx + Du, \qquad\qquad \text{(DLTI)}$$

one can conclude that the following result holds.

Theorem 9.6 (BIBO LTI conditions). *The following three statements are equivalent.*

1. *The system* (DLTI) *is uniformly BIBO stable.*

2. *For every entry $\bar{g}_{ij}(\rho)$ of $C A^\rho B$, we have*

$$\sum_{\rho=1}^{\infty} |\bar{g}_{ij}(\rho)| < \infty.$$

3. *Every pole of every entry of the transfer function of the system* (DLTI) *has magnitude strictly smaller than 1.* $\qquad \square$

9.6 EXERCISES

9.1. Consider the system

$$\dot{x} = \begin{bmatrix} -2 & 0 & 0 \\ 0 & 1 & 0 \\ 0 & 0 & -1 \end{bmatrix} x + := \begin{bmatrix} 1 \\ 0 \\ -1 \end{bmatrix} u, \qquad y = \begin{bmatrix} 1 & 1 & 0 \end{bmatrix} x + u.$$

(a) Compute the system's transfer function.

(b) Is the matrix A asymptotically stable, marginally stable, or unstable?

(c) Is this system BIBO stable? $\qquad \square$

LECTURE 10

Preview of Optimal Control

CONTENTS

This lecture provides a brief introduction to optimal control. Its main goal is to motivate several of the questions that will be addressed in subsequent lectures. This material is discussed in much greater detail in Lecture 20.

1. The Linear Quadratic Regulator Problem
2. Feedback Invariants
3. Feedback Invariants in Optimal Control
4. Optimal State Feedback
5. LQR with MATLAB®
6. Exercises

10.1 THE LINEAR QUADRATIC REGULATOR PROBLEM

Given a continuous-time LTI system

Note. For simplicity, here we assume that the D matrix is zero.

$$\dot{x} = Ax + Bu, \qquad\qquad y = Cx,$$

the *linear quadratic regulation (LQR)* problem consists of finding the control signal $u(t)$ that makes the following criterion as small as possible:

$$J_{\text{LQR}} := \int_0^\infty y(t)'Qy(t) + u(t)'Ru(t) \; dt, \tag{10.1}$$

where Q and R are positive-definite weighting matrices. The term

$$\int_0^\infty y(t)'Qy(t)dt$$

provides a measure of the *output energy*, and the term

$$\int_0^\infty u(t)'Ru(t)dt$$

provides a measure of the *control signal energy*. In LQR one seeks a controller that minimizes both energies. However, decreasing the energy of the output requires a large control signal, and a small control signal leads to large outputs. The role of the weighting matrices Q and R is to establish a trade-off between these conflicting goals.

1. When R is much larger than Q, the most effective way to decrease J_{LQR} is to employ a small control input, at the expense of a large output.

2. When R is much smaller than Q, the most effective way to decrease J_{LQR} is to obtain a very small output, even if this is achieved at the expense of employing a large control input.

10.2   FEEDBACK INVARIANTS

Given a continuous-time LTI system

$$\dot{x} = Ax + Bu, \qquad y = Cx, \qquad x \in \mathbb{R}^n,\ u \in \mathbb{R}^k,\ y \in \mathbb{R}^m, \qquad \text{(CLTI)}$$

Note. A *functional* maps functions (in this case signals; i.e., functions of time) to scalar values (in this case real numbers).

we say that a functional

$$H\big(x(\cdot); u(\cdot)\big)$$

that involves the system's input and state is a *feedback invariant* for the system (CLTI) if, when computed along a solution to the system, its value depends only on the initial condition $x(0)$ and not on the specific input signal $u(\cdot)$.

Proposition 10.1 (Feedback invariant). *For every symmetric matrix P, the functional*

$$H\big(x(\cdot); u(\cdot)\big) := -\int_0^\infty \big(Ax(t) + Bu(t)\big)' Px(t) + x(t)' P\big(Ax(t) + Bu(t)\big)\, dt$$

is a feedback invariant for (CLTI), as long as $\lim_{t\to\infty} x(t) = 0$.   □

Proof of Proposition 10.1. We can rewrite H as

$$H\big(x(\cdot); u(\cdot)\big) = -\int_0^\infty \dot{x}(t)' Px(t) + x(t)' P\dot{x}(t)\, dt$$

$$= -\int_0^\infty \frac{d\big(x(t)' Px(t)\big)}{dt}\, dt = x(0)' Px(0) - \lim_{t\to\infty} x(t)' Px(t) = x(0)' Px(0),$$

as long as $\lim_{t\to\infty} x(t) = 0$.   ∎

10.3   FEEDBACK INVARIANTS IN OPTIMAL CONTROL

Suppose that we are able to express a criterion J to be minimized by an appropriate choice of the input $u(\cdot)$ in the following form:

$$J = H\big(x(\cdot); u(\cdot)\big) + \int_0^\infty \Lambda\big(x(t), u(t)\big) dt, \qquad (10.2)$$

where H is a feedback invariant and the function $\Lambda(x, u)$ has the property that for every $x \in \mathbb{R}^n$

$$\min_{u \in \mathbb{R}^k} \Lambda(x, u) = 0.$$

In this case, the control

$$u(t) = \arg\min_{u \in \mathbb{R}^k} \Lambda(x, u),$$

Note. If one wants to restrict the optimization to solutions that lead to an asymptotically stable closed-loop system, then H needs to be a feedback invariant only for inputs that lead to $x(t) \to 0$ (as in Proposition 10.1).

will minimize the criterion J, and the optimal value of J is equal to the feedback invariant

$$J = H\big(x(\cdot); u(\cdot)\big).$$

Note that it is not possible to get a lower value for J since (1) the feedback invariant $H\big(x(\cdot); u(\cdot)\big)$ is never affected by u and (2) a smaller value for J would require the integral in the right-hand side of (10.2) to be negative, which is not possible, since $\Lambda\big(x(t), u(t)\big)$ can at best be as low as zero.

10.4 OPTIMAL STATE FEEDBACK

It turns out that the LQR criterion can be expressed as in (10.2) for an appropriate choice of feedback invariant. In fact, the feedback invariant in Proposition 10.1 will work, provided that we choose the matrix P appropriately. To check that this is so, we add and subtract this feedback invariant to the LQR criterion and conclude that

$$
\begin{aligned}
J_{\text{LQR}} &:= \int_0^\infty x'C'QC'x + u'Ru \; dt \\
&= H\big(x(\cdot); u(\cdot)\big) \\
&\quad + \int_0^\infty x'C'QC'x + u'Ru + (Ax + Bu)'Px + x'P(Ax + Bu) \; dt \\
&= H\big(x(\cdot); u(\cdot)\big) + \int_0^\infty x'(A'P + PA + C'QC')x + u'Ru + 2u'B'Px \; dt.
\end{aligned}
$$

By completing the squares as follows, we group the quadratic term in u with the cross-term in u times x:

$$(u' + x'K')R(u + Kx) = u'Ru + x'PBR^{-1}B'Px + 2u'B'Px, \quad K := R^{-1}B'P,$$

from which we conclude that

$$
\begin{aligned}
J_{\text{LQR}} = H\big(x(\cdot); u(\cdot)\big) + \int_0^\infty & x'(A'P + PA + C'QC' - PBR^{-1}B'P)x \\
& + (u' + x'K')R(u + Kx) \; dt.
\end{aligned}
$$

If we are able to select the matrix P so that

$$A'P + PA + C'QC - PBR^{-1}B'P = 0,$$

we obtain precisely an expression such as (10.2) with

$$\Lambda(x, u) := (u' + x'K')R(u + Kx),$$

which has a minimum equal to zero for

$$u = -Kx, \qquad\qquad K := R^{-1}B'P, \qquad\qquad (10.3)$$

leading to the following closed loop:

$$\dot{x} = Ax + BKx = (A - BR^{-1}B'P)x.$$

The following was proved:

Theorem 10.1. *Assume that there exists a symmetric solution P to the following* alge-braic Riccati equation (ARE)

$$A'P + PA + C'QC - PBR^{-1}B'P = 0 \qquad\qquad (10.4)$$

for which $A - BR^{-1}B'P$ is a stability matrix. Then the feedback law (10.3) stabilizes the closed-loop system while minimizing the LQR criterion (10.1). □

Attention! Several questions still remain open.

1. Under what conditions does the LQR problem have a solution?

   Intuitively, the answer to this question should be "as long as there exists at least one signal u that takes y to zero with finite energy."

2. Under what conditions does the ARE (10.4) have a symmetric solution that leads to an asymptotically stable closed-loop system?

   One would like the answer to this question to coincide with the answer to the previous one, because this would mean that one could always solve the LQR problem by solving an ARE. This is "almost" true . . . .

These questions will be resolved in Part VI, where we revisit the LQR problem in much more detail. □

Attention! The ARE itself already provides some clues about whether or not the closed-loop system is stable. Indeed, if we write the Lyapunov equation for the closed loop and use (10.4), we get

$$(A - BR^{-1}B'P)'P + P(A - BR^{-1}B'P) = A'P + PA - 2PBR^{-1}B'P$$
$$= -\bar{Q} \leq 0$$

for $\bar{Q} := C'QC + PBR^{-1}B'P \geq 0$. In case $P > 0$ and $\bar{Q} > 0$, we could immediately conclude that the closed loop was stable by the Lyapunov stability theorem (Theorem 8.2). □

MATLAB® Hint 28. `lqr` solves the ARE and computes the optimal state feedback controller gain K. ▶ p. 91

Note. Asymptotic stability of the closed-loop system is needed because we assumed that $\lim_{t \to \infty} x(t)Px(t) = 0$.

10.5 LQR WITH MATLAB®

MATLAB® Hint 28 (lqr).   The command [K,S,E]=lqr(A,B,QQ,RR,NN) computes the optimal state feedback LQR controller for the process

$$\dot{x} = Ax + Bu$$

with the criterion

$$J := \int_0^\infty x(t)'QQx(t) + u'(t)RRu(t) + 2x'(t)NNu(t)dt.$$

For the criterion in (10.1), one should select

$$QQ = C'QC, \qquad\qquad RR = R, \qquad\qquad NN = 0.$$

This command returns the optimal state feedback matrix K, the solution P to the corresponding algebraic Riccati equation, and the poles E of the closed-loop system. □

10.6 EXERCISES

10.1 (Hamiltonian).   Consider the following LTI SISO system

$$\dot{x} = Ax + bu, \qquad\qquad y = cx, \qquad\qquad x \in \mathbb{R}^n, \ u, y \in \mathbb{R}.$$

(a) Show that when the matrix

$$\mathcal{O} := \begin{bmatrix} c \\ cA \\ \vdots \\ cA^{n-1} \end{bmatrix} \in \mathbb{R}^{n \times n}$$

is nonsingular, then the null space of the matrix $\begin{bmatrix} A - \lambda I \\ c \end{bmatrix} \in \mathbb{R}^{(n+1) \times n}$ contains only the zero vector, for every $\lambda \in \mathbb{C}$.

Hint: Prove the statement by contradiction.

(b) Show that if $x := \begin{bmatrix} x_1 \\ x_2 \end{bmatrix}$, $x_1, x_2 \in \mathbb{C}^n$ is an eigenvector of a matrix $H \in \mathbb{R}^{2n \times 2n}$ associated with an eigenvalue $\lambda := j\omega$ over the imaginary axis, then

$$\begin{bmatrix} x_2^* & x_1^* \end{bmatrix} Hx + (Hx)^* \begin{bmatrix} x_2 \\ x_1 \end{bmatrix} = 0, \tag{10.5}$$

where $(\cdot)^*$ denotes the complex conjugate transpose.

Hint: Note that the order of the indexes of x_1 and x_2 in (10.5) is opposite to the order in the definition of x.

(c) Show that if $x := \begin{bmatrix} x_1 \\ x_2 \end{bmatrix}$, $x_1, x_2 \in \mathbb{C}^n$ is an eigenvector of

$$H := \begin{bmatrix} A & -bb^T \\ -c^T c & -A^T \end{bmatrix}$$

associated with an eigenvalue $\lambda := j\omega$ over the imaginary axis, then $b^T x_2 = 0$ and $cx_1 = 0$.

Hint: Use equation (10.5) in part (b) and do not get the indexes of x_1 and x_2 exchanged by mistake.

(d) Show that if for every $\lambda \in \mathbb{C}$ the null spaces of the matrices

$$\begin{bmatrix} A - \lambda I \\ c \end{bmatrix} \in \mathbb{R}^{(n+1)\times n} \quad \text{and} \quad \begin{bmatrix} A^T - \lambda I \\ b^T \end{bmatrix} \in \mathbb{R}^{(n+1)\times n}$$

contain only the zero vector, then

$$H := \begin{bmatrix} A & -bb^T \\ -c^T c & -A^T \end{bmatrix}$$

cannot have any eigenvalues over the imaginary axis.

Hint: Do the proof by contradiction, using the result stated in part (c).

It may make you happy to know that if you succeeded in solving the exercises above, you have figured out the main steps in the proof of the following important theorem in optimal control, a generalization of which we find in Lecture 21.

Theorem 10.2. *For every realization $A \in \mathbb{R}^{n\times n}$, $b \in \mathbb{R}^{n\times 1}$, $c \in \mathbb{R}^{1\times n}$, for which the matrices*

Note. In this case, we say that the triple (A, b, c) is a minimal realization (cf. Lecture 17).

$$\mathcal{C} := \begin{bmatrix} b & Ab & \cdots & A^{n-1}b \end{bmatrix}, \quad \mathcal{O} := \begin{bmatrix} c \\ cA \\ \vdots \\ cA^{n-1} \end{bmatrix} \in \mathbb{R}^{n\times n}$$

are nonsingular, the Hamiltonian matrix $H := \begin{bmatrix} A & -bb^T \\ -c^T c & -A^T \end{bmatrix}$ has no eigenvalues over the imaginary axis. □

PART III
CONTROLLABILITY AND STATE FEEDBACK

LECTURE 11

Controllable and Reachable Subspaces

CONTENTS

This lecture introduces the notions of controllability and reachability, which are the basis of all state-space control design methods.

1. Controllable and Reachable Subspaces
2. Physical Examples and System Interconnections
3. Fundamental Theorem of Linear Equations (review)
4. Reachability and Controllability Gramians
5. Open-Loop Minimum-Energy Control
6. Controllability Matrix (LTI)
7. Discrete-Time Case
8. MATLAB® Commands
9. Exercise

11.1 CONTROLLABLE AND REACHABLE SUBSPACES

Consider the continuous-time LTV system

$$\dot{x} = A(t)x + B(t)u, \quad y = C(t)x + D(t)u, \quad x \in \mathbb{R}^n,\ u \in \mathbb{R}^k,\ y \in \mathbb{R}^m. \quad \text{(CLTV)}$$

We saw in Lecture 5 that a given input $u(\cdot)$ transfers the state $x(t_0) := x_0$ at time t_0 to the state $x(t_1) := x_1$ at time t_1 given by the variation of constants formula,

$$x_1 = \Phi(t_1, t_0)x_0 + \int_{t_0}^{t_1} \Phi(t_1, \tau)B(\tau)u(\tau)d\tau,$$

where $\Phi(\cdot)$ denotes the system's state transition matrix. The following two definitions express how powerful the input is in terms of transferring the state between two given states.

Definition 11.1 (Reachable subspace). Given two times $t_1 > t_0 \geq 0$, the *reachable* or *controllable-from-the-origin on* $[t_0, t_1]$ subspace $\mathcal{R}[t_0, t_1]$ consists of all states x_1 for which there exists an input $u : [t_0, t_1] \to \mathbb{R}^k$ that transfers the state from $x(t_0) = 0$ to $x(t_1) = x_1$; i.e.,

$$\mathcal{R}[t_0, t_1] := \left\{ x_1 \in \mathbb{R}^n : \exists u(.), \ x_1 = \int_{t_0}^{t_1} \Phi(t_1, \tau) B(\tau) u(\tau) d\tau \right\}. \qquad \square$$

Definition 11.2 (Controllable subspace). Given two times $t_1 > t_0 \geq 0$, the *controllable* or *controllable-to-the-origin on* $[t_0, t_1]$ subspace $\mathcal{C}[t_0, t_1]$ consists of all states x_0 for which there exists an input $u : [t_0, t_1] \to \mathbb{R}^k$ that transfers the state from $x(t_0) = x_0$ to $x(t_1) = 0$; i.e.,

$$\mathcal{C}[t_0, t_1] := \left\{ x_0 \in \mathbb{R}^n : \exists u(.), \ 0 = \Phi(t_1, t_0) x_0 + \int_{t_0}^{t_1} \Phi(t_1, \tau) B(\tau) u(\tau) d\tau \right\}. \qquad \square$$

The matrices $C(\cdot)$ and $D(\cdot)$ play no role in these definitions; therefore, one often simply talks about the reachable or controllable subspaces of the system

$$\dot{x} = A(t) x + B(t) u, \qquad x \in \mathbb{R}^n, \ u \in \mathbb{R}^k, \qquad \text{(AB-CLTV)}$$

or of the pair $\big(A(\cdot), B(\cdot) \big)$.

Attention! Determining the reachable subspace amounts to finding for which vectors $x_1 \in \mathbb{R}^n$ the equation

$$x_1 = \int_{t_0}^{t_1} \Phi(t_1, \tau) B(\tau) u(\tau) d\tau \qquad (11.1)$$

Note 9. Note that the two equations (11.1) and (11.2) differ only by exchanging $\Phi(t_1, \tau)$ with $\Phi(t_0, \tau)$ and $u(\cdot)$ with $v(\cdot) := -u(\cdot)$.

has a solution $u(\cdot)$. Similarly, determining the controllable subspace amounts to finding for which vectors $x_1 \in \mathbb{R}^n$ the equation

$$0 = \Phi(t_1, t_0) x_0 + \int_{t_0}^{t_1} \Phi(t_1, \tau) B(\tau) u(\tau) d\tau \quad \Leftrightarrow \quad x_0 = \int_{t_0}^{t_1} \Phi(t_0, \tau) B(\tau) v(\tau) d\tau \tag{11.2}$$

has a solution $v(\cdot) := -u(\cdot)$. $\qquad \square$

11.2   PHYSICAL EXAMPLES AND SYSTEM INTERCONNECTIONS

Example 11.1 (Parallel RC network). The state-space model of the electrical network in Figure 11.1(a) is given by

$$\dot{x} = \begin{bmatrix} -\frac{1}{R_1 C_1} & 0 \\ 0 & -\frac{1}{R_2 C_2} \end{bmatrix} x + \begin{bmatrix} \frac{1}{R_1 C_1} \\ \frac{1}{R_2 C_2} \end{bmatrix} u.$$

The solution to this system is given by

$$x(t) = \begin{bmatrix} e^{-\frac{t}{R_1 C_1}} x_1(0) \\ e^{-\frac{t}{R_2 C_2}} x_2(0) \end{bmatrix} + \int_0^t \begin{bmatrix} \frac{e^{-\frac{t-\tau}{R_1 C_1}}}{R_1 C_1} \\ \frac{e^{-\frac{t-\tau}{R_2 C_2}}}{R_2 C_2} \end{bmatrix} u(\tau) d\tau.$$

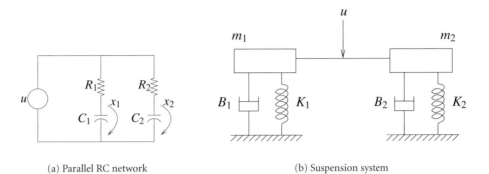

(a) Parallel RC network   (b) Suspension system

Figure 11.1. Uncontrollable systems.

When the two branches have the same time constant, i.e., $\frac{1}{R_1 C_1} = \frac{1}{R_2 C_2} = \omega$, we have

$$x(t) = e^{-\omega t} x(0) + \omega \int_0^t e^{-\omega(t-\tau)} u(\tau) d\tau \begin{bmatrix} 1 \\ 1 \end{bmatrix}.$$

This shows that if $x(0) = 0$, then $x(t)$ is always of the form

$$x(t) = \alpha(t) \begin{bmatrix} 1 \\ 1 \end{bmatrix}, \qquad \alpha(t) := \omega \int_0^t e^{-\omega(t-\tau)} u(\tau) d\tau,$$

and we cannot transfer the system from the origin to any state with $x_1(t) \neq x_2(t)$. The reachable subspace for this system is

$$\mathcal{R}[t_0, t_1] = \left\{ \alpha \begin{bmatrix} 1 \\ 1 \end{bmatrix} : \alpha \in \mathbb{R} \right\}, \quad \forall t_1 > t_0 \geq 0.$$

Suppose now that we want to transfer $x(0)$ to the origin. Then we need

$$0 = e^{-\omega t} x(0) + \alpha(t) \begin{bmatrix} 1 \\ 1 \end{bmatrix}, \qquad \alpha(t) := \omega \int_0^t e^{-\omega(t-\tau)} u(\tau) d\tau.$$

Clearly, this is possible only if $x(0)$ is aligned with $\begin{bmatrix} 1 & 1 \end{bmatrix}'$. The controllable subspace for this system is also

$$\mathcal{C}[t_0, t_1] = \left\{ \alpha \begin{bmatrix} 1 \\ 1 \end{bmatrix} : \alpha \in \mathbb{R} \right\}, \quad \forall t_1 > t_0 \geq 0.$$

However, we shall see shortly that when the time constants are different; i.e., $\frac{1}{R_1 C_1} \neq \frac{1}{R_2 C_2}$, any vector in $\mathbb{R}^2$ can be reached from the origin and the origin can be reached from any initial condition in $\mathbb{R}^2$; i.e.,

$$\mathcal{R}[t_0, t_1] = \mathcal{C}[t_0, t_1] = \mathbb{R}^2. \qquad \qquad \square$$

Example 11.2 (Suspension system). The state-space model of the mechanical suspension system in Figure 11.1(b) is given by

$$
\dot{x} = \begin{bmatrix} -\frac{b_1}{m_1} & -\frac{k_1}{m_1} & 0 & 0 \\ 1 & 0 & 0 & 0 \\ 0 & 0 & -\frac{b_2}{m_2} & -\frac{k_2}{m_2} \\ 0 & 0 & 1 & 0 \end{bmatrix} x + \begin{bmatrix} \frac{1}{2m_1} \\ 0 \\ \frac{1}{2m_2} \\ 0 \end{bmatrix} u,
$$

where $x := \begin{bmatrix} \dot{x}_1 & x_1 & \dot{x}_2 & x_2 \end{bmatrix}'$, and x_1 and x_2 are the spring displacements with respect to the equilibrium position. We assumed that the bar has negligible mass and therefore the force u is equally distributed between the two spring systems. □

This and the previous examples are special cases of the parallel connection in Figure 11.2(a), which is discussed next.

Example 11.3 (Parallel interconnection). Consider the parallel connection in Figure 11.2(a) of two systems with states $x_1, x_2 \in \mathbb{R}^n$. The overall system corresponds to the state-space model

$$
\dot{x} = \begin{bmatrix} A_1 & 0 \\ 0 & A_2 \end{bmatrix} x + \begin{bmatrix} B_1 \\ B_2 \end{bmatrix} u,
$$

where we chose for state $x := \begin{bmatrix} x_1' & x_2' \end{bmatrix}' \in \mathbb{R}^{2n}$. The solution to this system is given by

$$
x(t) = \begin{bmatrix} e^{A_1 t} x_1(0) \\ e^{A_2 t} x_2(0) \end{bmatrix} + \int_0^t \begin{bmatrix} e^{A_1(t-\tau)} B_1 \\ e^{A_2(t-\tau)} B_2 \end{bmatrix} u(\tau) d\tau.
$$

When $A_1 = A_2 = A$ and $B_1 = B_2 = B$, we have

$$
x(t) = \begin{bmatrix} e^{At} x_1(0) \\ e^{At} x_2(0) \end{bmatrix} + \begin{bmatrix} I \\ I \end{bmatrix} \int_0^t e^{A(t-\tau)} Bu(\tau) d\tau.
$$

This shows that if $x(0) = 0$, we cannot transfer the system from the origin to any state with $x_1(t) \neq x_2(t)$. Similarly, to transfer a state $x(t_0)$ to the origin, we must have $x_1(t_0) = x_2(t_0)$. □

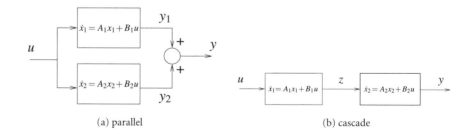

(a) parallel   (b) cascade

Figure 11.2. Block interconnections.

Attention! Parallel connections of similar systems are a common mechanism (but certainly not the only one) that leads to lack of reachability and controllability. Cascade connections, as in Figure 11.2(b), generally do not have this problem. However, they may lead to stability problems through resonance, as seen in Lecture 7. □

11.3 FUNDAMENTAL THEOREM OF LINEAR EQUATIONS (REVIEW)

Note. It is important to recognize Wx as a linear combination of the columns of W, with the coefficients of the linear combination given by the entries of the vector x.

Note. How to compute the image of a matrix W? A basis for the image is obtained by keeping only its linearly independent columns.

Note. How to compute the kernel of a matrix? Solve the equation $Wx = 0$ and find the vectors for which it has a solution.

MATLAB® Hint 29. svd(W) can be used to compute bases for ImW and kerW. ▶ p. 109

Given an $m \times n$ matrix W, the *range* or *image* is the set of vectors $y \in \mathbb{R}^m$ for which $y = Wx$ has a solution $x \in \mathbb{R}^n$; i.e.,

$$\text{Im } W := \left\{ y \in \mathbb{R}^m : \exists x \in \mathbb{R}^n, \ y = Wx \right\}.$$

The image of W is a linear subspace of $\mathbb{R}^m$, and its dimension is called the *rank* of the matrix W. The rank of W is equal to the number of linearly independent columns of W, which is also equal to the number of linearly independent rows of W.

The *kernel* or *null space* is the set

$$\ker W := \left\{ x \in \mathbb{R}^n : Wx = 0 \right\}.$$

The kernel of W is a linear subspace of $\mathbb{R}^n$, and its dimension is called the *nullity* of the matrix W. The following theorem relates the range and nullity of a matrix.

Theorem 11.1 (Fundamental theorem of linear equations). *For every $m \times n$ matrix W,*

$$\dim \ker W + \dim \text{Im } W = n. \qquad \square$$

The Fundamental theorem of linear equations is covered extensively, e.g., in [10, 14].

There exists a simple relationship between the kernel and image spaces. The *orthogonal complement* $\mathcal{V}^\perp$ of a linear subspace $\mathcal{V} \subset \mathbb{R}^n$ is the set of all vectors that are orthogonal to every vector in $\mathcal{V}$; i.e.,

$$\mathcal{V}^\perp = \left\{ x \in \mathbb{R}^n : x'z = 0, \ \forall z \in \mathcal{V} \right\}.$$

Lemma 11.1 (Range versus null space). *For every $m \times n$ matrix W,*

$$\text{Im } W = (\ker W')^\perp, \qquad \qquad \ker W = (\text{Im } W')^\perp. \qquad \square$$

Proof of Lemma 11.1. Assuming that $x \in \text{Im } W$,

$$\left. \begin{array}{ll} x \in \text{Im } W & \Rightarrow \quad \exists \eta : x = W\eta \\ z \in \ker W' & \Rightarrow \quad W'z = 0 \end{array} \right\} \quad \Rightarrow \quad z'x = z'W\eta = 0.$$

Note. For any subspace $\mathcal{V}$, one can show that $(\mathcal{V}^\perp)^\perp = \mathcal{V}$.

Therefore x is orthogonal to every vector in $\ker W'$, which means that $x \in (\ker W')^\perp$. We thus conclude that $\text{Im } W \subset (\ker W')^\perp$.

However,

$$\left. \begin{array}{l} \dim \ker W' + \dim(\ker W')^\perp = n \\ \dim \ker W' + \dim \text{Im } W' = n \end{array} \right\}$$

$$\Rightarrow \quad \dim \text{Im } W = \text{rank } W = \dim \text{Im } W' = \dim(\ker W')^\perp,$$

and therefore $\text{Im } W$ cannot be a strict subset of $(\ker W')^\perp$. ■

11.4 REACHABILITY AND CONTROLLABILITY GRAMIANS

The following definitions are useful in characterizing the reachable and controllable subspaces.

Definition 11.3 (Reachability and controllability Gramians). Given two times $t_1 > t_0 \geq 0$, the *reachability* and *controllability Gramians* of the system (AB-CLTV) are defined, respectively, by

$$W_R(t_0, t_1) := \int_{t_0}^{t_1} \Phi(t_1, \tau) B(\tau) B(\tau)' \Phi(t_1, \tau)' d\tau,$$

$$W_C(t_0, t_1) := \int_{t_0}^{t_1} \Phi(t_0, \tau) B(\tau) B(\tau)' \Phi(t_0, \tau)' d\tau. \qquad \square$$

As the name suggests, the reachability Gramian allows one to compute the reachable subspace.

Theorem 11.2 (Reachable subspace). *Given two times $t_1 > t_0 \geq 0$,*

$$\mathcal{R}[t_0, t_1] = \mathrm{Im}\, W_R(t_0, t_1),$$

Moreover, if $x_1 = W_R(t_0, t_1)\eta_1 \in \mathrm{Im}\, W_R(t_0, t_1)$, the control

$$u(t) = B(t)' \Phi(t_1, t)' \eta_1 \qquad t \in [t_0, t_1] \tag{11.3}$$

can be used to transfer the state from $x(t_0) = 0$ to $x(t_1) = x_1$. $\qquad \square$

Proof of Theorem 11.2. We start by showing that $x_1 \in \mathbf{Im}\, W_R(t_0, t_1) \Rightarrow x_1 \in \mathcal{R}[t_0, t_1]$. When $x_1 \in \mathrm{Im}\, W_R(t_0, t_1)$, there exists a vector $\eta_1 \in \mathbb{R}^n$ such that

$$x_1 = W_R(t_0, t_1)\eta_1.$$

To prove that $x_1 \in \mathcal{R}[t_0, t_1]$, it suffices to show that the input (11.3) does indeed transfer the state from $x(t_0) = 0$ to $x(t_1) = x_1$, and therefore $x_1 \in \mathcal{R}[t_0, t_1]$. To verify that this is so, we use the variation of constants formula for the input (11.3):

$$x(t_1) = \int_{t_0}^{t_1} \Phi(t_1, \tau) B(\tau) \underbrace{B(\tau)' \Phi(t_1, \tau)' \eta_1}_{u(\tau)} d\tau = W_R(t_0, t_1)\eta_1 = x_1.$$

We now show that $x_1 \in \mathcal{R}[t_0, t_1] \Rightarrow x_1 \in \mathbf{Im}\, W_R(t_0, t_1)$. When $x_1 \in \mathcal{R}[t_0, t_1]$, there exists an input $u(\cdot)$ for which

$$x_1 = \int_{t_0}^{t_1} \Phi(t_1, \tau) B(\tau) u(\tau) d\tau.$$

We show next that this leads to $x_1 \in \mathrm{Im}\, W_R(t_0, t_1) = (\ker W_R(t_0, t_1))^{\perp}$, which is to say that

$$x_1' \eta_1 = 0, \quad \forall \eta_1 \in \ker W_R(t_0, t_1). \tag{11.4}$$

Note. Both Gramians are symmetric positive-semidefinite $n \times n$ matrices; i.e., $W_R(t_0, t_1) = W_R(t_0, t_1)'$ and $x' W_R(t_0, t_1)x \geq 0$, $\forall x \in \mathbb{R}^n$. Similarly for W_C.

Note. To prove that two sets $\mathcal{A}$ and $\mathcal{B}$ are equal, one generally starts by showing that $\mathcal{A} \subset \mathcal{B}$ and then that $\mathcal{B} \subset \mathcal{A}$. The former amounts to showing that

$$x \in \mathcal{A} \Rightarrow x \in \mathcal{B}$$

and the latter that

$$x \in \mathcal{B} \Rightarrow x \in \mathcal{A}.$$

To verify that this is so, we pick some arbitrary vector $\eta_1 \in \ker W_R(t_0, t_1)$ and compute

$$x_1' \eta_1 = \int_{t_0}^{t_1} u(\tau)' B(\tau)' \Phi(t_1, \tau)' \eta_1 d\tau. \tag{11.5}$$

But since $\eta_1 \in \ker W_R(t_0, t_1)$, we have

$$\eta_1' W_R(t_0, t_1) \eta_1 = \int_{t_0}^{t_1} \eta_1' \Phi(t_1, \tau) B(\tau) B(\tau)' \Phi(t_1, \tau)' \eta_1 d\tau$$

$$= \int_{t_0}^{t_1} \| B(\tau)' \Phi(t_1, \tau)' \eta_1 \|^2 d\tau = 0,$$

which implies that

$$B(\tau)' \Phi(t_1, \tau)' \eta_1 = 0, \quad \forall \tau \in [t_0, t_1].$$

From this and (11.5), we conclude that (11.4) indeed holds.   ∎

A similar result can be proved for the controllable subspace.

Theorem 11.3 (Controllable subspace). *Given two times $t_1 > t_0 \geq 0$,*

$$\mathcal{C}[t_0, t_1] = \operatorname{Im} W_C(t_0, t_1).$$

Moreover, if $x_0 = W_C(t_0, t_1) \eta_0 \in \operatorname{Im} W_C(t_0, t_1)$, the control

Note. Cf. Note 9
(p. 96).
$$u(t) = -B(t)' \Phi(t_0, t)' \eta_0, \qquad t \in [t_0, t_1] \tag{11.6}$$

can be used to transfer the state from $x(t_0) = x_0$ to $x(t_1) = 0$.   □

11.5   OPEN-LOOP MINIMUM-ENERGY CONTROL

Suppose that a particular state x_1 belongs to the reachable subspace $\mathcal{R}[t_0, t_1]$ of the system (AB-CLTV). We saw in Theorem 11.2 that a specific control that can transfer the state from $x(t_0) = 0$ to $x(t_1) = x_1$ is given by

$$u(t) = B(t)' \Phi(t_1, t)' \eta_1, \qquad t \in [t_0, t_1], \tag{11.7}$$

where η_1 can be any vector for which

$$x_1 = W_R(t_0, t_1) \eta_1. \tag{11.8}$$

In general, there may be other controls that achieve the same goal, but controls of the form (11.7) are desirable because they *minimize control energy.*

To understand why this is so, suppose that $\bar{u}(\cdot)$ is another control that transfers the state to x_1 and therefore

$$x_1 = \int_{t_0}^{t_1} \Phi(t_1, \tau) B(\tau) u(\tau) d\tau = \int_{t_0}^{t_1} \Phi(t_1, \tau) B(\tau) \bar{u}(\tau) d\tau.$$

For this to hold, we must have

$$\int_{t_0}^{t_1} \Phi(t_1, \tau) B(\tau) v(\tau) d\tau = 0, \qquad (11.9)$$

where $v := \bar{u} - u$. The "energy" of $\bar{u}(\cdot)$ can be related to the energy of $u(\cdot)$ as follows:

$$\int_{t_0}^{t_1} \|\bar{u}(\tau)\|^2 d\tau$$

$$= \int_{t_0}^{t_1} \| \overbrace{B(t)' \Phi(t_1, \tau)' \eta_1}^{u(\tau)} + v(\tau)\|^2 d\tau$$

$$= \eta_1' W_R(t_0, t_1) \eta_1 + \int_{t_0}^{t_1} \|v(\tau)\|^2 d\tau + 2\eta_1' \int_{t_0}^{t_1} \Phi(t_1, \tau) B(\tau) v(\tau) d\tau.$$

Because of (11.9), the last term is equal to zero, and we conclude that the energy of $\bar{u}$ is minimized for $v(\cdot) = 0$; i.e., for $\bar{u} = u$. Moreover, for $v(\cdot) = 0$, we conclude that the energy required for the optimal control $u(\cdot)$ in (11.7) is given by

$$\int_{t_0}^{t_1} \|u(\tau)\|^2 d\tau = \eta_1' W_R(t_0, t_1) \eta_1.$$

These observations are summarized in the following theorem.

Note. There may be several vectors η_1 for which (11.8) holds, but they all differ by vectors in $\ker W_R(t_0, t_1)$, so they all lead to the same control energy and, in fact, to the same control, i.e., $v = 0$.

Note. Controls such as (11.3) and (11.6) are called *open loop* because $u(t)$ is precomputed and is not expressed as a function of the current state.

Theorem 11.4 (Minimum-energy control). *Given two times $t_1 > t_0 \geq 0$,*

1. *when $x_1 \in \mathcal{R}[t_0, t_1]$, the control (11.3) transfers the state from $x(t_0) = 0$ to $x(t_1) = x_1$ with the smallest amount of control energy, which is given by*

$$\int_{t_0}^{t_1} \|u(\tau)\|^2 d\tau = \eta_1' W_R(t_0, t_1) \eta_1, \qquad and$$

2. *when $x_1 \in \mathcal{C}[t_0, t_1]$, the control (11.6) transfers the state from $x(t_0) = x_0$ to $x(t_1) = 0$ with the smallest amount of control energy, which is given by*

$$\int_{t_0}^{t_1} \|u(\tau)\|^2 d\tau = \eta_0' W_C(t_0, t_1) \eta_0. \qquad \square$$

11.6   CONTROLLABILITY MATRIX (LTI)

Consider now the continuous-time LTI system

$$\dot{x} = Ax + Bu, \qquad\qquad x \in \mathbb{R}^n, \ u \in \mathbb{R}^k. \qquad \text{(AB-CLTI)}$$

For this system, the reachability and controllability Gramians are given, respectively, by

$$W_R(t_0, t_1) := \int_{t_0}^{t_1} e^{A(t_1-\tau)} B B' e^{A'(t_1-\tau)} d\tau = \int_0^{t_1-t_0} e^{At} B B' e^{A't} dt,$$

$$W_C(t_0, t_1) := \int_{t_0}^{t_1} e^{A(t_0-\tau)} B B' e^{A'(t_0-\tau)} d\tau = \int_0^{t_1-t_0} e^{-At} B B' e^{-A't} dt.$$

MATLAB® Hint 30.
`ctrb(sys)`
computes the
controllability matrix
of the state-space
system `sys`.
Alternatively, one can
use `ctrb(A,B)`
directly.   ▶ p. 109

The *controllability matrix* of the time-invariant system (AB-CLTI) is defined to be

$$\mathcal{C} := \begin{bmatrix} B & AB & A^2B & \cdots & A^{n-1}B \end{bmatrix}_{n \times (kn)}$$

and provides a particularly simple method to compute the reachable and controllable subspaces.

Theorem 11.5. *For any two times t_0, t_1, with $t_1 > t_0 \geq 0$, we have*

$$\mathcal{R}[t_0, t_1] = \operatorname{Im} W_R(t_0, t_1) = \operatorname{Im} \mathcal{C} = \operatorname{Im} W_C(t_0, t_1) = \mathcal{C}[t_0, t_1]. \qquad \square$$

Attention! This result has several important implications.

1. *Time reversibility.* The notions of controllable and reachable subspaces coincide for continuous-time LTI systems, which means that if one can go from the origin to some state x_1, then one can also go from x_1 to the origin.

   Because of this, for continuous-time LTI systems *one simply studies controllability and neglects reachability.*

2. *Time scaling.* The notions of controllable and reachable subspaces do not depend on the time interval considered. This means that if it is possible to transfer the state from the origin to some state x_1 in a finite interval $[t_0, t_1]$, then it is possible to do the same transfer in any other time finite interval $[\bar{t}_0, \bar{t}_1]$. Similarly for the controllable subspace.

Note. However, time
scaling does not
extend from infinite to
finite time intervals.
E.g., for the system

$$\dot{x} = \begin{bmatrix} 0 & 0 \\ 0 & -1 \end{bmatrix} x + \begin{bmatrix} 1 \\ 0 \end{bmatrix} u,$$

it is possible to
transfer the state from
$\begin{bmatrix} 1 & 1 \end{bmatrix}'$ to the origin
in "infinite time" but
not in finite time.

   Because of this, for continuous-time LTI systems *one generally does not specify the time interval $[t_0, t_1]$ under consideration.* $\qquad \square$

Proof of Theorem 11.5. The first and last equalities have already been proved, so it remains to prove the middle ones. We start with the second equality.

We start by showing that $x_1 \in \mathcal{R}[t_0, t_1] = \operatorname{Im} W_R(t_0, t_1) \Rightarrow x_1 \in \operatorname{Im} \mathcal{C}$. When $x_1 \in \mathcal{R}[t_0, t_1]$, there exists an input $u(\cdot)$ that transfers the state from $x(t_0) = 0$ to $x(t_1) = x_1$, and therefore

$$x_1 = \int_{t_0}^{t_1} e^{A(t_1-\tau)} B u(\tau) d\tau.$$

But we saw in Lecture 6 that, using the Cayley-Hamilton theorem, we can write

$$e^{At} = \sum_{i=0}^{n-1} \alpha_i(t) A^i, \qquad \forall t \in \mathbb{R}$$

for appropriately defined scalar functions $\alpha_0(t), \alpha_1(t), \ldots, \alpha_{n-1}(t)$ [cf. (6.5)]. Therefore

$$x_1 = \sum_{i=0}^{n-1} A^i B \left(\int_{t_0}^{t_1} \alpha_i(t_1 - \tau) u(\tau) d\tau \right) = \mathcal{C} \begin{bmatrix} \int_{t_0}^{t_1} \alpha_0(t_1 - \tau) u(\tau) d\tau \\ \vdots \\ \int_{t_0}^{t_1} \alpha_{n-1}(t_1 - \tau) u(\tau) d\tau \end{bmatrix},$$

which shows that $x_1 \in \mathrm{Im}\,\mathcal{C}$.

We show next that $x_1 \in \mathbf{Im}\,\mathcal{C} \Rightarrow x_1 \in \mathcal{R}[t_0, t_1] = \mathbf{Im}\,W_R(t_0, t_1)$. When $x_1 \in \mathrm{Im}\,\mathcal{C}$, there exists a vector $\nu \in \mathbb{R}^{kn}$ for which

$$x_1 = \mathcal{C}\nu. \tag{11.10}$$

We show next that this leads to $x_1 \in \mathrm{Im}\,W_R(t_0, t_1) = (\ker W_R(t_0, t_1))^\perp$, which is to say that

$$\eta_1' x_1 = \eta_1' \mathcal{C}\nu = 0, \quad \forall \eta_1 \in \ker W_R(t_0, t_1). \tag{11.11}$$

To verify that this is so, we pick an arbitrary vector $\eta_1 \in \ker W_R(t_0, t_1)$. We saw in the proof of Theorem 11.2 that such vector η_1 has the property that

$$\eta_1' e^{A(t_1 - \tau)} B = 0, \quad \forall \tau \in [t_0, t_1].$$

Note. Check that

$$\frac{d^k(\eta_1' e^{A(t_1-\tau)} B)}{d\tau^k}$$
$$= (-1)^k \eta_1' A^k e^{A(t_1-\tau)} B.$$

(Cf. Exercise 11.1.)

Taking k time derivatives with respect to τ, we further conclude that

$$(-1)^k \eta_1' A^k e^{A(t_1 - \tau)} B = 0, \quad \forall \tau \in [t_0, t_1], \; k \geq 0, \tag{11.12}$$

and in particular for $\tau = t_1$, we obtain

$$\eta_1' A^k B = 0, \quad \forall k \geq 0.$$

It follows that $\eta_1' \mathcal{C} = 0$ and therefore (11.11) indeed holds.

Since the corresponding proofs for the controllable subspace are analogous, we do not present them. ∎

Example 11.4 (Parallel RC network, continued). The controllability matrix for the electrical network in Figure 11.1(a) and Example 11.1 is given by

$$\mathcal{C} = \begin{bmatrix} B & AB \end{bmatrix} = \begin{bmatrix} \frac{1}{R_1 C_1} & \frac{-1}{R_1^2 C_1^2} \\ \frac{1}{R_2 C_2} & \frac{-1}{R_2^2 C_2^2} \end{bmatrix}.$$

When the two branches have the same time constant, i.e., $\frac{1}{R_1 C_1} = \frac{1}{R_2 C_2} =: \omega$, we have

$$\mathcal{C} = \begin{bmatrix} \omega & -\omega^2 \\ \omega & -\omega^2 \end{bmatrix},$$

and therefore

$$\mathcal{R}[t_0, t_1] = \mathcal{C}[t_0, t_1] = \mathrm{Im}\,\mathcal{C} = \left\{ \alpha \begin{bmatrix} 1 \\ 1 \end{bmatrix} : \alpha \in \mathbb{R} \right\}, \quad \forall t_1 > t_0 \geq 0.$$

However, when the time constants are different, i.e., $\frac{1}{R_1 C_1} \neq \frac{1}{R_2 C_2}$,

$$\det \mathcal{C} = \frac{1}{R_1^2 C_1^2 R_2 C_2} - \frac{1}{R_1 C_1 R_2^2 C_2^2} = \frac{1}{R_1 C_1 R_2 C_2} \left(\frac{1}{R_1 C_1} - \frac{1}{R_2 C_2} \right) \neq 0,$$

which means that $\mathcal{C}$ is nonsingular, and therefore

$$\mathcal{R}[t_0, t_1] = \mathcal{C}[t_0, t_1] = \operatorname{Im} \mathcal{C} = \mathbb{R}^2. \qquad \square$$

11.7   DISCRETE-TIME CASE

Consider the discrete-time LTV system

$$x(t + 1) = A(t)x(t) + B(t)u(t), \qquad x \in \mathbb{R}^n, \ u \in \mathbb{R}^k. \qquad \text{(AB-DLTV)}$$

We saw in Lecture 5 that a given input $u(\cdot)$ transfers the state $x(t_0) := x_0$ at time t_0 to the state $x(t_1) := x_1$ at time t_1 given by the variation of constants formula,

$$x_1 = \Phi(t_1, t_0)x_0 + \sum_{\tau = t_0}^{t_1 - 1} \Phi(t_1, \tau + 1)B(\tau)u(\tau),$$

where $\Phi(\cdot)$ denotes the system's state transition matrix.

Definition 11.4 (Reachable and controllable subspaces). Given two times $t_1 > t_0 \geq 0$, the *reachable* or *controllable-from-the-origin* on $[t_0, t_1]$ subspace $\mathcal{R}[t_0, t_1]$ consists of all states x_1 for which there exists an input $u : \{t_0, t_0 + 1, \ldots, t_1 - 1\} \to \mathbb{R}^k$ that transfers the state from $x(t_0) = 0$ to $x(t_1) = x_1$; i.e.,

$$\mathcal{R}[t_0, t_1] := \left\{ x_1 \in \mathbb{R}^n : \exists u(.), \ x_1 = \sum_{\tau = t_0}^{t_1 - 1} \Phi(t_1, \tau + 1)B(\tau)u(\tau) \right\}.$$

The *controllable* or *controllable-to-the-origin* on $[t_0, t_1]$ subspace $\mathcal{C}[t_0, t_1]$ consists of all states x_0 for which there exists an input $u : \{t_0, t_0 + 1, \ldots, t_1 - 1\} \to \mathbb{R}^k$ that transfers the state from $x(t_0) = x_0$ to $x(t_1) = 0$; i.e.,

$$\mathcal{C}[t_0, t_1] := \left\{ x_0 \in \mathbb{R}^n : \exists u(.), \ 0 = \Phi(t_1, t_0)x_0 + \sum_{\tau = t_0}^{t_1 - 1} \Phi(t_1, \tau + 1)B(\tau)u(\tau) \right\}. \quad \square$$

Definition 11.5 (Reachability and controllability Gramians). Given two times $t_1 > t_0 \geq 0$, the *reachability* and *controllability Gramians* of the system (AB-DLTV) are defined, respectively, by

$$W_R(t_0, t_1) := \sum_{\tau = t_0}^{t_1 - 1} \Phi(t_1, \tau + 1)B(\tau)B(\tau)'\Phi(t_1, \tau + 1)',$$

$$W_C(t_0, t_1) := \sum_{\tau = t_0}^{t_1 - 1} \Phi(t_0, \tau + 1)B(\tau)B(\tau)'\Phi(t_0, \tau + 1)'. \qquad \square$$

Attention! The definition of the discrete-time controllability Gramian requires a backward-in-time state transition matrix $\Phi(t_0, \tau + 1)$ from time $\tau + 1$ to time $t_0 \leq \tau < \tau + 1$. This matrix is well defined only when

$$x(\tau + 1) = A(\tau)A(\tau - 1) \cdots A(t_0)x(t_0), \quad t_0 \leq \tau \leq t_1 - 1$$

can be solved for $x(t_0)$, i.e., when all the matrices $A(t_0)$, $A(t_0 + 1)$, ..., $A(t_1 - 1)$ are nonsingular. When this does not happen, *the controllability Gramian cannot be defined.* □

These Gramians allow us to determine exactly what the reachable and controllable spaces are.

Theorem 11.6 (Reachable and controllable subspaces). *Given two times $t_1 > t_0 \geq 0$,*

$$\mathcal{R}[t_0, t_1] = \operatorname{Im} W_R(t_0, t_1), \qquad \mathcal{C}[t_0, t_1] = \operatorname{Im} W_C(t_0, t_1).$$

Moreover,

1. if $x_1 = W_R(t_0, t_1)\eta_1 \in \operatorname{Im} W_R(t_0, t_1)$, the control

$$u(t) = B(t)'\Phi(t_1, t + 1)'\eta_1, \quad t \in [t_0, t_1 - 1] \tag{11.13}$$

can be used to transfer the state from $x(t_0) = 0$ to $x(t_1) = x_1$, and

2. if $x_0 = W_C(t_0, t_1)\eta_0 \in \operatorname{Im} W_C(t_0, t_1)$, the control

$$u(t) = -B(t)'\Phi(t_0, t + 1)'\eta_0, \quad t \in [t_0, t_1 - 1]$$

can be used to transfer the state from $x(t_0) = x_0$ to $x(t_1) = 0$. □

Consider now the discrete-time LTI system

$$x^+ = Ax + Bu, \qquad x \in \mathbb{R}^n,\ u \in \mathbb{R}^k. \tag{AB-DLTI}$$

For this system, the reachability and controllability Gramians are given, respectively, by

$$W_R(t_0, t_1) := \sum_{\tau=t_0}^{t_1-1} A^{t_1-1-\tau} BB'(A')^{t_1-1-\tau},$$

$$W_C(t_0, t_1) := \sum_{\tau=t_0}^{t_1-1} A^{t_0-1-\tau} BB'(A')^{t_0-1-\tau},$$

and the *controllability matrix of* (AB-DLTI) is given by

$$\mathcal{C} := \begin{bmatrix} B & AB & A^2B & \cdots & A^{n-1}B \end{bmatrix}_{n \times (kn)}.$$

Theorem 11.7. *For any two times $t_1 > t_0 \geq 0$, with $t_1 \geq t_0 + n$, we have*

$$\mathcal{R}[t_0, t_1] = \operatorname{Im} W_R(t_0, t_1) = \operatorname{Im} \mathcal{C} = \operatorname{Im} W_C(t_0, t_1) = \mathcal{C}[t_0, t_1].$$ □

Attention! This result differs from the continuous-time counterparts in two significant ways.

1. *Time reversibility.* In discrete time, the notions of controllable and reachable subspaces coincide only when the matrix A is nonsingular. Otherwise, we have

$$\mathcal{R}[t_0, t_1] = \operatorname{Im} \mathcal{C} \subset \mathcal{C}[t_0, t_1],$$

but the reverse inclusion does not hold; i.e., there are states x_1 that can be transferred to the origin, but it is not possible to find an input to transfer the origin to these states.

Because of this, when A is singular, one must study reachability and controllability of discrete-time systems separately.

2. *Time scaling.* In discrete time, the notions of controllable and reachable subspaces do not depend on the time interval only when the intervals have length larger than or equal to n time steps. When $t_1 - t_0 < n$, we have

$$\mathcal{R}[t_0, t_1] \subset \operatorname{Im} \mathcal{C},$$

but the reverse inclusion does not hold; i.e., there are states x_1 than can be reached in n time steps, but not in $t_1 - t_0 < n$ time steps.

In discrete-time systems, when one omits the interval under consideration, it is implicitly assumed that it has length no smaller than n, in which case we have time scaling. $\qquad\Box$

DISCRETE-TIME CASE PROOFS

Proof of Theorem 11.6. We start by showing that $x_1 \in \mathbf{Im}\, W_R(t_0, t_1) \Rightarrow x_1 \in \mathcal{R}[t_0, t_1]$. When $x_1 \in \operatorname{Im} W_R(t_0, t_1)$, there exists a vector $\eta_1 \in \mathbb{R}^n$ such that

$$x_1 = W_R(t_0, t_1)\eta_1.$$

To prove that $x_1 \in \mathcal{R}[t_0, t_1]$, it suffices to show that the input (11.13) does indeed transfer the state from $x(t_0) = 0$ to $x(t_1) = x_1$, and therefore $x_1 \in \mathcal{R}[t_0, t_1]$. To verify that this is so, we use the variation of constants formula for the input (11.13):

$$x(t_1) = \sum_{\tau=t_0}^{t_1-1} \Phi(t_1, \tau + 1)B(\tau) \underbrace{B(\tau)'\Phi(t_1, \tau + 1)'\eta_1}_{u(\tau)} = W_R(t_0, t_1)\eta_1 = x_1.$$

We show next that $x_1 \in \mathcal{R}[t_0, t_1] \Rightarrow x_1 \in \mathbf{Im}\, W_R(t_0, t_1)$. To prove by contradiction, assume that there exists an input $u(\cdot)$ for which

$$x_1 = \sum_{\tau=t_0}^{t_1-1} \Phi(t_1, \tau + 1)B(\tau)u(\tau), \tag{11.14}$$

but $x_1 \notin \operatorname{Im} W_R(t_0, t_1) = (\ker W_R(t_0, t_1))^\perp$. Since $x_1 \notin (\ker W_R(t_0, t_1))^\perp$, there must be a vector η_1 in $\ker W_R(t_0, t_1)$ that is not orthogonal to x_1; i.e.,

$$W_R(t_0, t_1)\eta_1 = 0, \qquad\qquad \eta_1'x_1 \neq 0.$$

But then

$$\eta_1' W_R(t_0, t_1)\eta_1 = \sum_{\tau=t_0}^{t_1-1} \eta_1' \Phi(t_1, \tau+1) B(\tau) B(\tau)' \Phi(t_1, \tau+1)' \eta_1$$

$$= \sum_{\tau=t_0}^{t_1-1} \| B(\tau)' \Phi(t_1, \tau+1)' \eta_1 \| = 0,$$

which implies that

$$B(\tau)' \Phi(t_1, \tau+1)' \eta_1 = 0, \qquad \forall \tau \in \{t_0, t_0+1, \ldots, t_1-1\}.$$

From this and (11.14), we conclude that

$$\eta_1' x_1 = \sum_{\tau=t_0}^{t_1-1} \eta_1' \Phi(t_1, \tau+1) B(\tau) u(\tau) = 0,$$

which contradicts the fact that η_1 is not orthogonal to x_1. ∎

Proof of Theorem 11.7. The first and last equalities have already been proved, so it remains to prove the middle ones. We start with the second equality.

We start by showing that $x_1 \in \mathcal{R}[t_0, t_1] = \mathbf{Im}\, W_R(t_0, t_1) \Rightarrow x_1 \in \mathbf{Im}\,\mathcal{C}$. When $x_1 \in \mathcal{R}[t_0, t_1]$, there exists an input $u(\cdot)$ that transfers the state from $x(t_0) = 0$ to $x(t_1) = x_1$, and therefore

$$x_1 = \sum_{\tau=t_0}^{t_1-1} A^{t_1-1-\tau} B u(\tau).$$

But we saw in Lecture 6 that, using the Cayley-Hamilton theorem, we can write

$$A^t = \sum_{i=0}^{n-1} \alpha_i(t) A^i, \qquad \forall t \in \mathbb{R}$$

for appropriately defined scalar functions $\alpha_0(t), \alpha_1(t), \ldots, \alpha_{n-1}(t)$ [cf. (6.6)]. Therefore

$$x_1 = \sum_{i=0}^{n-1} A^i B \Big(\sum_{\tau=t_0}^{t_1-1} \alpha_i(t_1-1-\tau) u(\tau) \Big) = \mathcal{C} \begin{bmatrix} \sum_{i=0}^{n-1} \alpha_0(t_1-1-\tau)u(\tau) \\ \vdots \\ \sum_{i=0}^{n-1} \alpha_{n-1}(t_1-1-\tau)u(\tau) \end{bmatrix},$$

which shows that $x_1 \in \mathbf{Im}\,\mathcal{C}$.

We show next that $x_1 \in \mathbf{Im}\,\mathcal{C} \Rightarrow x_1 \in \mathcal{R}[t_0, t_1] = \mathbf{Im}\, W_R(t_0, t_1)$. When $x_1 \in \mathbf{Im}\,\mathcal{C}$, there exists a vector $v \in \mathbb{R}^{kn}$ such that

$$x_1 = \mathcal{C} v = \begin{bmatrix} B & AB & A^2 B & \cdots & A^{n-1}B \end{bmatrix} \begin{bmatrix} v_0 \\ \vdots \\ v_{n-1} \end{bmatrix}' = \sum_{i=0}^{n-1} A^i B v_i,$$

where ν is broken into n k-vectors ν_i. We show that $x_1 \in \mathcal{R}[t_0, t_1]$ because the n-step control

$$u(\tau) = \begin{cases} 0 & t_0 \le t < t_1 - n \\ \nu_{t_1-1-\tau} & t_1 - n \le t \le t_1 - 1 \end{cases}$$

transfers the system from the origin to

$$x(t_1) = \sum_{\tau=t_0}^{t_1-1} A^{t_1-1-\tau} B u(\tau) = \sum_{\tau=t_1-n}^{t_1-1} A^{t_1-1-\tau} B \nu_{t_1-1-\tau} = \sum_{i=0}^{n-1} A^i B \nu_i = x_1.$$

Note that this control requires $t_1 - n \ge t_0$.

Since the corresponding proofs for the controllable subspace are analogous, we do not present them. ∎

11.8 MATLAB® COMMANDS

MATLAB® Hint 29 (svd). The command [U,S,V]=svd(W) can be used to compute a basis for the image and kernel of the $n \times m$ matrix W. This command computes a *singular value decomposition* of W, i.e., (square) orthogonal matrices $U_{n \times n}$, $V_{m \times m}$, and a (real) diagonal matrix $S_{m \times m}$ such that W = USV'.

Notation. A square matrix U is called *orthogonal* if its inverse exists and is equal to its transpose, i.e., $UU' = U'U = I$.

1. The columns of U corresponding to nonzero rows of S are an orthonormal basis for Im W.

2. The columns of V (rows of V') corresponding to the zero columns of S are an orthonormal basis for ker W. □

MATLAB® Hint 30 (ctrb). The function ctrb(sys) computes the controllability matrix of the system sys. The system must be specified by a state-space model using, e.g., sys=ss(A,B,C,D), where A,B,C,D are a realization of the system. Alternatively, one can use ctrb(A,B) directly. □

11.9 EXERCISE

11.1. Verify that

$$\frac{d^k(\eta_1' e^{A(t_1-\tau)} B)}{d\tau^k} = (-1)^k \eta_1' A^k e^{A(t_1-\tau)} B.$$

□

LECTURE 12

Controllable Systems

CONTENTS

This lecture introduces the notion of a controllable system and presents several tests to determine whether a system is controllable.

1. Controllable Systems
2. Eigenvector Test for Controllability
3. Lyapunov Test for Controllability
4. Feedback Stabilization Based on the Lyapunov Test
5. Exercises

12.1 CONTROLLABLE SYSTEMS

Notation. In this section, we jointly present the results for continuous and discrete time and use a slash to separate the two cases.

Consider the following continuous- and discrete-time LTV systems

$$\dot{x} = A(t)x + B(t)u \quad / \quad x(t+1) = A(t)x(t) + B(t)u(t), \quad x \in \mathbb{R}^n, \ u \in \mathbb{R}^k.$$
$$\text{(AB-LTV)}$$

Definition 12.1 (Reachable system). Given two times $t_1 > t_0 \geq 0$, the system (AB-LTV), or simply the pair $\big(A(\cdot), B(\cdot)\big)$, is *(completely state-) reachable on* $[t_0, t_1]$ if $\mathcal{R}[t_0, t_1] = \mathbb{R}^n$, i.e., if the origin can be transferred to every state. □

Definition 12.2 (Controllable system). Given two times $t_1 > t_0 \geq 0$, the system (AB-LTV), or simply the pair $\big(A(\cdot), B(\cdot)\big)$, is *(completely state-) controllable on* $[t_0, t_1]$ if $\mathcal{C}[t_0, t_1] = \mathbb{R}^n$, i.e., if every state can be transferred to the origin. □

Note. For continuous-time LTI systems $\mathcal{R}[t_0, t_1] = \mathcal{C}[t_0, t_1]$, and therefore one often talks about only controllability.

Notation. A system that is not controllable is called *uncontrollable*.

Consider now the LTI systems

$$\dot{x} = Ax + Bu \quad / \quad x^+ = Ax + Bu, \qquad x \in \mathbb{R}^n, \ u \in \mathbb{R}^k. \quad \text{(AB-LTI)}$$

We saw in Theorem 11.5 that

$$\text{Im} \, \mathcal{C} = \mathcal{R}[t_0, t_1] = \mathcal{C}[t_0, t_1].$$

Note. In discrete time, this holds for $t_1 - t_0 \geq n$, and nonsingular A.

Since $\mathcal{C}$ has n rows, $\text{Im} \, \mathcal{C}$ is a subspace of $\mathbb{R}^n$, so its dimension can be at most n. For controllability, $\text{Im} \, \mathcal{C} = \mathbb{R}^n$, and therefore the dimension of $\text{Im} \, \mathcal{C}$ must be exactly n. This reasoning leads to the following theorem.

Note. In discrete time, when A is singular, we simply have

$$\text{Im}\,\mathcal{C} = \mathcal{R}[t_0, t_1] \subset \mathcal{C}[t_0, t_1].$$

In this case rank $\mathcal{C} = n$ implies that $\mathcal{R}[t_0, t_1] = \mathcal{C}[t_0, t_1] = \mathbb{R}^n$. However, one could have rank $\mathcal{C} < n$. In this case, $\text{Im}\,\mathcal{C} = \mathcal{R}[t_0, t_1] \subset \mathbb{R}^n$ (strict inclusion) and yet $\mathcal{C}[t_0, t_1] = \mathbb{R}^n$.

Theorem 12.1 (Controllability matrix test). *The LTI system* (AB-LTI) *is controllable if and only if*

$$\text{rank}\,\mathcal{C} = n. \qquad \Box$$

Although Theorem 12.1 provides a simple test for controllability, there are a few other useful tests that we introduce next. Some of these actually lead to feedback control design methods.

12.2   EIGENVECTOR TEST FOR CONTROLLABILITY

Given an $n \times n$ matrix A, a linear subspace $\mathcal{V}$ of $\mathbb{R}^n$ is said to be *A-invariant* if for every vector $v \in \mathcal{V}$ we have $Av \in \mathcal{V}$. The following properties of invariant subspaces will be used.

Properties. Given an $n \times n$ matrix A and a nonzero A-invariant subspace $\mathcal{V} \subset \mathbb{R}^n$, the following statements are true.

P12.1 If one constructs an $n \times k$ matrix V whose columns form a basis for $\mathcal{V}$, there exists a $k \times k$ matrix $\bar{A}$ such that

Note. For $k = 1$, this means that the (only) column of V is an eigenvector of A.

$$AV = V\bar{A}. \tag{12.1}$$

P12.2 $\mathcal{V}$ contains at least one eigenvector of A.

Proof. Let A and $\mathcal{V}$ be as in the statement of the proposition.

P12.1   Since the ith column v_i of the matrix V belongs to $\mathcal{V}$ and $\mathcal{V}$ is A-invariant, $Av_i \in \mathcal{V}$. This means that it can be written as a linear combination of the columns of V; i.e., there exists a column vector $\bar{a}_i$ such that

$$Av_i = V\bar{a}_i, \qquad \forall i \in \{1, 2, \ldots, k\}.$$

Putting all these equations together, we conclude that

$$\begin{bmatrix} Av_1 & Av_2 & \cdots & Av_k \end{bmatrix} = \begin{bmatrix} V\bar{a}_1 & V\bar{a}_2 & \cdots & V\bar{a}_k \end{bmatrix} \quad \Leftrightarrow \quad AV = V\bar{A},$$

where all the $\bar{a}_i$ are used as columns for $\bar{A}$.

P12.2   Let $\bar{v}$ be an eigenvector of the matrix $\bar{A}$ in (12.1) corresponding to the eigenvalue λ. Then

$$AV\bar{v} = V\bar{A}\bar{v} = \lambda V\bar{v},$$

and therefore $v := V\bar{v}$ is an eigenvector of A. Moreover, since v is a linear combination of the columns of V, it must belong to $\mathcal{V}$. ∎

Theorem 12.2 (Eigenvector test for controllability). *The LTI system* (AB-LTI) *is controllable if and only if there is no eigenvector of A' in the kernel of B'.* □

Proof of Theorem 12.2. We start by proving that if the system (AB-LTI) is controllable, then every eigenvector of A' is not in the kernel of B'. To prove by contradiction, assume that there exists an eigenvalue $A'x = \lambda x$, with $x \neq 0$, for which $B'x = 0$. Then

$$\mathcal{C}'x = \begin{bmatrix} B' \\ B'A' \\ \vdots \\ B'(A')^{n-1} \end{bmatrix} x = \begin{bmatrix} B'x \\ \lambda B'x \\ \vdots \\ \lambda^{n-1} B'x \end{bmatrix} = 0. \tag{12.2}$$

This means that the null space of $\mathcal{C}$ has at least one nonzero vector, and therefore, from the fundamental theorem of linear equations, we conclude that

$$\dim \ker \mathcal{C}' \geq 1 \quad \Rightarrow \quad \operatorname{rank}\mathcal{C} = \operatorname{rank}\mathcal{C}' = n - \dim \ker \mathcal{C}' < n,$$

which contradicts the controllability of (AB-LTI).

Conversely, suppose now that (AB-LTI) is not controllable, and therefore that

$$\operatorname{rank}\mathcal{C} = \operatorname{rank}\mathcal{C}' < n \quad \Rightarrow \quad \dim \ker \mathcal{C}' = n - \operatorname{rank}\mathcal{C}' \geq 1.$$

It turns out that $\ker \mathcal{C}'$ is A'-invariant. Indeed, if $x \in \ker \mathcal{C}'$, then (12.2) holds, and therefore

$$x \in \ker \mathcal{C}' \quad \Rightarrow \quad \mathcal{C}'A'x = \begin{bmatrix} B'A' \\ B'(A')^2 \\ \vdots \\ B'(A')^n \end{bmatrix} x = \begin{bmatrix} 0 \\ 0 \\ \vdots \\ B'(A')^n x \end{bmatrix}.$$

But by the Cayley-Hamilton theorem, A^n can be written as a linear combination of the lower powers of A', and therefore $B'(A')^n x$ can be written as a linear combination of the terms

$$B'x, A'B'x, \ldots, (A')^{n-1} Bx,$$

which are all zero because of (12.2). We therefore conclude that

$$x \in \ker \mathcal{C}' \quad \Rightarrow \quad \mathcal{C}'A'x = 0 \quad \Rightarrow \quad A'x \in \ker \mathcal{C}',$$

which confirms that $\ker \mathcal{C}'$ is A'-invariant.

From Property P12.2, we then conclude that $\ker \mathcal{C}'$ must contain at least one eigenvector x of A'. But since $\mathcal{C}'x = 0$, we necessarily have $B'x = 0$. This concludes the proof, since we also showed that if the system (AB-LTI) is not controllable, then there must exist an eigenvector of A' in the kernel of B'. ■

The following test is essentially an elegant restatement of the eigenvector test.

Theorem 12.3 (Popov-Belevitch-Hautus [PBH] test for controllability). *The LTI system* (AB-LTI) *is controllable if and only if*

$$\text{rank} \begin{bmatrix} A - \lambda I & B \end{bmatrix} = n, \qquad \forall \lambda \in \mathbb{C}. \tag{12.3}$$

Proof of Theorem 12.3. From the Fundamental theorem of linear equations, we conclude that

$$\dim \ker \begin{bmatrix} A' - \lambda I \\ B' \end{bmatrix} = n - \text{rank} \begin{bmatrix} A - \lambda I & B \end{bmatrix}, \quad \forall \lambda \in \mathbb{C},$$

and therefore the condition (12.3) can also be rewritten as

$$\dim \ker \begin{bmatrix} A' - \lambda I \\ B' \end{bmatrix} = 0, \qquad \forall \lambda \in \mathbb{C}, \tag{12.4}$$

which means that the kernel of $\begin{bmatrix} A' - \lambda I \\ B' \end{bmatrix}$ can contain only the zero vector. This means that (12.3) is also equivalent to

$$\ker \begin{bmatrix} A' - \lambda I \\ B' \end{bmatrix} = \left\{ x \in \mathbb{R}^n : A'x = \lambda x, \quad B'x = 0 \right\} = \{0\}, \qquad \forall \lambda \in \mathbb{C},$$

which is precisely equivalent to the statement that there can be no eigenvector of A' in the kernel of B'. ∎

12.3 LYAPUNOV TEST FOR CONTROLLABILITY

Consider again the LTI systems

$$\dot{x} = Ax + Bu \quad / \quad x^+ = Ax + Bu, \qquad x \in \mathbb{R}^n, \ u \in \mathbb{R}^k. \tag{AB-LTI}$$

Theorem 12.4 (Lyapunov test for controllability). *Assume that A is a stability matrix/Schur stable. The LTI system* (AB-LTI) *is controllable if and only if there is a unique positive-definite solution W to the following Lyapunov equation*

$$AW + WA' = -BB' \quad / \quad AWA' - W = -BB' \tag{12.5}$$

Moreover, the unique solution to (12.5) is equal to

Note. Opposite to what happens in the Lyapunov stability theorem (Theorem 8.2), A' appears now to the right of W instead of to the left.

$$W = \int_0^\infty e^{A\tau} BB' e^{A'\tau} d\tau = \lim_{t_1 - t_0 \to \infty} W_R(t_0, t_1)$$

$$/ \quad W = \sum_{\tau=0}^\infty A^\tau BB'(A')^\tau = \lim_{t_1 - t_0 \to \infty} W_R(t_0, t_1). \tag{12.6}$$

Attention! Controllability and reachability are *finite-time concepts*; e.g., controllability means that the origin can be reached from any state in finite time. However, there are uncontrollable systems for which the origin can be reached in infinite time from any state. In view of this, one should emphasize that *the "infinite time" Gramian in (12.5)–(12.6) still provides information only about (finite time) controllability*. Note that the system

$$\dot{x} = -x + 0 \cdot u$$

can be transferred to the origin in infinite time (due to asymptotic stability). However, the "infinite time" Gramian is still equal to zero and therefore is not positive-definite. This is consistent with the fact that this system is not controllable.   □

Proof of Theorem 12.4. We do the proof for continuous time and start by showing that if (12.5) has a positive-definite solution W, then the system (AB-LTI) is controllable. The simplest way to do this is by using the eigenvector test. To do this, assume that (12.5) holds, and let $x \neq 0$ be an eigenvector of A' associated with the eigenvalue λ, i.e., $A'x = \lambda x$. Then

$$x^*(AW + WA')x = -x^*BB'x = -\|B'x\|^2, \qquad (12.7)$$

where $(\cdot)^*$ denotes the complex conjugate transpose. But the left-hand side of this equation is equal to

$$(A'x^*)'Wx + x^*WA'x = \lambda^* x^*Wx + \lambda x^*Wx = 2\Re[\lambda]x^*Wx. \qquad (12.8)$$

Since W is positive-definite, this expression must be strictly negative (note that $\Re[\lambda] < 0$ because A is a stability matrix), and therefore $B'x \neq 0$. We conclude that every eigenvalue of A' is not in the kernel of B', which implies controllability by the eigenvector test.

To prove the converse, we assume that (AB-LTI) is controllable. Equation (12.5) can be written as

$$\bar{A}'W + W\bar{A} = -Q, \qquad \bar{A} := A', \qquad Q := BB',$$

which was the equation that we analyzed in the proof of the Lyapunov stability theorem (Theorem 8.2). Since A is a stability matrix, $\bar{A} := A'$ is also a stability matrix, and therefore we can reuse the proof of the Lyapunov stability theorem (Theorem 8.2) to conclude that (12.6) is the unique solution to (12.5).

The only issue that needs special attention is that in Theorem 8.2 we used the fact that $Q = BB'$ was positive-definite to show that the solution W was also positive-definite. Here, $Q = BB'$ may not be positive-definite, but it turns out that controllability of the pair (A, B) suffices to establish that W is positive-definite, even if Q is not. Indeed, given an arbitrary vector $x \neq 0$,

$$x'Wx = x'\Big(\int_0^\infty e^{A\tau}BB'e^{A'\tau}d\tau\Big)x \geq x'\Big(\int_0^1 e^{A\tau}BB'e^{A'\tau}d\tau\Big)x$$
$$= x'W_R(0,1)x > 0,$$

because $W_R(0,1) > 0$, due to controllability.   ■

Note. This is because A and A' have the same eigenvalues: $\det(\lambda I - A) = \det(\lambda I - A')$.

Note 6. This reasoning allows us to add a sixth equivalent condition to the Lyapunov stability theorem (Theorem 8.2).   ▶ p. 114

Note 6 (Controllability condition in the Lyapunov stability theorem). The results in Theorem 12.4 allow us to add a sixth equivalent condition to the Lyapunov stability theorem (Theorem 8.2). The full theorem with the additional condition is reproduced below.

Theorem 12.5 (Lyapunov stability, updated). *The following six conditions are equivalent.*

1. *The system (H-CLTI) is asymptotically stable.*

2. *The system (H-CLTI) is exponentially stable.*

3. *All the eigenvalues of A have strictly negative real parts.*

4. *For every symmetric positive-definite matrix Q, there exists a unique solution P to the Lyapunov equation*

$$A'P + PA = -Q. \tag{12.9}$$

Moreover, P is symmetric, positive-definite, and equal to $P := \int_0^\infty e^{A't} Q e^{At} dt$.

Note. The inequality (12.10) is called a *linear matrix inequality (LMI)*. The term "linear" comes from the linearity of the left-hand side in P and $<$ refers to the fact that the left-hand side must be *negative-definite*.

5. *There exists a symmetric positive-definite matrix P for which the following Lyapunov matrix inequality holds:*

$$A'P + PA < 0. \tag{12.10}$$

Note. Opposite to what happens in the Lyapunov stability theorem (Theorem 8.2), A' appears now on the right of P instead of on the left.

MATLAB® Hint 26. `P=lyap(A,B*B')` solves the Lyapunov equation (12.11).   ▶ p. 77

6. *For every matrix B for which the pair (A, B) is controllable, there exists a unique solution P to the Lyapunov equation*

$$AP + PA' = -BB'. \tag{12.11}$$

Moreover, P is symmetric, positive-definite, and equal to $P = \int_0^\infty e^{A\tau} BB' e^{A'\tau} d\tau$. $\qquad\square$

Proof of Theorem 12.5. Theorem 12.4 actually states only that if A is a stability matrix, then (12.11) has a unique symmetric positive-definite solution P. To show that condition 6 is indeed equivalent to asymptotic stability, one still needs to show that when (12.11) holds for a symmetric positive-definite matrix P, then the matrix A must be a stability matrix. To show this, assume that (12.11) holds, and let λ be an eigenvector of A' and let $x \neq 0$ be the corresponding eigenvector; i.e., $A'x = \lambda x$. Then

$$x^*(AP + PA')x = -x^* BB' x = -\|B'x\|^2,$$

where $(\cdot)^*$ denotes the complex conjugate transpose. Expanding the left-hand side of this equation, we obtain

$$(A'x^*)'Px + x^* PA'x = \lambda^* x^* Px + \lambda x^* Px = 2\Re[\lambda]x^* Px = -\|B'x\|^2.$$

Note. This is because A and A' have the same eigenvalues, since $\det(\lambda I - A) = \det(\lambda I - A')$.

Because of controllability, x cannot belong to the kernel of B', and therefore $2\Re[\lambda]x^* Px$ is strictly negative. Since P is positive-definite, $x^* Px$ is strictly positive, and we conclude that $\Re[\lambda] < 0$. This shows that A' is a stability matrix, which also means that A is a stability matrix. ∎

12.4 FEEDBACK STABILIZATION BASED ON THE LYAPUNOV TEST

Assume that the continuous-time LTI system

$$\dot{x} = Ax + Bu, \qquad\qquad x \in \mathbb{R}^n,\ u \in \mathbb{R}^k \qquad\qquad \text{(AB-CLTI)}$$

is controllable. Controllability of the pair (A, B) guarantees that the pair $(-\mu I - A, B)$ is controllable for every $\mu \in \mathbb{R}$. This is a consequence of the eigenvector test, because A' and $-\mu I - A'$ have exactly the same eigenvectors:

$$A'x = \lambda x \quad \Leftrightarrow \quad (-\mu I - A')x = (-\mu - \lambda)x$$

and therefore (λ, x) is an eigenvalue-eigenvector pair for A' if and only if $(-\mu - \lambda, x)$ is an eigenvalue-eigenvector pair for $-\mu I - A'$. From this, we also conclude that by making μ sufficiently large we can always make $-\mu I - A'$ a stability matrix.

Suppose that we indeed choose μ sufficiently large so that $-\mu I - A$ is a stability matrix. From the Lyapunov test, we conclude that there must exist a positive-definite matrix W such that

MATLAB® Hint 31.
Equation (12.12) can be solved using
`W=lyap(-mu*eye(n)-A,B*B')`
(cf. MATLAB® Hint 26, p. 77).

$$(-\mu I - A)W + W(-\mu I - A)' = -BB' \quad \Leftrightarrow \quad AW + WA' - BB' = -2\mu W. \tag{12.12}$$

Multiplying the right-hand side equation on both sides by $P := W^{-1} > 0$, we obtain

$$PA + A'P - PBB'P = -2\mu P,$$

which can be further rewritten as

$$P(A - BK) + (A - BK)'P = -2\mu P, \qquad K := \frac{1}{2}B'P. \tag{12.13}$$

Since $P > 0$ and $2\mu P > 0$, we conclude from the Lyapunov stability theorem (Theorem 8.2) that $A - BK$ must be a stability matrix. This means that the state feedback control

$$u = -Kx$$

Note. This is because if we define $v := x'Px$, we conclude from (12.13) that $\dot{v} = -2\mu v$. Therefore v converges to zero as fast as $e^{-2\mu t}$. Since $P > 0$, this means that $\|x\|^2$ converges to zero at the same rate. Therefore $\|x\|$ must converge to zero as fast as $\sqrt{e^{-2\mu t}} = e^{-\mu t}$. See also Exercise 8.6.

asymptotically stabilizes the system (AB-CLTI). It turns out that all the eigenvalues of the closed-loop system actually have their real parts smaller than or equal to $-\mu$. Since we could have chosen μ arbitrarily large, we conclude that the following result is true.

Theorem 12.6. *When the system* (AB-CLTI) *is controllable, for every* $\mu > 0$*, it is possible to find a state feedback controller* $u = -Kx$ *that places all eigenvalues of the closed-loop system* $\dot{x} = (A - BK)x$ *on the complex semiplane* $\Re[s] \leq -\mu$. $\qquad\square$

The discrete-time equivalent of this result can be stated as follows.

Theorem 12.7. *When the system*

$$x^+ = Ax + Bu, \qquad\qquad x \in \mathbb{R}^n,\ u \in \mathbb{R}^k \qquad\qquad \text{(AB-DLTI)}$$

is controllable, for every $\mu > 0$*, it is possible to find a state feedback controller* $u = -Kx$ *that places all eigenvalues of the closed-loop system* $x^+ = (A - BK)x$ *in the complex plane disk* $|s| \leq \mu$. $\qquad\square$

Attention! The conditions in Theorems 12.6 and 12.7 are actually necessary and sufficient for controllability. In particular, for example in continuous time, one can also show that if for every $\mu > 0$ it is possible to find a state feedback controller $u = -Kx$ that places all eigenvalues of the closed-loop system $\dot{x} = (A - BK)x$ on the complex semiplane $\Re[s] \leq \mu$, then the pair (A, B) must be controllable. $\qquad\square$

12.5 EXERCISES

12.1 (*A*-invariance and controllability). Consider the LTI systems

$$\dot{x} = Ax + Bu \quad / \quad x^+ = Ax + Bu, \qquad x \in \mathbb{R}^n,\; u \in \mathbb{R}^k. \qquad \text{(AB-LTI)}$$

Prove the following two statements:

(a) The controllable subspace $\mathcal{C}$ of the system (AB-LTI) is A-invariant.

(b) The controllable subspace $\mathcal{C}$ of the system (AB-LTI) contains $\mathrm{Im}\, B$. $\qquad\square$

12.2 (Satellite). The equations of motion of a satellite, linearized around a steady-state solution, are given by $\dot{x} = Ax + Bu$, where x_1 and x_2 denote the perturbations in the radius and the radial velocity, respectively, x_3 and x_4 denote the perturbations in the angle and the angular velocity, and

$$A := \begin{bmatrix} 0 & 1 & 0 & 0 \\ 3\omega^2 & 0 & 0 & 2\omega \\ 0 & 0 & 0 & 1 \\ 0 & -2\omega & 0 & 1 \end{bmatrix}, \qquad B := \begin{bmatrix} 0 & 0 \\ 1 & 0 \\ 0 & 0 \\ 0 & 1 \end{bmatrix}.$$

The input vector consists of a radial thruster u_1 and a tangential thruster u_2.

(a) Show that the system is controllable from u.

(b) Can the system still be controlled if the radial thruster fails? What if the tangential thruster fails?

12.3 (Controllable canonical form). Consider a system in controllable canonical form

$$A = \begin{bmatrix} -\alpha_1 I_{k\times k} & -\alpha_2 I_{k\times k} & \cdots & -\alpha_{n-1} I_{k\times k} & -\alpha_n I_{k\times k} \\ I_{k\times k} & 0_{k\times k} & \cdots & 0_{k\times k} & 0_{k\times k} \\ 0_{k\times k} & I_{k\times k} & \cdots & 0_{k\times k} & 0_{k\times k} \\ \vdots & \vdots & \ddots & \vdots & \vdots \\ 0_{k\times k} & 0_{k\times k} & \cdots & I_{k\times k} & 0_{k\times k} \end{bmatrix}_{nk\times nk},$$

$$B = \begin{bmatrix} I_{k\times k} \\ 0_{k\times k} \\ \vdots \\ 0_{k\times k} \\ 0_{k\times k} \end{bmatrix}_{nk\times k}, \qquad C = \begin{bmatrix} N_1 & N_2 & \cdots & N_{n-1} & N_n \end{bmatrix}_{m\times nk}.$$

Show that such a system is always controllable. $\qquad\square$

LECTURE 13

Controllable Decompositions

CONTENTS

This lecture introduces a family of state-space similarity transformations that highlight the system's controllability (or lack thereof).

1. Invariance with Respect to Similarity Transformations
2. Controllable Decomposition
3. Block Diagram Interpretation
4. Transfer Function
5. MATLAB® Commands
6. Exercise

13.1 INVARIANCE WITH RESPECT TO SIMILARITY TRANSFORMATIONS

Consider the LTI systems

$$\dot{x} = Ax + Bu \quad / \quad x^+ = Ax + Bu, \qquad x \in \mathbb{R}^n, \ u \in \mathbb{R}^k \qquad \text{(AB-LTI)}$$

and a similarity transformation $\bar{x} := T^{-1}x$, leading to

$$\dot{\bar{x}} = \bar{A}\bar{x} + \bar{B}u, \qquad \bar{A} := T^{-1}AT, \qquad \bar{B} := T^{-1}B. \qquad (13.1)$$

The controllability matrices $\mathcal{C}$ and $\bar{\mathcal{C}}$ of the systems (AB-LTI) and (13.1), respectively, are related by

$$\bar{\mathcal{C}} = \begin{bmatrix} \bar{B} & \bar{A}\bar{B} & \cdots & \bar{A}^{n-1}\bar{B} \end{bmatrix} = \begin{bmatrix} T^{-1}B & T^{-1}AB & \cdots & T^{-1}A^{n-1}B \end{bmatrix} = T^{-1}\mathcal{C}.$$

Note. Why? because multiplication by a nonsingular matrix does not change the rank of a matrix.

Therefore

$$\text{rank}\,\bar{\mathcal{C}} = \text{rank}\,T^{-1}\mathcal{C} = \text{rank}\,\mathcal{C},$$

Note. Similarity transformations actually preserve the dimension of the controllable subspace, even when the system is not controllable.

because T^{-1} is nonsingular. Since the controllability of a system is determined by the rank of its controllability matrix, we conclude that controllability is preserved through similarity transformations, as formally stated in the following result.

Theorem 13.1. *The pair (A, B) is controllable if and only if the pair $(\bar{A}, \bar{B}) = (T^{-1}AT, T^{-1}B)$ is controllable.* □

13.2 CONTROLLABLE DECOMPOSITION

Consider again the LTI systems

$$\dot{x} = Ax + Bu \quad / \quad x^+ = Ax + Bu, \qquad x \in \mathbb{R}^n, \ u \in \mathbb{R}^k. \qquad \text{(AB-LTI)}$$

We saw in Exercise 12.1 that the controllable subspace $\mathcal{C}$ of the system (AB-LTI) is A-invariant and contains Im B. Because of A-invariance, by constructing an $n \times \bar{n}$ matrix V whose columns form a basis for $\mathcal{C}$, there exists an $\bar{n} \times \bar{n}$ matrix A_c such that

$$AV = V A_\text{c}.$$

Note. The number of columns of V is $\bar{n}$, and therefore $\bar{n}$ is also the dimension of the controllable subspace.

Moreover, since Im $B \subset \mathcal{C}$, the columns of B can be written as a linear combination of the columns of V, and therefore there exists an $\bar{n} \times k$ matrix B_c such that

$$B = V B_\text{c}.$$

When the system (AB-LTI) is controllable, $\bar{n} = \dim \mathcal{C} = n$, and the matrix V is square and nonsingular. Otherwise, let U be an $n \times (n - \bar{n})$ matrix whose columns are linearly independent of each other and also linearly independent of the columns of V.

Note. The columns of U are vectors that complete the columns of V to form a basis for $\mathbb{R}^n$.

Suppose that we define a nonsingular matrix T by combining V and U side by side:

$$T := \begin{bmatrix} V_{n \times \bar{n}} & U_{n \times (n - \bar{n})} \end{bmatrix}_{n \times n}.$$

We then conclude that

$$AT = A \begin{bmatrix} V & U \end{bmatrix} = \begin{bmatrix} AV & AU \end{bmatrix} = \begin{bmatrix} V A_\text{c} & T T^{-1} AU \end{bmatrix} = \begin{bmatrix} T \begin{bmatrix} A_\text{c} \\ 0 \end{bmatrix} & T T^{-1} AU \end{bmatrix}.$$

By partitioning the $n \times (n - \bar{n})$ matrix $T^{-1}AU$ as

$$T^{-1} AU = \begin{bmatrix} A_{12} \\ A_\text{u} \end{bmatrix},$$

we further obtain

Notation. This form is often called the *standard form for uncontrollable systems.*

$$AT = T \begin{bmatrix} A_\text{c} & A_{12} \\ 0 & A_\text{u} \end{bmatrix}, \qquad\qquad B = V B_\text{c} = T \begin{bmatrix} B_\text{c} \\ 0 \end{bmatrix},$$

which can be rewritten as

MATLAB® Hint 32.
`[Abar,Bbar,Cbar,T]`
`= ctrbf(A,B,C)`
computes the controllable decomposition of the system with realization
A,B,C. ▶ p. 122

$$\begin{bmatrix} A_\text{c} & A_{12} \\ 0 & A_\text{u} \end{bmatrix} = T^{-1} AT, \qquad\qquad \begin{bmatrix} B_\text{c} \\ 0 \end{bmatrix} = T^{-1} B. \qquad (13.2)$$

The similarity transformation constructed using this procedure is called a *controllable decomposition* and has several interesting properties, as stated in the following theorem.

Theorem 13.2 (Controllable decomposition). *For every LTI system* (AB-LTI), *there is a similarity transformation that takes the system to the form* (13.2), *for which*

1. *the controllable subspace of the transformed system* (13.2) *is given by*

$$\bar{\mathcal{C}} = \mathrm{Im} \begin{bmatrix} I_{\bar{n} \times \bar{n}} \\ 0 \end{bmatrix}$$

and

2. *the pair* (A_c, B_c) *is controllable.*   □

Proof of Theorem 13.2. To compute the controllable subspace of the transformed system, we compute its controllability matrix

$$\bar{C} = \begin{bmatrix} \begin{bmatrix} B_c \\ 0 \end{bmatrix} & \begin{bmatrix} A_c & A_{12} \\ 0 & A_u \end{bmatrix} \begin{bmatrix} B_c \\ 0 \end{bmatrix} & \cdots & \begin{bmatrix} A_c & A_{12} \\ 0 & A_u \end{bmatrix}^{n-1} \begin{bmatrix} B_c \\ 0 \end{bmatrix} \end{bmatrix}$$
$$= \begin{bmatrix} B_c & A_c B_c & \cdots & A_c^{n-1} B_c \\ 0 & 0 & \cdots & 0 \end{bmatrix}.$$

Since similarity transformations preserve the dimension of the controllable subspace, which was $\bar{n}$ for the original system,

$$\mathrm{rank}\, \bar{C} = \bar{n}.$$

Since the number of nonzero rows of $\bar{C}$ is exactly $\bar{n}$, all these rows must be linearly dependent. Therefore

$$\mathrm{Im}\, \bar{C} = \mathrm{Im} \begin{bmatrix} I_{\bar{n} \times \bar{n}} \\ 0 \end{bmatrix}.$$

Moreover,

$$\mathrm{rank} \begin{bmatrix} B_c & A_c B_c & \cdots & A_c^{n-1} B_c \end{bmatrix} = \bar{n}.$$

But since A_c is $\bar{n} \times \bar{n}$, by the Cayley-Hamilton theorem,

$$\mathrm{rank} \begin{bmatrix} B_c & A_c B_c & \cdots & A_c^{n-1} B_c \end{bmatrix} = \mathrm{rank} \begin{bmatrix} B_c & A_c B_c & \cdots & A_c^{\bar{n}-1} B_c \end{bmatrix} = \bar{n},$$

which proves that the pair (A_c, B_c) is controllable.   ■

13.3   BLOCK DIAGRAM INTERPRETATION

Consider now LTI systems with outputs

$$\dot{x}/x^+ = Ax + Bu, \qquad y = Cx + Du, \qquad x \in \mathbb{R}^n,\ u \in \mathbb{R}^k,\ y \in \mathbb{R}^m \qquad \text{(LTI)}$$

and let T be the similarity transformation that leads to the controllable decomposition

$$\begin{bmatrix} A_c & A_{12} \\ 0 & A_u \end{bmatrix} = T^{-1} A T, \qquad \begin{bmatrix} B_c \\ 0 \end{bmatrix} = T^{-1} B, \qquad \begin{bmatrix} C_c & C_u \end{bmatrix} = CT.$$

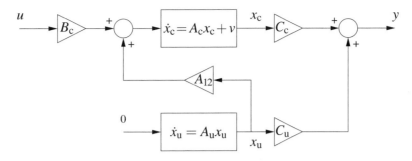

Figure 13.1. Controllable decomposition. The direct feed-through term D was omitted to simplify the diagram.

In general, the transformed output matrix CT has no particular structure, but for convenience we partition it into the first $\bar{n}$ columns and the remaining ones.

Partitioning the state of the transformed system as

Notation. The vectors x_c and x_u are called the *controllable* and the *uncontrollable* components of the state, respectively.

$$\bar{x} = T^{-1}x = \begin{bmatrix} x_c \\ x_u \end{bmatrix}, \qquad x_c \in \mathbb{R}^{\bar{n}}, \; x_u \in \mathbb{R}^{n-\bar{n}},$$

its state-space model can be written as

$$\begin{bmatrix} \dot{x}_c \\ \dot{x}_u \end{bmatrix} = \begin{bmatrix} A_c & A_{12} \\ 0 & A_u \end{bmatrix} \begin{bmatrix} x_c \\ x_u \end{bmatrix} + \begin{bmatrix} B_c \\ 0 \end{bmatrix} u, \qquad y = \begin{bmatrix} C_c & C_u \end{bmatrix} \begin{bmatrix} x_c \\ x_u \end{bmatrix} + Du.$$

Note. This is consistent with statement 1 in Theorem 13.2.

Figure 13.1 shows a block representation of this system, which highlights the fact that the input u cannot affect the x_u component of the state. Moreover, the controllability of the pair (A_c, B_c) means that the x_c component of the state can always be taken to the origin by an appropriate choice of $u(\cdot)$.

13.4 TRANSFER FUNCTION

Since similarity transformations do not change the system's transfer function, we can use the state-space model for the transformed system to compute the transfer function $T(s)$ of the original system

Note. This could have been deduced directly from the block diagram representation in Figure 13.1. In computing the transfer function we can ignore initial conditions, and, in this case, the x_u component of the state plays no role because it is identically zero.

$$T(s) = \begin{bmatrix} C_c & C_u \end{bmatrix} \begin{bmatrix} sI - A_c & -A_{12} \\ 0 & sI - A_u \end{bmatrix}^{-1} \begin{bmatrix} B_c \\ 0 \end{bmatrix} + D.$$

Since the matrix that needs to be inverted is upper triangular, its inverse is also upper triangular, and the diagonal blocks of the inverse are the inverses of the diagonal block of the matrix. Therefore

$$T(s) = \begin{bmatrix} C_c & C_u \end{bmatrix} \begin{bmatrix} (sI - A_c)^{-1} & \times \\ 0 & s(I - A_u)^{-1} \end{bmatrix} \begin{bmatrix} B_c \\ 0 \end{bmatrix} = C_c(sI - A_c)^{-1}B_c + D.$$

This shows that the transfer function of the system (LTI) is equal to the transfer function of its controllable part.

13.5 MATLAB® COMMANDS

MATLAB® Hint 32 (`ctrbf`). The command

$$[\texttt{Abar,Bbar,Cbar,T}] \ = \ \texttt{ctrbf(A,B,C)}$$

computes the controllable decomposition of the system with realization A, B, C. The matrices returned are such that

$$\texttt{Abar} = \begin{bmatrix} A_u & 0 \\ A_{21} & A_c \end{bmatrix} = \text{TAT}', \quad \texttt{Bbar} = \begin{bmatrix} 0 \\ B_c \end{bmatrix} = \text{TB}, \quad \texttt{Cbar} = \text{CT}', \quad \text{T}' = \text{T}^{-1}.$$

This decomposition places the uncontrollable modes *on top of* the controllable ones, opposite to what happens in (13.2). Moreover, the nonsingular matrix T is chosen to be orthogonal.

The command `[Abar,Bbar,Cbar,T] = ctrbf(A,B,C,tol)` further specifies the tolerance `tol` used to select the uncontrollable modes. □

13.6 EXERCISE

13.1 (Controllable decomposition). Consider an LTI system with realization

$$A = \begin{bmatrix} -1 & 0 \\ 0 & -1 \end{bmatrix}, \qquad B = \begin{bmatrix} -1 \\ 1 \end{bmatrix}, \qquad C = \begin{bmatrix} 1 & 0 \\ 0 & 1 \end{bmatrix}, \qquad D = \begin{bmatrix} 2 \\ 1 \end{bmatrix}.$$

Is this realization controllable? If not, perform a controllable decomposition to obtain a controllable realization of the same transfer function. □

LECTURE 14

Stabilizability

CONTENTS

This lecture introduces the concept of stabilizability.

14.1 STABILIZABLE SYSTEM

We saw in Lecture 13 that any LTI system is algebraically equivalent to a system in the following standard form for uncontrollable systems:

$$
\begin{bmatrix} \dot{x}_c/x_c^+ \\ \dot{x}_u/x_u^+ \end{bmatrix} = \begin{bmatrix} A_c & A_{12} \\ 0 & A_u \end{bmatrix} \begin{bmatrix} x_c \\ x_u \end{bmatrix} + \begin{bmatrix} B_c \\ 0 \end{bmatrix} u, \qquad x_c \in \mathbb{R}^{\bar{n}}, \ x_c \in \mathbb{R}^{n-\bar{n}}, \qquad (14.1a)
$$

$$
y = \begin{bmatrix} C_c & C_u \end{bmatrix} \begin{bmatrix} x_c \\ x_u \end{bmatrix} + Du, \qquad u \in \mathbb{R}^k, m \in \mathbb{R}^m. \qquad (14.1b)
$$

Definition 14.1 (Stabilizable system). The pair (A, B) is *stabilizable* if it is algebraically equivalent to a system in the standard form for uncontrollable systems (14.1) with $n = \bar{n}$ (i.e., A_u nonexistent) or with A_u a stability matrix. ☐

Since for stabilizable systems we have

$$
\dot{x}_u/x_u^+ = A_u x_u,
$$

with A_u a stability matrix, x_u converges to zero exponentially fast, and therefore we have

$$
\dot{x}_c/x_c^+ = A_c x_c + B_c u + d, \qquad\qquad y = C_c x_c + Du + n,
$$

Note. Any *controllable* system is stabilizable, because in this case $\bar{n} = n$ and the matrix A_u does not exist. Also, any *asymptotically stable* system is stabilizable, because in this case both A_c and A_u are stability matrices.

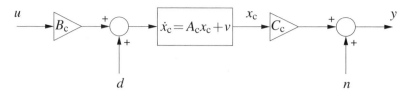

Figure 14.1. Controllable part of a stabilizable system. The direct feed-through term D was omitted to simplify the diagram.

where

$$d(t) := A_{12}x_{\mathrm{u}}(t), \qquad\qquad n(t) := C_{\mathrm{u}}x_{\mathrm{u}}(t), \qquad\qquad \forall t \geq 0$$

can be viewed as disturbance and noise terms, respectively, that converge to zero exponentially fast (cf. Figure 14.1).

Attention! Stabilizability can be viewed as an infinite-time version of controllability in the sense that if a system is stabilizable, then its state can be transferred to the origin from any initial state, but this may require *infinite time*. In particular, if the system is not controllable, then x_{u} will indeed "reach" the origin only as $t \to \infty$. $\square$

14.2 EIGENVECTOR TEST FOR STABILIZABILITY

Investigating the stabilizability of the LTI systems

$$\dot{x} = Ax + Bu \quad / \quad x^{+} = Ax + Bu \qquad x \in \mathbb{R}^n, \ u \in \mathbb{R}^k \qquad\qquad \text{(AB-LTI)}$$

from the definition requires the computation of their controllable decompositions. However, there are alternative tests that avoid this intermediate step.

Theorem 14.1 (Eigenvector test for stabilizability).

1. *The continuous-time LTI system* (AB-LTI) *is stabilizable if and only if every eigenvector of A' corresponding to an eigenvalue with a positive or zero real part is not in the kernel of B'.*

2. *The discrete-time LTI system* (AB-LTI) *is stabilizable if and only if every eigenvector of A' corresponding to an eigenvalue with magnitude larger or equal to 1 is not in the kernel of B'.* $\square$

Proof of Theorem 14.1. Let T be the similarity transformation that leads the system (AB-LTI) to the controllable decomposition, and let (14.1) be the corresponding standard form; i.e.,

$$\bar{A} := \begin{bmatrix} A_{\mathrm{c}} & A_{12} \\ 0 & A_{\mathrm{u}} \end{bmatrix} = T^{-1}AT, \qquad\qquad \bar{B} := \begin{bmatrix} B_{\mathrm{c}} \\ 0 \end{bmatrix} = T^{-1}B.$$

We start by proving that if the system (AB-LTI) is stabilizable, then every "unstable" eigenvector of A' is not in the kernel of B'. To prove by contradiction, assume that

Note. The term "unstable" should be understood in the appropriate sense, depending on whether we are considering continuous or discrete time. In either case, for the purposes of stabilizability, eigenvalues on the "boundary" are considered unstable.

there exists an "unstable" eigenvalue-eigenvector pair (λ, x) for which

$$A'x = \lambda x, \quad B'x = 0 \quad \Leftrightarrow \quad (T\bar{A}T^{-1})'x = \lambda x, \quad (T\bar{B})'x = 0$$

$$\Leftrightarrow \quad \begin{bmatrix} A'_{\mathrm{c}} & 0 \\ A'_{12} & A'_{\mathrm{u}} \end{bmatrix} T'x = \lambda T'x, \quad \begin{bmatrix} B'_{\mathrm{c}} & 0 \end{bmatrix} T'x = 0$$

$$\Leftrightarrow \quad \begin{bmatrix} A'_{\mathrm{c}} & 0 \\ A'_{12} & A'_{\mathrm{u}} \end{bmatrix} \begin{bmatrix} x_{\mathrm{c}} \\ x_{\mathrm{u}} \end{bmatrix} = \lambda \begin{bmatrix} x_{\mathrm{c}} \\ x_{\mathrm{u}} \end{bmatrix}, \quad \begin{bmatrix} B'_{\mathrm{c}} & 0 \end{bmatrix} \begin{bmatrix} x_{\mathrm{c}} \\ x_{\mathrm{u}} \end{bmatrix} = 0,$$

$$(14.2)$$

where $\begin{bmatrix} x'_{\mathrm{c}} & x'_{\mathrm{u}} \end{bmatrix}' := T'x \neq 0$. Since the pair $(A_{\mathrm{c}}, B_{\mathrm{c}})$ is controllable and

$$A'_{\mathrm{c}} x_{\mathrm{c}} = \lambda x_{\mathrm{c}}, \qquad\qquad B'_{\mathrm{c}} x_{\mathrm{c}} = 0,$$

we must have $x_{\mathrm{c}} = 0$ (and consequently $x_{\mathrm{u}} \neq 0$), since otherwise this would violate the eigenvector test for controllability. This means that λ must be an eigenvalue of A_{u} because

Note. This is because A_{u} and A'_{u} have the same eigenvalues: $\det(\lambda I - A_{\mathrm{u}}) = \det(\lambda I - A'_{\mathrm{u}})$.

$$A'_{\mathrm{u}} x_{\mathrm{u}} = \lambda x_{\mathrm{u}},$$

which contradicts the stabilizability of the system (AB-LTI) because λ is "unstable."

Conversely, suppose now that the system (AB-LTI) is not stabilizable. Therefore A'_{u} has an "unstable" eigenvalue-eigenvector pair

$$A'_{\mathrm{u}} x_{\mathrm{u}} = \lambda x_{\mathrm{u}}, \qquad x_{\mathrm{u}} \neq 0.$$

Then

$$\bar{A}' \begin{bmatrix} 0 \\ x_{\mathrm{u}} \end{bmatrix} = \begin{bmatrix} A'_{\mathrm{c}} & 0 \\ A'_{12} & A'_{\mathrm{u}} \end{bmatrix} \begin{bmatrix} 0 \\ x_{\mathrm{u}} \end{bmatrix} = \lambda \begin{bmatrix} 0 \\ x_{\mathrm{u}} \end{bmatrix},$$

$$\bar{B}' \begin{bmatrix} 0 \\ x_{\mathrm{u}} \end{bmatrix} = \begin{bmatrix} B'_{\mathrm{c}} & 0 \end{bmatrix} \begin{bmatrix} 0 \\ x_{\mathrm{u}} \end{bmatrix} = 0, \quad \begin{bmatrix} 0 \\ x_{\mathrm{u}} \end{bmatrix} \neq 0.$$

We have thus far found an "unstable" eigenvector of $\bar{A}'$ in the kernel of $\bar{B}'$, so $(\bar{A}, \bar{B})$ cannot be stabilizable. To conclude that the original pair (A, B) is also not stabilizable, we use the equivalences in (14.2) to conclude that

$$x := (T')^{-1} \begin{bmatrix} 0 \\ x_{\mathrm{u}} \end{bmatrix} \quad \Leftrightarrow \quad \begin{bmatrix} 0 \\ x_{\mathrm{u}} \end{bmatrix} := T'x$$

is an "unstable" eigenvector of A' in the kernel of B'.   ∎

14.3 POPOV-BELEVITCH-HAUTUS (PBH) TEST FOR STABILIZABILITY

For stabilizability, one can also reformulate the eigenvector test as a rank condition, as was done in Theorem 12.3 for controllability.

Theorem 14.2 (Popov-Belevitch-Hautus [PBH] test for stabilizability).

1. *The continuous-time LTI system* (AB-LTI) *is stabilizable if and only if*

$$\text{rank} \begin{bmatrix} A - \lambda I & B \end{bmatrix} = n, \qquad \forall \lambda \in \mathbb{C} : \Re[\lambda] \geq 0. \tag{14.3}$$

2. *The discrete-time LTI system* (AB-LTI) *is stabilizable if and only if*

$$\text{rank} \begin{bmatrix} A - \lambda I & B \end{bmatrix} = n, \qquad \forall \lambda \in \mathbb{C} : |\lambda| \geq 1. \qquad \square$$

The proof of this theorem is analogous to the proof of Theorem 12.3, except that now we need to restrict our attention to only the "unstable" portion of $\mathbb{C}$.

14.4 LYAPUNOV TEST FOR STABILIZABILITY

Consider again the LTI systems

$$\dot{x} = Ax + Bu \quad / \quad x^+ = Ax + Bu, \qquad x \in \mathbb{R}^n, \ u \in \mathbb{R}^k. \tag{AB-LTI}$$

Note. The term BB' in (14.4) appears with opposite sign with respect to the Lyapunov test for controllability, where the Lyapunov equality was
$AP + PA' + BB' = 0.$

Note. Equation (14.4) is known as a *linear matrix inequality (LMI)*. The term "linear" comes from the fact that the left-hand side is linear in P, and $<$ refers to the fact that the left-hand side must be *negative-definite*.

Theorem 14.3 (Lyapunov test for stabilizability).   *The LTI system* (AB-LTI) *is stabilizable if and only if there is a positive-definite solution P to the following Lyapunov matrix inequality*

$$AP + PA' - BB' < 0 \quad / \quad APA' - P - BB' < 0. \tag{14.4}$$

Proof of Theorem 14.3. We do the proof for continuous time and start by showing that if (14.4) has a positive-definite solution P, then the system (AB-LTI) is stabilizable. The simplest way to do this is by using the eigenvector test. Assume that (14.4) holds, and let $x \neq 0$ be an eigenvector of A' associated with the "unstable" eigenvalue λ; i.e., $A'x = \lambda x$. Then

$$x^*(AP + PA')x < x^*BB'x = \|B'x\|^2,$$

where $(\cdot)^*$ denotes the complex conjugate transpose. But the left-hand side of this equation is equal to

$$(A'x^*)'Px + x^*PA'x = \lambda^* x^* Px + \lambda x^* Px = 2\Re[\lambda]x^*Px.$$

Since P is positive-definite and $\Re[\lambda] \geq 0$, we conclude that

$$0 \leq 2\Re[\lambda]x^*Px < \|B'x\|^2,$$

and therefore x must not belong to the kernel of B'.

To prove the converse, we assume that the system (AB-LTI) is stabilizable. Let T be the similarity transformation that leads the system (AB-LTI) to the controllable decomposition, and let (14.1) be the corresponding standard form; i.e.,

$$\bar{A} := \begin{bmatrix} A_c & A_{12} \\ 0 & A_u \end{bmatrix} = T^{-1}AT, \qquad\qquad \bar{B} := \begin{bmatrix} B_c \\ 0 \end{bmatrix} = T^{-1}B,$$

We saw in Section 12.4 (regarding feedback stabilization based on the Lyapunov test) that controllability of the pair (A_c, B_c) guarantees the existence of a positive-definite matrix P_c such that

$$A_c P_c + P_c A_c' - B_c B_c' = -Q_c < 0$$

[cf. equation (12.12)]. On the other hand, since A_u is a stability matrix, we conclude from the Lyapunov stability theorem (Theorem 8.2) that there exists a positive-definite matrix P_u such that

$$A_u P_u + P_u A_u' = -Q_u < 0.$$

Defining

$$\bar{P} = \begin{bmatrix} P_c & 0 \\ 0 & \rho P_u \end{bmatrix}$$

for some scalar $\rho > 0$ to be determined shortly, we conclude that

$$
\begin{aligned}
\bar{A}\bar{P} &+ \bar{P}\bar{A}' - \bar{B}\bar{B}' \\
&= \begin{bmatrix} A_c & A_{12} \\ 0 & A_u \end{bmatrix} \begin{bmatrix} P_c & 0 \\ 0 & \rho P_u \end{bmatrix} + \begin{bmatrix} P_c & 0 \\ 0 & \rho P_u \end{bmatrix} \begin{bmatrix} A_c' & 0 \\ A_{12}' & A_u \end{bmatrix} - \begin{bmatrix} B_c \\ 0 \end{bmatrix} \begin{bmatrix} B_c' & 0 \end{bmatrix} \\
&= -\begin{bmatrix} Q_c & -\rho A_{12} P_u \\ -\rho P_u A_{12}' & \rho Q_u \end{bmatrix}.
\end{aligned}
$$

It turns out that by making ρ positive, but sufficiently small, the right-hand side can be made negative-definite . The proof is completed by verifying that the matrix

Note. This can be proved by completing the square. See Exercise 14.1.

$$P = T \begin{bmatrix} P_c & 0 \\ 0 & \rho P_u \end{bmatrix} T'$$

satisfies (14.4)   ∎

14.5 FEEDBACK STABILIZATION BASED ON THE LYAPUNOV TEST

Assume that the continuous-time LTI system

$$\dot{x} = Ax + Bu, \qquad\qquad x \in \mathbb{R}^n,\ u \in \mathbb{R}^k \qquad\qquad \text{(AB-CLTI)}$$

is stabilizable. We saw in the Lyapunov test for stabilizability (Theorem 14.3) that this guarantees the existence of a positive-definite solution P for which

$$AP + PA' - BB' < 0.$$

Defining $K := \frac{1}{2} B' P^{-1}$, this inequality can be rewritten as

$$\left(A - \frac{1}{2}BB'P^{-1}\right)P + P\left(A - \frac{1}{2}BB'P^{-1}\right)' = (A - BK)P + P(A - BK)' < 0.$$

Multiplying this inequality on the left and right by $Q := P^{-1}$, we obtain

$$Q(A - BK) + (A - BK)'Q < 0.$$

Since $Q > 0$, we conclude from the Lyapunov stability theorem (Theorem 8.2) that $A - BK$ must be a stability matrix. This means that the state feedback control

$$u = -Kx$$

asymptotically stabilizes the system (AB-CLTI).

Note. This result justifies the name "stabilizable."

Theorem 14.4. *When the system (AB-CLTI) is stabilizable, it is always possible to find a state feedback controller $u = -Kx$ that makes the closed-loop system $\dot{x} = (A - BK)x$ asymptotically stable.* ☐

The discrete-time equivalent of this result is as follows.

Note. As opposed to the analogous result for controllability, we now cannot make the closed-loop eigenvalues arbitrarily fast (cf. Theorem 12.6).

Theorem 14.5. *When the system*

$$x^+ = Ax + Bu, \qquad\qquad x \in \mathbb{R}^n,\ u \in \mathbb{R}^k \qquad\qquad \text{(AB-DLTI)}$$

is stabilizable, it is always possible to find a state feedback controller $u = -Kx$ that makes the closed-loop system $x^+ = (A - BK)x$ asymptotically stable. ☐

Attention! The conditions in Theorems 14.4 and 14.5 are actually necessary and sufficient for stabilizability. In particular, one can also show that if it is possible to find a state feedback controller $u = -Kx$ that makes the closed-loop system $\dot{x}/x^+ = (A - BK)x$ asymptotically stable, then the pair (A, B) must be stabilizable. ☐

14.6 EIGENVALUE ASSIGNMENT

We saw in Section 14.5 that when a system is stabilizable, it is possible to find a feedback controller that makes the closed loop asymptotically stable.

When the system is not only stabilizable, but also controllable, we saw in Sections 12.4 that one can actually make the closed-loop eigenvalues arbitrarily fast. It turn out that for controllable systems, one has complete freedom to select the closed-loop eigenvalues.

MATLAB® Hint 33. `K=place(A,B,P)` computes a matrix K such that the eigenvalues of A-B K are those specified in the vector P. This command should be used with great caution and generally avoided because it is numerically badly conditioned. ▶ p. 128

Theorem 14.6 (Eigenvalue assignment). *Assume that the system*

$$\dot{x}/x^+ = Ax + Bu, \qquad\qquad x \in \mathbb{R}^n,\ u \in \mathbb{R}^k \qquad\qquad \text{(AB-CLTI)}$$

is controllable. Given any set of n complex numbers $\lambda_1, \lambda_2, \ldots, \lambda_n$, there exists a state feedback matrix K such that the closed-loop system $\dot{x}/x^+ = (A - BK)x$ has eigenvalues equal to the λ_i. ☐

The proof of this theorem can be found in [1, Section 4.2 B]. The special case of a SISO system in controllable canonical form is proved in Exercise 14.2.

14.7 MATLAB® COMMANDS

MATLAB® Hint 33 (`place`). The command K=place(A,B,P) computes a matrix K such that the eigenvalues of A-B K are those specified in the vector P. The pair (A,B) should be controllable, and the vector P should have no repeated eigenvalues. This command should be used with great caution (and generally avoided), because it is numerically badly conditioned. □

14.8 EXERCISES

14.1 (Positive definiteness of a partitioned matrix). Consider a symmetric matrix P that can be partitioned as follows:

$$P = \begin{bmatrix} Q & \rho S \\ \rho S' & \rho R \end{bmatrix},$$

where Q and R are both square symmetric and positive-definite matrices and ρ is a positive scalar. Show that the matrix P is positive-definite for a sufficiently small, but positive, ρ.

Hint: Show that we can pick $\rho > 0$ so that $x'Px > 0$ for every nonzero vector x, by completing the squares. □

14.2 (Eigenvalue assignment). Consider the SISO LTI system in controllable canonical form

$$\dot{x} = Ax + Bu, \qquad x \in \mathbb{R}^n,\ u \in \mathbb{R}^1, \qquad \text{(AB-DLTI)}$$

where

$$A = \begin{bmatrix} -\alpha_1 & -\alpha_2 & \cdots & -\alpha_{n-1} & -\alpha_n \\ 1 & 0 & \cdots & 0 & 0 \\ 0 & 1 & \cdots & 0 & 0 \\ \vdots & \vdots & \ddots & \vdots & \vdots \\ 0 & 0 & \cdots & 1 & 0 \end{bmatrix}_{n \times n}, \qquad B = \begin{bmatrix} 1 \\ 0 \\ \vdots \\ 0 \\ 0 \end{bmatrix}_{n \times 1}.$$

(a) Compute the characteristic polynomial of the closed-loop system for

$$u = -Kx, \qquad K := \begin{bmatrix} k_1 & k_2 & \cdots & k_n \end{bmatrix}.$$

Hint: Compute the determinant of $(sI - A + BK)$ by doing a Laplacian expansion along the first line of this matrix.

(b) Suppose you are given n complex numbers $\lambda_1, \lambda_2, \ldots, \lambda_n$ as desired locations for the closed-loop eigenvalues. Which characteristic polynomial for the closed-loop system would lead to these eigenvalues?

(c) Based on the answers to parts (a) and (b), propose a procedure to select K that would result in the desired values for the closed-loop eigenvalues.

(d) Suppose that

$$A = \begin{bmatrix} 1 & 2 & 3 \\ 1 & 0 & 0 \\ 0 & 1 & 0 \end{bmatrix}, \qquad B = \begin{bmatrix} 1 \\ 0 \\ 0 \end{bmatrix}.$$

Find a matrix K for which the closed-loop eigenvalues are $\{-1, -1, -2\}$. □

14.3 (Transformation to controllable canonical form). Consider the following third-order SISO LTI system

$$\dot{x} = Ax + Bu, \qquad x \in \mathbb{R}^3, \ u \in \mathbb{R}^1. \qquad \text{(AB-CLTI)}$$

Assume that the characteristic polynomial of A is given by

$$\det(sI - A) = s^3 + \alpha_1 s^2 + \alpha_2 s + \alpha_3$$

and consider the 3×3 matrix

$$T := C \begin{bmatrix} 1 & \alpha_1 & \alpha_2 \\ 0 & 1 & \alpha_1 \\ 0 & 0 & 1 \end{bmatrix}, \qquad (14.5)$$

where C is the system's controllability matrix.

(a) Show that the following equality holds:

$$B = T \begin{bmatrix} 1 \\ 0 \\ 0 \end{bmatrix}.$$

(b) Show that the following equality holds:

$$AT = T \begin{bmatrix} -\alpha_1 & -\alpha_2 & -\alpha_3 \\ 1 & 0 & 0 \\ 0 & 1 & 0 \end{bmatrix}.$$

Hint: Compute separately the left- and right-hand side of the equation above and then show that the two matrices are equal with the help of the Cayley-Hamilton theorem.

(c) Show that if the system (AB-CLTI) is controllable, then T is a nonsingular matrix.

(d) Combining parts (a)–(c), you showed that, if the system (AB-CLTI) is controllable, then the matrix T given by equation (14.5) can be viewed as a similarity transformation that transforms the system into the controllable canonical form

$$T^{-1}AT = \begin{bmatrix} -\alpha_1 & -\alpha_2 & -\alpha_3 \\ 1 & 0 & 0 \\ 0 & 1 & 0 \end{bmatrix}, \qquad T^{-1}B = \begin{bmatrix} 1 \\ 0 \\ 0 \end{bmatrix}.$$

Use this to find the similarity transformation that transforms the following pair into the controllable canonical form

$$A := \begin{bmatrix} 6 & 4 & 1 \\ -5 & -4 & 0 \\ -4 & -3 & -1 \end{bmatrix}, \qquad B := \begin{bmatrix} 1 \\ -1 \\ -1 \end{bmatrix}.$$

Hint: You may use the MATLAB® functions `poly(A)` *to compute the characteristic polynomial of* A *and* `ctrb(A,B)` *to compute the controllability matrix of the pair* `(A,B)`. □

PART IV
Observability and Output Feedback

LECTURE 15

Observability

Contents

This lecture introduces the notions of observability and constructibility.

1. Motivation: Output Feedback
2. Unobservable Subspace
3. Unconstructible Subspace
4. Physical Examples
5. Observability and Constructibility Gramians
6. Gramian-based Reconstruction
7. Discrete-Time Case
8. Duality (LTI)
9. Observability Tests

15.1 MOTIVATION: OUTPUT FEEDBACK

Consider the continuous-time LTI system

$$\dot{x} = Ax + Bu, \qquad y = Cx + Du, \qquad x \in \mathbb{R}^n,\ u \in \mathbb{R}^k,\ y \in \mathbb{R}^m. \qquad \text{(CLTI)}$$

We saw in Lecture 14 (Section 14.5) that if the pair (A, B) is stabilizable, then there exists a state feedback control law

$$u = -Kx \qquad (15.1)$$

that asymptotically stabilizes the system (CLTI), i.e., for which $A - BK$ is a stability matrix. However, when only the output y can be measured (as opposed to the whole state x), the control law (15.1) cannot be implemented. In principle, this difficulty can be overcome if it is possible to reconstruct the state of the system based on its measured output and perhaps also on the control input that is applied.

When the C matrix is invertible, instantaneous reconstruction of x from y and u is possible by solving the output equation for x:

$$x(t) = C^{-1}\big(y(t) - Du(t)\big).$$

However, this would be possible only if the number of outputs was equal to the number of states (C is square). When the number of outputs is strictly smaller than the number of states, instantaneous reconstruction of x from y and u is not possible, but it may still be possible to reconstruct the state from the input $u(t)$ and output $y(t)$ over an interval $[t_0, t_1]$. Two formulations are usually considered.

1. *Observability* refers to determining $x(t_0)$ from the *future* inputs and outputs $u(t)$ and $y(t)$, $t \in [t_0, t_1]$.

2. *Constructibility* refers to determining $x(t_1)$ from the *past* inputs and outputs $u(t)$ and $y(t)$, $t \in [t_0, t_1]$.

15.2 UNOBSERVABLE SUBSPACE

Consider the continuous-time LTV system

$$\dot{x} = A(t)x + B(t)u, \quad y = C(t)x + D(t)u, \quad x \in \mathbb{R}^n, \ u \in \mathbb{R}^k, \ y \in \mathbb{R}^m. \quad \text{(CLTV)}$$

We have seen in Lecture 5 that the system's state $x_0 := x(t_0)$ at time t_0 is related to its input and output on the interval $[t_0, t_1]$ by the variation of constants formula:

$$y(t) = C(t)\Phi(t, t_0)x_0 + \int_{t_0}^{t} C(t)\Phi(t, \tau)B(\tau)u(\tau)d\tau + D(t)u(t), \qquad \forall t \in [t_0, t_1],$$

$$\tag{15.2}$$

where $\Phi(\cdot)$ denotes the system's state transition matrix. To study the system's observability, we need to determine under which conditions we can solve

$$\tilde{y}(t) = C(t)\Phi(t, t_0)x_0, \qquad \forall t \in [t_0, t_1] \tag{15.3}$$

Notation. Given an input/output pair $u(t)$, $y(t)$, $t \in [t_0, t_1]$, we say that it is *compatible with an initial state x_0* if (15.2) [or equivalently (15.3)] holds.

for the unknown $x_0 \in \mathbb{R}^n$, where

$$\tilde{y}(t) = y(t) - \int_{t_0}^{t} C(t)\Phi(t, \tau)B(\tau)u(\tau)d\tau - D(t)u(t), \qquad \forall t \in [t_0, t_1].$$

This motivates the following definition.

Definition 15.1 (Unobservable subspace). Given two times $t_1 > t_0 \geq 0$, the *unobservable subspace on $[t_0, t_1]$* $\mathcal{UO}[t_0, t_1]$ consists of all states $x_0 \in \mathbb{R}^n$ for which

$$C(t)\Phi(t, t_0)x_0 = 0, \qquad \forall t \in [t_0, t_1]. \qquad \qquad \square$$

The importance of the unobservable subspace stems from the following properties.

Properties (Unobservable subspace). Suppose we are given two times $t_1 > t_0 \geq 0$ and an input/output pair $u(t)$, $y(t)$, $t \in [t_0, t_1]$.

P15.1 When a particular initial state $x_0 = x(t_0)$ is compatible with the input/output pair, then every initial state of the form

$$x_0 + x_{\mathrm{u}}, \qquad x_{\mathrm{u}} \in \mathcal{UO}[t_0, t_1]$$

is also compatible with the same input/output pair.

This is because

$$\begin{cases} \tilde{y}(t) = C(t)\Phi(t, t_0)x_0, & \forall t \in [t_0, t_1] \\ 0 = C(t)\Phi(t, t_0)x_u, & \forall t \in [t_0, t_1] \end{cases}$$

$$\Rightarrow \tilde{y}(t) = C(t)\Phi(t, t_0)(x_0 + x_u), \qquad \forall t \in [t_0, t_1].$$

P15.2 When the unobservable subspace contains only the zero vector, then there exists at most one initial state that is compatible with the input/output pair.

This is because if two different states $x_0, \bar{x}_0 \in \mathbb{R}^n$ were compatible with the same input/output pair, we would have

$$\begin{cases} \tilde{y}(t) = C(t)\Phi(t, t_0)x_0, & \forall t \in [t_0, t_1] \\ \tilde{y}(t) = C(t)\Phi(t, t_0)\bar{x}_0, & \forall t \in [t_0, t_1] \end{cases}$$

$$\Rightarrow 0 = C(t)\Phi(t, t_0)(x_0 - \bar{x}_0), \qquad \forall t \in [t_0, t_1],$$

and therefore $x_0 - \bar{x}_0 \neq 0$ would have to belong to the unobservable subspace. □

These properties motivate the following definition.

Definition 15.2 (Observable system). Given two times $t_1 > t_0 \geq 0$, the system (CLTV) is *observable* if its unobservable subspace contains only the zero vector; i.e., $\mathcal{UO}[t_0, t_1] = 0$. □

Note. Because of property P15.2, it is possible to uniquely reconstruct the state of an observable system from (future) inputs/outputs.

The matrices $B(\cdot)$ and $D(\cdot)$ play no role in the definition of the unobservable subspace; therefore one often simply talks about the unobservable subspace or the observability of the system

$$\dot{x} = A(t)x, \qquad y = C(t)x, \qquad x \in \mathbb{R}^n, \ y \in \mathbb{R}^m \qquad \text{(AC-CLTV)}$$

or simply of the pair $\big(A(\cdot), C(\cdot)\big)$.

15.3 UNCONSTRUCTIBLE SUBSPACE

The "future" system's state $x_1 := x(t_1)$ at time t_1 can also be related to the system's input and output on the interval $[t_0, t_1]$ by the variation of constants formula:

$$y(t) = C(t)\Phi(t, t_1)x_1 + \int_{t_1}^{t} C(t)\Phi(t, \tau)B(\tau)u(\tau)d\tau + D(t)u(t), \quad \forall t \in [t_0, t_1].$$

This formula motivates the following definition.

Definition 15.3 (Unconstructible subspace). Given two times $t_1 > t_0 \geq 0$, the *unconstructible subspace on* $[t_0, t_1]$ $\mathcal{UC}[t_0, t_1]$ consists of all states x_1 for which

$$C(t)\Phi(t, t_1)x_1 = 0, \qquad \forall t \in [t_0, t_1]. \qquad □$$

The importance of the unconstructible subspace stems from the following simple properties.

Properties (Unconstructible subspace). Suppose we are given two times $t_1 > t_0 \geq 0$ and an input/output pair $u(t)$, $y(t)$, $t \in [t_0, t_1]$.

P15.3 When a particular final state $x_1 = x(t_1)$ is compatible with the input/output pair, then every final state of the form

$$x_1 + x_u, \qquad x_u \in \mathcal{UC}[t_0, t_1]$$

is also compatible with the same input/output pair.

P15.4 When the unconstructible subspace contains only the zero vector, then there exists at most one final state that is compatible with the input/output pair. $\square$

Note. Because of property P15.4, it is possible to uniquely reconstruct the state of a constructible system from (past) inputs/outputs.

Definition 15.4 (Constructible system). Given two times $t_1 > t_0 \geq 0$, the system (CLTV) is *constructible* if its unconstructible subspace contains only the zero vector, i.e., $\mathcal{UC}[t_0, t_1] = 0$. $\square$

15.4 PHYSICAL EXAMPLES

Example 15.1 (Parallel interconnection). Consider the parallel interconnection in Figure 15.1 of two systems with states $x_1, x_2 \in \mathbb{R}^n$. The overall system corresponds to the following state-space model

$$\dot{x} = \begin{bmatrix} A_1 & 0 \\ 0 & A_2 \end{bmatrix} x + \begin{bmatrix} B_1 \\ B_2 \end{bmatrix} u, \qquad\qquad y = \begin{bmatrix} C_1 & C_2 \end{bmatrix} x,$$

where we chose for state $x := \begin{bmatrix} x_1' & x_2' \end{bmatrix}' \in \mathbb{R}^{2n}$. The output to this system is given by

$$y(t) = C_1 e^{A_1 t} x_1(0) + C_1 e^{A_2 t} x_2(0) + \int_0^t \left(C_1 e^{A_1(t-\tau)} B_1 + C_2 e^{A_2(t-\tau)} B_2 \right) u(\tau) d\tau.$$

When $A_1 = A_2 = A$ and $C_1 = C_2 = C$, we have

$$y(t) = C e^{At} \left(x_1(0) + x_2(0) \right) + \int_0^t C e^{A(t-\tau)} (B_1 + B_2) u(\tau) d\tau.$$

This shows that, solely by knowing the input and output of the system, we cannot distinguish between initial states for which $x_1(0) + x_2(0)$ is the same. $\square$

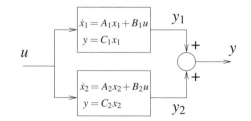

Figure 15.1. Parallel interconnections.

15.5 OBSERVABILITY AND CONSTRUCTIBILITY GRAMIANS

The following definitions are useful to characterize the unobservable and unconstructible subspaces.

Definition 15.5 (Observability and constructibility Gramians). Given two times $t_1 > t_0 \geq 0$, the *observability* and *constructibility Gramians* of the system (CLTV) are defined by

$$W_O(t_0, t_1) := \int_{t_0}^{t_1} \Phi(\tau, t_0)' C(\tau)' C(\tau) \Phi(\tau, t_0) d\tau,$$

$$W_{Cn}(t_0, t_1) := \int_{t_0}^{t_1} \Phi(\tau, t_1)' C(\tau)' C(\tau) \Phi(\tau, t_1) d\tau. \qquad \square$$

As the names suggest, these Gramians allow one to compute the unobservable and the unconstructible spaces.

Theorem 15.1 (Unobservable and unconstructible subspaces). *Given two times $t_1 > t_0 \geq 0$,*

$$\mathcal{UO}[t_0, t_1] = \ker W_O(t_0, t_1), \qquad \mathcal{UC}[t_0, t_1] = \ker W_{Cn}(t_0, t_1). \qquad \square$$

Proof of Theorem 15.1. From the definition of the observability Gramian, for every $x_0 \in \mathbb{R}^n$, we have

$$x_0' W_O(t_0, t_1) x_0 = \int_{t_0}^{t_1} x_0' \Phi(\tau, t_0)' C(\tau)' C(\tau) \Phi(\tau, t_0) x_0 \, d\tau$$

$$= \int_{t_0}^{t_1} \| C(\tau) \Phi(\tau, t_0) x_0 \|^2 d\tau.$$

Therefore

$$x_0 \in \ker W_O(t_0, t_1) \quad \Rightarrow \quad C(\tau)\Phi(\tau, t_0)x_0 = 0, \ \forall \tau \in [t_0, t_1]$$
$$\Rightarrow \quad x_0 \in \mathcal{UO}[t_0, t_1].$$

Conversely,

$$x_0 \in \mathcal{UO}[t_0, t_1] \quad \Rightarrow \quad C(\tau)\Phi(\tau, t_0)x_0 = 0, \ \forall \tau \in [t_0, t_1]$$
$$\Rightarrow \quad x_0 \in \ker W_O(t_0, t_1).$$

For the second implication, we are using the fact that, for any given positive-semidefinite matrix W, $x'Wx = 0$ implies that $Wx = 0$. This implication is not true for nonsemidefinite matrices. A similar argument can be made for the unconstructible subspace. ∎

This result provides a first method to determine whether a system is observable or constructible, because the kernel of a square matrix contains only the zero vector when the matrix is nonsingular .

Corollary 15.1 (Observable and constructible systems). *Suppose we are given two times $t_1 > t_0 \geq 0$.*

1. *The system* (CLTV) *is observable if and only if* rank $W_O(t_0, t_1) = n$.

2. *The system* (CLTV) *is constructible if and only if* rank $W_{Cn}(t_0, t_1) = n$. $\qquad \square$

Note. Both Gramians are symmetric positive-semidefinite $n \times n$ matrices.

Note. These Gramians are very similar to the controllability and reachability Gramians, except that now the transposes appear on the left and the $B(\cdot)$ matrix has been replaced by the $C(\cdot)$ matrix.

Note. Recall from the fundamental theorem of linear equations that for an $m \times n$ matrix W, dim ker W + rank $W = n$.

15.6 GRAMIAN-BASED RECONSTRUCTION

Consider the continuous-time LTV system

$$\dot{x} = A(t)x + B(t)u, \quad y = C(t)x + D(t)u, \quad x \in \mathbb{R}^n, \ u \in \mathbb{R}^k, \ y \in \mathbb{R}^m. \quad \text{(CLTV)}$$

We have seen that the system's state $x_0 := x(t_0)$ at time t_0 is related to its input and output on the interval $[t_0, t_1]$ by

$$\tilde{y}(t) = C(t)\Phi(t, t_0)x_0, \qquad \forall t \in [t_0, t_1], \qquad (15.4)$$

where

$$\tilde{y}(t) = y(t) - \int_{t_0}^{t} C(t)\Phi(t, \tau)B(\tau)u(\tau)d\tau - D(t)u(t), \qquad \forall t \in [t_0, t_1]. \quad (15.5)$$

Premultiplying (15.4) by $\Phi(t, t_0)'C(t)'$ and integrating between t_0 and t_1 yields

$$\int_{t_0}^{t_1} \Phi(t, t_0)'C(t)'\tilde{y}(t)dt = \int_{t_0}^{t_1} \Phi(t, t_0)'C(t)'C(t)\Phi(t, t_0)x_0 dt,$$

which can be written as

$$W_O(t_0, t_1)x_0 = \int_{t_0}^{t_1} \Phi(t, t_0)'C(t)'\tilde{y}(t)dt.$$

If the system is observable, $W_O(t_0, t_1)$ is invertible, and we conclude that

$$x_0 = W_O(t_0, t_1)^{-1} \int_{t_0}^{t_1} \Phi(t, t_0)'C(t)'\tilde{y}(t)dt,$$

which allows us to reconstruct $x(t_0)$ from the future inputs and outputs on $[t_0, t_1]$. A similar construction can be carried out to reconstruct $x(t_1)$ from past inputs and outputs for reconstructible systems. This is summarized in the following statement.

Theorem 15.2 (Gramian-based reconstruction). *Suppose we are given two times $t_1 > t_0 \geq 0$ and an input/output pair $u(t), y(t), t \in [t_0, t_1]$.*

1. When the system (CLTV) is observable

$$x(t_0) = W_O(t_0, t_1)^{-1} \int_{t_0}^{t_1} \Phi(t, t_0)'C(t)'\tilde{y}(t)dt,$$

where $\tilde{y}(t)$ is defined by (15.5).

2. When the system (CLTV) is constructible

$$x(t_1) = W_{Cn}(t_0, t_1)^{-1} \int_{t_0}^{t_1} \Phi(t, t_1)'C(t)'\bar{y}(t)dt,$$

where

$$\bar{y}(t) := y(t) - \int_{t_1}^{t} C(t)\Phi(t, \tau)B(\tau)u(\tau)d\tau - D(t)u(t), \qquad \forall t \in [t_0, t_1].$$

$\square$

15.7 DISCRETE-TIME CASE

Consider the discrete-time LTV system

$$x(t+1) = A(t)x(t) + B(t)u(t), \qquad y(t) = C(t)x(t) + D(t)u(t), \qquad \text{(DLTV)}$$

for which the system's state $x_0 := x(t_0)$ at time t_0 is related to its input and output on the interval $t_0 \leq t < t_1$ by the variation of constants formula,

$$y(t) = C(t)\Phi(t, t_0)x_0 + \sum_{\tau=t_0}^{t-1} C(t)\Phi(t, \tau)B(\tau)u(\tau)d\tau + D(t)u(t), \quad \forall t_0 \leq t < t_1.$$

Definition 15.6 (Unobservable and unconstructible subspaces). Given two times $t_1 > t_0 \geq 0$, the *unobservable subspace on* $[t_0, t_1)$ $\mathcal{UO}[t_0, t_1)$ consists of all states x_0 for which

$$C(t)\Phi(t, t_0)x_0 = 0, \qquad \forall t_0 \leq t < t_1. \qquad \square$$

The *unconstructible subspace on* $[t_0, t_1)$ $\mathcal{UC}[t_0, t_1)$ consists of all states x_1 for which

$$C(t)\Phi(t, t_1)x_1 = 0, \qquad \forall t_0 \leq t < t_1. \qquad \square$$

Attention! The definition of the discrete-time unconstructible subspace requires a backward-in-time state transition matrix $\Phi(t, t_1)$ from time t_1 to time $t \leq t_1 - 1 < t_1$. This matrix is well defined only when

$$x(t_1) = A(t_1 - 1)A(t_1 - 2) \cdots A(\tau)x(t), \quad t_0 \leq \tau \leq t_1 - 1$$

can be solved for $x(t)$, i.e., when all the matrices $A(t_0)$, $A(t_0 + 1)$, ... , $A(t_1 - 1)$ are nonsingular. When this does not happen, *the unconstructibility subspace cannot be defined.* $\qquad \square$

Note. It is possible to uniquely reconstruct the state of an observable system from (future) inputs/outputs.

Note. It is possible to uniquely reconstruct the state of a constructible system from (past) inputs/outputs.

Definition 15.7 (Observable and constructible systems). Given two times $t_1 > t_0 \geq 0$, the system (DLTV) is *observable* if its unobservable subspace contains only the zero vector, and it is *constructible* if its unconstructible subspace contains only the zero vector. $\qquad \square$

The matrices $B(\cdot)$ and $D(\cdot)$ play no role in the definition of the unobservable subspace, therefore one often simply talks about the unobservable subspace or the observability of the system

$$x(t+1) = A(t)x(t), \qquad y(t) = C(t)x(t), \qquad x \in \mathbb{R}^n, \ y \in \mathbb{R}^m \qquad \text{(AC-DLTV)}$$

or simply of the pair $\big(A(\cdot), C(\cdot)\big)$.

Definition 15.8 (Observability and constructibility Gramians). Given two times $t_1 > t_0 \geq 0$, the *observability* and *constructibility Gramians* of the system (AC-DLTV) are defined by

$$W_O(t_0, t_1) := \sum_{\tau=t_0}^{t_1-1} \Phi(\tau, t_0)' C(\tau)' C(\tau)\Phi(\tau, t_0),$$

$$W_{Cn}(t_0, t_1) := \sum_{\tau=t_0}^{t_1-1} \Phi(\tau, t_1)' C(\tau)' C(\tau)\Phi(\tau, t_1). \qquad \square$$

Theorem 15.3 (Unobservable and unconstructible subspaces). *Given two times* $t_1 > t_0 \geq 0$,

$$\mathcal{UO}[t_0, t_1] = \ker W_O(t_0, t_1), \qquad \mathcal{UC}[t_0, t_1] = \ker W_{Cn}(t_0, t_1). \qquad \Box$$

Theorem 15.4 (Gramian-based reconstruction). *Suppose we are given two times* $t_1 > t_0 \geq 0$ *and an input/output pair* $u(t)$, $y(t)$, $t_0 \leq t < t_1$.

1. *When the system (DLTV) is observable*

$$x(t_0) = W_O(t_0, t_1)^{-1} \sum_{t=t_0}^{t_1-1} \Phi(t, t_0)' C(t)' \tilde{y}(t),$$

   where

$$\tilde{y}(t) := y(t) - \sum_{\tau=t_0}^{t-1} C(t)\Phi(t, \tau)B(\tau)u(\tau)d\tau - D(t)u(t), \qquad \forall t_0 \leq t < t_1.$$

2. *When the system (DLTV) is constructible*

$$x(t_1) = W_{Cn}(t_0, t_1)^{-1} \sum_{t=t_0}^{t_1-1} \Phi(t, t_1)' C(t)' \bar{y}(t),$$

   where

$$\bar{y}(t) := y(t) - \sum_{\tau=t_1}^{t-1} C(t)\Phi(t, \tau)B(\tau)u(\tau)d\tau - D(t)u(t), \qquad \forall t_0 \leq t < t_1.$$

$$\Box$$

15.8   DUALITY (LTI)

Consider the continuous-time LTI system

$$\dot{x} = Ax + Bu, \qquad y = Cx + Du, \qquad x \in \mathbb{R}^n, \ u \in \mathbb{R}^k, \ y \in \mathbb{R}^m. \qquad \text{(CLTI)}$$

So far we have shown the following:

Note. Why can we omit "on $[t_0, t_1]$" for controllability? Because for time-invariant systems, we have time scalability; i.e., controllability does not depend on the interval under consideration.

The system (CLTI) is controllable $\quad \Leftrightarrow \quad \text{rank } W_C(t_0, t_1) = n, \quad$ (15.6a)

where $W_C(t_0, t_1) := \displaystyle\int_{t_0}^{t_1} e^{A(\tau-t_0)} B B' e^{A'(\tau-t_0)} d\tau.$

The system (CLTI) is observable on $[t_0, t_1] \quad \Leftrightarrow \quad \text{rank } W_O(t_0, t_1) = n, \quad$ (15.6b)

where $W_O(t_0, t_1) := \displaystyle\int_{t_0}^{t_1} e^{A'(\tau-t_0)} C' C e^{A(\tau-t_0)} d\tau.$

Suppose that we construct the following *dual system*

Note. All matrices were replaced by their transposes, the B and C matrices were exchanged, and the dimensions of the input and the output were also exchanged.

$$\dot{\bar{x}} = A'\bar{x} + C'\bar{u}, \qquad \bar{y} = B'\bar{x} + D'\bar{u}, \qquad x \in \mathbb{R}^n, \ \bar{u} \in \mathbb{R}^m, \ \bar{y} \in \mathbb{R}^k. \qquad (15.7)$$

For this system we have the following.

The system (15.7) is controllable $\qquad \Leftrightarrow \qquad$ rank $\bar{W}_C(t_0, t_1) = n, \qquad (15.8a)$

where $\bar{W}_C(t_0, t_1) := \displaystyle\int_{t_0}^{t_1} e^{A'(\tau - t_0)} C'C e^{A(\tau - t_0)} d\tau$.

The system (15.7) is observable on $[t_0, t_1]$ $\qquad \Leftrightarrow \qquad$ rank $\bar{W}_O(t_0, t_1) = n, \qquad (15.8b)$

where $\bar{W}_O(t_0, t_1) := \displaystyle\int_{t_0}^{t_1} e^{A(\tau - t_0)} BB' e^{A'(\tau - t_0)} d\tau$.

By matching the conditions (15.6) for the original system (CLTI) with the conditions (15.8) for the dual system (15.7), we obtain the following result.

Theorem 15.5 (Duality controllability/observability). *Suppose we are given two times* $t_1 > t_0 \geq 0$.

 1. *The system* (CLTI) *is controllable if and only if the system* (15.7) *is observable on* $[t_0, t_1]$.

 2. *The system* (CLTI) *is observable on* $[t_0, t_1]$ *if and only if the system* (15.7) *is controllable.* $\qquad\qquad\square$

A similar result can be obtained for reachability/constructibility.

Theorem 15.6 (Duality reachability/constructability). *Suppose we are given two times* $t_1 > t_0 \geq 0$.

 1. *The system* (CLTI) *is reachable if and only if the system* (15.7) *is constructible on* $[t_0, t_1]$.

 2. *The system* (CLTI) *is constructible on* $[t_0, t_1]$ *if and only if the system* (15.7) *is reachable.* $\qquad\qquad\square$

Attention! This result has several important implications.

Note. Why? Because we have already established time scaling for controllability and reachability of LTI systems.

 1. *Time scaling.* The notions of observability and constructibility do not depend on the time interval considered. This means that if it is possible to uniquely reconstruct the state $x(t_0)$ from (future) inputs and outputs on an interval $[t_0, t_1]$, then it is also possible to uniquely reconstruct the state $x(\bar{t}_0)$ from inputs and outputs on any other interval $[\bar{t}_0, \bar{t}_1]$, and similarly for constructibility.

 Because of this, for continuous-time LTI systems, *one generally does not specify the time interval* $[t_0, t_1]$ *under consideration.*

Note. Why? Because we have already established that controllability and reachability coincide for continuous-time LTI systems.

2. *Time reversibility.* The notions of observability and constructibility coincide for continuous-time LTI systems, which means that if one can reconstruct the state from future inputs/outputs, one can also reconstruct it from past inputs/outputs.

Because of this, for continuous-time LTI systems, *one simply studies observability and omits constructibility.*   □

Attention! For time-varying systems, duality is more complicated, because the state transition matrix of the dual system must be the transpose of the state transition matrix of the original system, but this is not obtained by simply transposing $A(t)$. □

15.9 OBSERVABILITY TESTS

Notation. In this section, we jointly present the results for continuous and discrete time and use a slash to separate the two cases.

Consider the LTI system

$$\dot{x}/x^{+} = Ax, \qquad y = Cx, \qquad x \in \mathbb{R}^n, \ y \in \mathbb{R}^m. \qquad \text{(AC-LTI)}$$

From the duality theorems in Section 15.8, we can conclude that a pair (A, C) is observable if and only if the pair (A', C') is controllable. This allows us to use all previously discussed tests for controllability to determine whether or not a system is observable.

To apply the controllability matrix test to the pair (A', C'), we construct the corresponding controllability matrix

$$\mathcal{C} = \begin{bmatrix} C' & A'C' & (A')^2 C' & \cdots & (A')^{n-1}C' \end{bmatrix}_{(kn) \times n} = \mathcal{O}',$$

where $\mathcal{O}$ denotes the *observability matrix* of the system (AC-LTI) , which is defined by

MATLAB® Hint 34. `obsv(sys)` computes the observability matrix of the state-space system `sys`. Alternatively, one can use `obsv(A,C)` directly.   ▶ p. 145

$$\mathcal{O} := \begin{bmatrix} C \\ C A \\ C A^2 \\ \vdots \\ C A^{n-1} \end{bmatrix}_{(kn) \times n}.$$

Since $\operatorname{rank}\mathcal{C} = \operatorname{rank}\mathcal{O}' = \operatorname{rank}\mathcal{O}$, we obtain the following test.

Note. Now, we do not need to work with left-eigenvectors nor left-kernels.

Theorem 15.7 (Observability matrix test). *The system* (AC-LTI) *is observable if and only if* $\operatorname{rank} \mathcal{O} = n$.   □

All other controllability tests also have observability counterparts.

Notation. The eigenvalues corresponding to eigenvectors of A in the kernel of C are called the *unobservable modes*, and the remaining ones are called the *observable modes*.

Theorem 15.8 (Eigenvector test for observability). *The system* (AC-LTI) *is observable if and only if no eigenvector of A is in the kernel of C.*   □

Theorem 15.9 (Popov-Belevitch-Hautus [PBH] test for observability). *The system* (AC-LTI) *is observable if and only if*

$$\operatorname{rank}\begin{bmatrix} A - \lambda I \\ C \end{bmatrix} = n, \qquad \forall \lambda \in \mathbb{C}.$$ □

Theorem 15.10 (Lyapunov test for observability). *Assume that A is a stability matrix/Schur stable. The system* (AC-LTI) *is observable if and only if there is a unique positive-definite solution W to the Lyapunov equation*

$$A'W + WA = -C'C \quad / \quad A'WA - W = -C'C. \tag{15.9}$$

Moreover, the unique solution to (15.9) *is*

Note. Now equation (15.9) very much resembles the one in the Lyapunov stability theorem (Theorem 8.2).

$$W = \int_0^\infty e^{A'\tau} C'C e^{A\tau} d\tau = \lim_{t_1 - t_0 \to \infty} W_O(t_0, t_1)$$

$$/ \quad W = \sum_{\tau=0}^\infty (A')^\tau C'C A^\tau d\tau = \lim_{t_1 - t_0 \to \infty} W_O(t_0, t_1). \quad □$$

Table 15.1 summarizes these results and contrasts them with the corresponding controllability tests.

15.10   MATLAB® COMMANDS

MATLAB® Hint 34 (`obsv`). The function `obsv(sys)` computes the observability matrix of the state-space system `sys`. The system must be specified by a state-space model using, e.g., `sys=ss(A,B,C,D)`, where `A,B,C,D` are a realization of the system. Alternatively, one can use `obsv(A,C)` directly.   □

15.11   EXERCISES

15.1 (Diagonal Systems). Consider the following system

$$\dot{x} = \begin{bmatrix} 1 & 0 & 0 \\ 0 & 0 & 0 \\ 0 & 0 & -1 \end{bmatrix} x, \qquad\qquad y = \begin{bmatrix} c_1 & c_2 & c_3 \end{bmatrix} u,$$

where $c_1, c_2,$ and c_3 are unknown scalars.

(a) Provide an example of values for c_1, c_2, and c_3 for which the system is not observable.

(b) Provide an example of values for c_1, c_2, and c_3 for which the system is observable.

(c) Provide a necessary and sufficient condition on the c_i so that the system is observable.

Table 15.1. Controllability and Observability Tests for LTI Systems

	Controllability	Stabilizability	Observability	Detectability
Definition	Every initial state can be taken to the origin in finite time.	Asymptotically stable uncontrollable component of the state in standard form for uncontrollable systems.	Every nonzero initial state results in a nonzero output.	Asymptotically stable unobservable component of the state in standard form for unobservable systems.
Rank test	$\operatorname{rank}\underbrace{\left[B\ AB \cdots A^{n-1}B \right]}_{\substack{\text{controllability}\\\text{matrix}}} = n$		$\operatorname{rank}\underbrace{\left[\begin{array}{c} C \\ CA \\ \vdots \\ CA^{n-1} \end{array}\right]}_{\substack{\text{observability}\\\text{matrix}}} = n$	
Eigenvector test	No eigenvector of A' in the kernel of B'.	No unstable or marginally stable eigenvector of A' in the kernel of B'.	No eigenvector of A in the kernel of C.	No unstable or marginally stable eigenvector of A in the kernel of C.
Popov-Belevitch-Hautus (PBH) test	$\operatorname{rank}\left[A - \lambda I\ B \right] = n,$ $\forall \lambda \in \mathbb{C}.$	$\operatorname{rank}\left[A - \lambda I\ B \right] = n,$ $\forall \lambda \in \mathbb{C}: \Re[\lambda] \geq 0$ (continuous time) or $\operatorname{rank}\left[A - \lambda I\ B \right] = n,$ $\forall \lambda \in \mathbb{C}: \lvert \lambda \rvert \geq 1$ (discrete time).	$\operatorname{rank}\left[\begin{array}{c} A - \lambda I \\ C \end{array}\right] = n,$ $\forall \lambda \in \mathbb{C}.$	$\operatorname{rank}\left[\begin{array}{c} A - \lambda I \\ C \end{array}\right] = n,$ $\forall \lambda \in \mathbb{C}: \Re[\lambda] \geq 0$ (continuous time) or $\operatorname{rank}\left[\begin{array}{c} A - \lambda I \\ C \end{array}\right] = n,$ $\forall \lambda \in \mathbb{C}: \lvert \lambda \rvert \geq 1$ (discrete time).
Lyapunov test (A asymptotically stable)	$\exists W > 0:$ $AW + WA' + BB' = 0$ (continuous time) or $\exists W > 0:$ $AWA' - W + BB' = 0$ (discrete time).	$\exists P > 0:$ $AP + PA' - BB' < 0$ (continuous time) or $\exists P > 0:$ $APA' - P - BB' < 0$ (discrete time).	$\exists W > 0:$ $WA + A'W + C'C = 0$ (continuous time) or $\exists W > 0:$ $A'WA - W + C'C = 0$ (discrete time).	$\exists P > 0:$ $PA + A'P - C'C < 0$ (continuous time) or $\exists P > 0:$ $A'PA - P - C'C < 0$ (discrete time).
Eigenvalue assignment	$\forall \{\lambda_i\},\ \exists K:$ $\lambda[A - BK] = \{\lambda_i\}.$	$\exists K : A - BK$ is asymptotically stable.	$\forall \{\lambda_i\},\ \exists L:$ $\lambda[A - LC] = \{\lambda_i\}.$	$\exists L : A - LC$ is asymptotically stable.

Hint: Use the eigenvector test. Make sure that you provide a condition that when true the system is guaranteed to be observable, but when false the system is guaranteed to not be observable.

(d) Generalize the previous result for an arbitrary system with a single output and diagonal matrix A. □

LECTURE 16

Output Feedback

CONTENTS

This lecture addresses the output feedback problem.

1. Observable Decomposition
2. Kalman Decomposition Theorem
3. Detectability
4. Detectability Tests
5. State Estimation
6. Eigenvalue Assignment by Output Injection
7. Stabilization through Output Feedback
8. MATLAB® Commands
9. Exercises

16.1 OBSERVABLE DECOMPOSITION

Notation. In this lecture, we jointly present the results for continuous and discrete time and use a slash to separate the two cases.

Consider the LTI system

$$\dot{x}/x^+ = Ax + Bu, \qquad y = Cx + Du, \qquad x \in \mathbb{R}^n, \ y \in \mathbb{R}^m \qquad \text{(AC-LTI)}$$

and a similarity transformation $\bar{x} := T^{-1}x$, leading to

$$\dot{\bar{x}}/\bar{x}^+ = \bar{A}\bar{x} + \bar{B}u, \quad y = \bar{C}\bar{x} + Du, \quad \bar{A} := T^{-1}AT, \quad \bar{B} := T^{-1}B, \quad \bar{C} := CT. \tag{16.1}$$

The observability matrices $\mathcal{O}$ and $\bar{\mathcal{O}}$ of the systems (AC-LTI) and (16.1), respectively, are related by

$$\bar{\mathcal{O}} = \begin{bmatrix} \bar{C} \\ \bar{C}\bar{A} \\ \vdots \\ \bar{C}\bar{A}^{n-1} \end{bmatrix} = \begin{bmatrix} C \\ CA \\ \vdots \\ CA^{n-1} \end{bmatrix} T = \mathcal{O}T.$$

Since the observability of a system is determined by the rank of its observability matrix, which does not change by multiplication by a nonsingular matrix, we obtain the following result.

Note. Similarity
transformations
actually preserve the
dimension of the
unobservable
subspace.

Theorem 16.1 (Invariance with respect to similarity transformations). *The pair* (A, C) *is observable if and only if the pair* $(\bar{A}, \bar{C}) = (T^{-1}AT, CT)$ *is observable.* □

As with controllability, it is possible to find similarity transformations that highlight the unobservable subspace. In fact, by applying the controllable decomposition theorem (Theorem 13.2) to the pair (A', C'), we obtain the following result.

Notation. This form
is often called the
*standard form for
unobservable systems.*

Theorem 16.2 (Observable decomposition). *For every LTI system (AC-LTI), there is a similarity transformation that takes the system to the form*

$$\begin{bmatrix} A_{\mathrm{o}} & 0 \\ A_{21} & A_{\mathrm{u}} \end{bmatrix} = T^{-1}AT, \qquad \begin{bmatrix} B_{\mathrm{o}} & B_{\mathrm{u}} \end{bmatrix} = T^{-1}B, \qquad \begin{bmatrix} C_{\mathrm{o}} & 0 \end{bmatrix} = CT, \qquad (16.2)$$

for which

MATLAB® Hint 35.
[Abar,Bbar,Cbar,T]
= obsvf(A,B,C)
computes the
observable
decomposition of the
system with
realization
A,B,C.   ▶ p. 156

1. *the unobservable subspace of the transformed system (16.2) is given by*

$$\mathcal{U\bar{O}} = \mathrm{Im} \begin{bmatrix} 0 \\ I_{\bar{n} \times \bar{n}} \end{bmatrix},$$

   where $\bar{n}$ *denotes the dimension of the unobservable subspace* $\mathcal{UO}$ *of the original system, and*

2. *the pair* $(A_{\mathrm{o}}, C_{\mathrm{o}})$ *is observable.* □

By partitioning the state of the transformed system as

Notation. The vectors
x_{o} and x_{u} are called
the *observable* and the
unobservable
components of the
state, respectively.

$$\bar{x} = T^{-1}x = \begin{bmatrix} x_{\mathrm{o}} \\ x_{\mathrm{u}} \end{bmatrix}, \qquad\qquad x_{\mathrm{o}} \in \mathbb{R}^{n-\bar{n}}, \ x_{\mathrm{u}} \in \mathbb{R}^{\bar{n}},$$

its state-space model can be written as follows

$$\begin{bmatrix} \dot{x}_{\mathrm{o}} \\ \dot{x}_{\mathrm{u}} \end{bmatrix} = \begin{bmatrix} A_{\mathrm{o}} & 0 \\ A_{21} & A_{\mathrm{u}} \end{bmatrix} \begin{bmatrix} x_{\mathrm{o}} \\ x_{\mathrm{u}} \end{bmatrix} + \begin{bmatrix} B_{\mathrm{o}} \\ B_{\mathrm{u}} \end{bmatrix} u, \qquad y = \begin{bmatrix} C_{\mathrm{o}} & 0 \end{bmatrix} \begin{bmatrix} x_{\mathrm{o}} \\ x_{\mathrm{u}} \end{bmatrix} + Du.$$

Note. This is
consistent with
statement 1 in
Theorem 16.2.

Figure 16.1 shows a block representation of this system, which highlights the fact that the x_{u} component of the state cannot be reconstructed from the output. Moreover, the observability of the pair $(A_{\mathrm{o}}, C_{\mathrm{o}})$ means that the x_{o} component of the state can be uniquely reconstructed from the input and output.

16.2 KALMAN DECOMPOSITION THEOREM

Consider the LTI system

$$\dot{x}/x^{+} = Ax + Bu, \qquad y = Cx + Du, \qquad x \in \mathbb{R}^{n}, \ u \in \mathbb{R}^{k}, \ y \in \mathbb{R}^{m}. \qquad \text{(LTI)}$$

In Lecture 13, we saw that every LTI system can be transformed through a similarity transformation into the following standard form for uncontrollable systems:

$$\begin{bmatrix} \dot{x}_{\mathrm{c}}/x_{\mathrm{c}}^{+} \\ \dot{x}_{\bar{\mathrm{c}}}/x_{\bar{\mathrm{c}}}^{+} \end{bmatrix} = \begin{bmatrix} A_{\mathrm{c}} & A_{12} \\ 0 & A_{\bar{\mathrm{c}}} \end{bmatrix} \begin{bmatrix} x_{\mathrm{c}} \\ x_{\bar{\mathrm{c}}} \end{bmatrix} + \begin{bmatrix} B_{\mathrm{c}} \\ 0 \end{bmatrix} u, \qquad y = \begin{bmatrix} C_{\mathrm{c}} & C_{\bar{\mathrm{c}}} \end{bmatrix} \begin{bmatrix} x_{\mathrm{c}} \\ x_{\bar{\mathrm{c}}} \end{bmatrix} + Du,$$

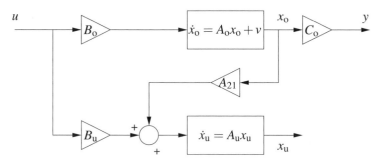

Figure 16.1. Observable decomposition. The direct feed-through term D was omitted to simplify the diagram.

in which the pair (A_c, B_c) is controllable [cf. Figure 16.2(a)]. This was obtained by choosing a similarity transformation

$$\begin{bmatrix} x_c \\ x_{\bar{c}} \end{bmatrix} := T^{-1}x, \qquad\qquad T := \begin{bmatrix} V_c & V_{\bar{c}} \end{bmatrix},$$

whose *leftmost* columns V_c form a basis for the (A-invariant) controllable subspace $\mathcal{C}$ of the pair (A, B). Using duality, we further concluded that every LTI system can also be transformed into the following form standard form for unobservable systems:

$$\begin{bmatrix} \dot{x}_o/x_o^+ \\ \dot{x}_{\bar{o}}/x_{\bar{o}}^+ \end{bmatrix} = \begin{bmatrix} A_o & 0 \\ A_{21} & A_{\bar{o}} \end{bmatrix} \begin{bmatrix} x_o \\ x_{\bar{o}} \end{bmatrix} + \begin{bmatrix} B_o \\ B_{\bar{o}} \end{bmatrix} u, \qquad y = \begin{bmatrix} C_o & 0 \end{bmatrix} \begin{bmatrix} x_o \\ x_{\bar{o}} \end{bmatrix} + Du,$$

in which the pair (A_o, C_o) is observable [cf. Figure 16.2(b)]. This is obtained by choosing a similarity transformation

$$\begin{bmatrix} x_o \\ x_{\bar{o}} \end{bmatrix} := T^{-1}x, \qquad\qquad T := \begin{bmatrix} V_o & V_{\bar{o}} \end{bmatrix},$$

whose *rightmost* columns $V_{\bar{o}}$ form a basis for the (A-invariant) unobservable subspace $\mathcal{UO}$ of the pair (A, C).

Suppose now that we choose a similarity transformation

$$\bar{x} := T^{-1}x, \qquad\qquad T := \begin{bmatrix} V_{co} & V_{c\bar{o}} & V_{\bar{c}o} & V_{\bar{c}\bar{o}} \end{bmatrix}$$

such that

1. the columns of $V_{c\bar{o}}$ form a basis for the (A-invariant) subspace $\mathcal{C} \cap \mathcal{UO}$,

2. the columns of $\begin{bmatrix} V_{co} & V_{c\bar{o}} \end{bmatrix}$ form a basis for the (A-invariant) controllable subspace $\mathcal{C}$ of the pair (A, B), and

3. the columns of $\begin{bmatrix} V_{c\bar{o}} & V_{\bar{c}\bar{o}} \end{bmatrix}$ form a basis for the (A-invariant) unobservable subspace $\mathcal{UO}$ of the pair (A, C).

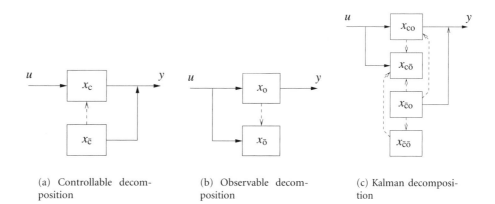

(a) Controllable decomposition

(b) Observable decomposition

(c) Kalman decomposition

Figure 16.2. Schematic representation of the structural decompositions.

This similarity transformation leads to a system in the form

$$
\begin{bmatrix} \dot{x}_{\mathrm{co}}/x_{\mathrm{co}}^+ \\ \dot{x}_{\mathrm{c\bar{o}}}/x_{\mathrm{c\bar{o}}}^+ \\ \dot{x}_{\mathrm{\bar{c}o}}/x_{\mathrm{\bar{c}o}}^+ \\ \dot{x}_{\mathrm{\bar{c}\bar{o}}}/x_{\mathrm{\bar{c}\bar{o}}}^+ \end{bmatrix} = \begin{bmatrix} A_{\mathrm{co}} & 0 & A_{\times o} & 0 \\ A_{\mathrm{c\times}} & A_{\mathrm{c\bar{o}}} & A_{\times\times} & A_{\times\bar{o}} \\ 0 & 0 & A_{\mathrm{\bar{c}o}} & 0 \\ 0 & 0 & A_{\mathrm{\bar{c}\times}} & A_{\mathrm{\bar{c}\bar{o}}} \end{bmatrix} \begin{bmatrix} x_{\mathrm{co}} \\ x_{\mathrm{c\bar{o}}} \\ x_{\mathrm{\bar{c}o}} \\ x_{\mathrm{\bar{c}\bar{o}}} \end{bmatrix} + \begin{bmatrix} B_{\mathrm{co}} \\ B_{\mathrm{c\bar{o}}} \\ 0 \\ 0 \end{bmatrix} u, \tag{16.3a}
$$

$$
y = \begin{bmatrix} C_{\mathrm{co}} & 0 & C_{\mathrm{\bar{c}o}} & 0 \end{bmatrix} \begin{bmatrix} x_{\mathrm{co}} \\ x_{\mathrm{c\bar{o}}} \\ x_{\mathrm{\bar{c}o}} \\ x_{\mathrm{\bar{c}\bar{o}}} \end{bmatrix} + Du. \tag{16.3b}
$$

MATLAB® Hint 36.
`[msys,T] =`
`minreal(sys)`
returns an orthogonal
matrix `T` such that
$(\mathtt{TAT}^{-1}, \mathtt{TB}, \mathtt{CT}^{-1})$
is a Kalman
decomposition of
`(A, B, C)`. ▶ p. 163

This similarity transformation is called a *canonical Kalman decomposition*, and it is represented schematically in Figure 16.2(c). This decomposition has several important properties as stated in the following theorem.

Theorem 16.3 (Kalman decomposition). *For every LTI system (AB-LTI), there is a similarity transformation that takes it to the form* (16.3), *for which*

1. *the pair* $\left(\begin{bmatrix} A_{\mathrm{co}} & 0 \\ A_{\mathrm{c\times}} & A_{\mathrm{c\bar{o}}} \end{bmatrix}, \begin{bmatrix} B_{\mathrm{co}} \\ B_{\mathrm{c\bar{o}}} \end{bmatrix} \right)$ *is controllable,*

2. *the pair* $\left(\begin{bmatrix} A_{\mathrm{co}} & A_{\times o} \\ 0 & A_{\mathrm{\bar{c}o}} \end{bmatrix}, \begin{bmatrix} C_{\mathrm{co}} & C_{\mathrm{\bar{c}o}} \end{bmatrix} \right)$ *is observable,*

3. *the triple* $(A_{\mathrm{co}}, B_{\mathrm{co}}, C_{\mathrm{co}})$ *is both controllable and observable, and*

Note. In discrete time, the Laplace transform variable s should be replaced by the $\mathcal{Z}$-transform variable z.

4. *the transfer function $C(sI - A)^{-1}B + D$ of the original system is the same as the transfer function $C_{co}(sI - A_{co})^{-1}B_{co} + D$ of the controllable and observable system.* □

Note. Statement 4 of Theorem 16.3 can be concluded directly from Figure 16.2(c), since for zero initial conditions $x_{\bar{c}o}$ and $x_{\bar{c}\bar{o}}$ remain identically zero and $x_{c\bar{o}}$ never affects the output.

16.3 DETECTABILITY

We just saw that any LTI system is algebraically equivalent to a system in the following standard form for unobservable systems:

$$\begin{bmatrix} \dot{x}_o/x_o^+ \\ \dot{x}_u/x_u^+ \end{bmatrix} = \begin{bmatrix} A_o & 0 \\ A_{21} & A_u \end{bmatrix} \begin{bmatrix} x_o \\ x_u \end{bmatrix} + \begin{bmatrix} B_o \\ B_u \end{bmatrix} u, \qquad x_o \in \mathbb{R}^{\bar{n}}, \ x_o \in \mathbb{R}^{n-\bar{n}}, \qquad (16.4a)$$

$$y = \begin{bmatrix} C_o & 0 \end{bmatrix} \begin{bmatrix} x_o \\ x_u \end{bmatrix} + Du, \qquad\qquad u \in \mathbb{R}^k, m \in \mathbb{R}^m. \qquad (16.4b)$$

Note. Any *observable* system is detectable, because in this case $\bar{n} = n$ and the matrix A_u does not exist. Also, any *asymptotically stable* system is detectable, because in this case both A_o and A_u are stability matrices.

Definition 16.1 (Detectable system). The pair (A, C) is *detectable* if it is algebraically equivalent to a system in the standard form for unobservable systems (16.4) with $n = \bar{n}$ (i.e., A_u nonexistent) or with A_u a stability matrix. □

For a continuous-time system, the evolution of the unobservable component of the state is determined by

$$\dot{x}_u = A_u x_u + A_{21} x_o + B_u u.$$

Regarding $A_{21}x_o + B_u u$ as the input, we can use the variation of constants formula to conclude that

$$x_u(t) = e^{A_u(t-t_0)} x_u(t_0) + \int_{t_0}^t e^{A_u(t-\tau)}\big(A_{21}x_o(\tau) + B_u u(\tau)\big)d\tau.$$

Since the pair (A_o, C_o) is observable, it is possible to reconstruct x_o from the input and output, and therefore the integral term can be perfectly reconstructed. For detectable systems, the term $e^{A_u(t-t_0)}x_u(t_0)$ eventually converges to zero, and therefore one can guess $x_u(t)$ up to an error that converges to zero exponentially fast.

16.4 DETECTABILITY TESTS

Investigating the detectability of an LTI system

$$\dot{x}/x^+ = Ax, \qquad y = Cx, \qquad x \in \mathbb{R}^n, \ y \in \mathbb{R}^m, \qquad (\text{AC-LTI})$$

from the definition requires the computation of the observable decomposition. However, there are alternative tests that avoid this intermediate step. These tests can be deduced by duality from the stabilizability tests.

Theorem 16.4 (Eigenvector test for detectability).

1. *The continuous-time LTI system (AC-LTI) is detectable if and only if every eigenvector of A corresponding to an eigenvalue with a positive or zero real part is not in the kernel of C.*

2. *The discrete-time LTI system (AC-LTI) is detectable if and only if every eigenvector of A corresponding to an eigenvalue with magnitude larger than or equal to 1 is not in the kernel of C.* □

Theorem 16.5 (Popov-Belevitch-Hautus [PBH] test for detectability).

1. *The continuous-time LTI system (AC-LTI) is detectable if and only if*

$$\text{rank} \begin{bmatrix} A - \lambda I \\ C \end{bmatrix} = n, \qquad \forall \lambda \in \mathbb{C} : \Re[\lambda] \geq 0.$$

2. *The discrete-time LTI system (AC-LTI) is detectable if and only if*

$$\text{rank} \begin{bmatrix} A - \lambda I \\ C \end{bmatrix} = n, \qquad \forall \lambda \in \mathbb{C} : |\lambda| \geq 1.$$ □

Theorem 16.6 (Lyapunov test for detectability).

1. *The continuous-time LTI system (AC-LTI) is detectable if and only if there is a positive-definite solution P to the Lyapunov matrix inequality*

$$A'P + PA - C'C < 0.$$

2. *The discrete-time LTI system (AC-LTI) is detectable if and only if there is a positive-definite solution P to the Lyapunov matrix inequality*

$$A'PA - P - C'C < 0.$$ □

Table 15.1 summarizes these results and contrasts them with the corresponding stabilizability tests.

16.5 STATE ESTIMATION

Consider the continuous-time LTI system

$$\dot{x} = Ax + Bu, \qquad y = Cx + Du, \qquad x \in \mathbb{R}^n, \ u \in \mathbb{R}^k, \ y \in \mathbb{R}^m, \qquad \text{(CLTI)}$$

and let

$$u = -Kx \qquad (16.5)$$

be a state feedback control law that asymptotically stabilizes (CLTI), i.e., for which $A - BK$ is a stability matrix. When only the output y can be measured, the control law (16.5) cannot be implemented, but if the pair (A, C) is detectable, it should be possible to estimate x from the system's output up to an error that vanishes as $t \to \infty$.

In Lecture 15, we saw that the state of an observable system can be reconstructed from the its input and output over an interval $[t_0, t_1]$ using the observability or constructibility Gramians. However, the formulas derived provide only the value of the state at a particular instant of time, instead of the continuous estimate required to implement (16.5).

The simplest state estimator consists of a copy of the original system,

Notation. This is called an *open-loop state estimator.*

$$\dot{\hat{x}} = A\hat{x} + Bu. \tag{16.6}$$

To study the performance of this state estimator, we define the *state estimation error*

$$e := \hat{x} - x.$$

Taking derivatives, we conclude that

$$\dot{e} = A\hat{x} + Bu - (A\hat{x} + Bu) = Ae.$$

Therefore, when A is a stability matrix, the open-loop state estimator (16.6) results in an error that converges to zero exponentially fast, *for every input signal u.*

When the matrix A is not a stability matrix, it is still possible to construct an asymptotically correct state estimate, but to achieve this we need a closed-loop estimator of the form

Note. The term $L(\hat{y} - y)$ is used to correct any deviations of $\hat{x}$ from the true value x. When $\hat{x} = x$, we have $\hat{y} = y$ and this term disappears.

$$\dot{\hat{x}} = A\hat{x} + Bu - L(\hat{y} - y), \qquad \hat{y} = C\hat{x} + Du, \tag{16.7}$$

for some *output injection matrix gain $L \in \mathbb{R}^{n \times m}$.* Now the state estimation error evolves according to

Note. In Lecture 23, we shall see that this type of state estimator can actually be optimal.

$$\dot{e} = A\hat{x} + Bu - L(\hat{y} - y) - (A\hat{x} + Bu) = (A - LC)e.$$

Theorem 16.7. *Consider the closed-loop state estimator* (16.7). *If the output injection matrix gain $L \in \mathbb{R}^{n \times m}$ makes $A - LC$ a stability matrix, then the state estimation error e converges to zero exponentially fast, for every input signal u.* $\square$

16.6 EIGENVALUE ASSIGNMENT BY OUTPUT INJECTION

The following results can also be obtained by duality from the eigenvalue assignment results that we proved for controllable and stabilizable systems.

Theorem 16.8. *When the system pair (A, C) is detectable, it is always possible to find a matrix gain $L \in \mathbb{R}^{n \times m}$ such that $A - LC$ is a stability matrix.* $\square$

Theorem 16.9. *Assume that the pair (A, C) is observable. Given any set of n complex numbers $\lambda_1, \lambda_2, \dots, \lambda_n$, there exists a state feedback matrix $L \in \mathbb{R}^{n \times m}$ such that $A - LC$ has eigenvalues equal to the λ_i.* $\square$

Attention! The condition in Theorem 16.8 is actually necessary and sufficient for detectability. In particular, one can also show that if it is possible to find a matrix gain L that makes $A - LC$ a stability matrix, then the pair (A, C) must be detectable. $\square$

Attention! The condition in Theorem 16.9 is actually necessary and sufficient for observability. In particular, one can also show that if it is possible to arbitrarily assign all eigenvalues of $A - LC$ by choosing L, then the pair (A, C) must be observable.

$\square$

16.7 STABILIZATION THROUGH OUTPUT FEEDBACK

Consider again the following LTI system

$$\dot{x}/x^+ = Ax + Bu, \qquad y = Cx + Du, \qquad x \in \mathbb{R}^n,\ u \in \mathbb{R}^k,\ y \in \mathbb{R}^m, \qquad \text{(LTI)}$$

that is asymptotically stabilized by the state feedback control law

$$u = -Kx, \tag{16.8}$$

and let

$$\dot{\hat{x}}/\hat{x}^+ = A\hat{x} + Bu - L(\hat{y} - y), \qquad \hat{y} = C\hat{x} + Du$$

be a state estimator for which $A - LC$ is a stability matrix. If the state x cannot be measured, one may be tempted to use the state estimate $\hat{x}$ instead of the actual state x in (16.8),

$$u = -K\hat{x}.$$

This results in a controller with the following state-space model

$$\dot{\hat{x}}/\hat{x}^+ = A\hat{x} + Bu - L(C\hat{x} + Du - y), \qquad u = -K\hat{x}$$

$$\Leftrightarrow \quad \dot{\hat{x}}/\hat{x}^+ = (A - LC - BK + LDK)\hat{x} + Ly, \qquad u = -K\hat{x} \tag{16.9}$$

and transfer function

$$\hat{C}(s) := -K(sI - A + LC + BK - LDK)^{-1}L$$
$$/ \quad -K(zI - A + LC + BK - LDK)^{-1}L.$$

Note. One could have obtained a state-space model for the closed-loop system using the state $[x'\ \bar{x}']'$. However, this would not lead to the simple diagonal structure in (16.10), and it would make it more difficult to establish the asymptotic stability of the closed-loop system.

To study whether the resulting closed-loop system is stable, we construct a state-space model for the closed-loop system with state $\bar{x} := \begin{bmatrix} x' & e' \end{bmatrix}'$. To do this, we recall that

$$\dot{e}/e^+ = (A - LC)e$$

and also that

$$\dot{x}/x^+ = Ax + Bu = Ax - BK\hat{x} = Ax - BK(e + x) = (A - BK)x - BKe.$$

From these two equations, we conclude that

$$\begin{bmatrix} \dot{x}/x^+ \\ \dot{e}/e^+ \end{bmatrix} = \begin{bmatrix} A - BK & -BK \\ 0 & A - LC \end{bmatrix} \begin{bmatrix} \dot{x} \\ \dot{e} \end{bmatrix}. \tag{16.10}$$

The following theorem results from the triangular structure of this matrix.

Theorem 16.10 (Separation). *The closed loop of the process* (LTI) *with the output feedback controller* (16.9) *results in a system whose eigenvalues are the union of the eigenvalues of the state feedback closed-loop matrix* $A - BK$ *with the eigenvalues of the state estimator matrix* $A - LC$. □

This is called the *separation theorem*, because one can design the state feedback gain K and the output injection gain L independently.

16.8 MATLAB® COMMANDS

MATLAB® Hint 35 (obsvf). The command [Abar,Bbar,Cbar,T] = obsvf(A,B,C) computes the observable decomposition of the system with realization A,B,C. The matrices returned are such that

$$\text{Abar} = \begin{bmatrix} A_u & A_{12} \\ 0 & A_o \end{bmatrix} = \text{T A T}', \quad \text{Bbar} = \text{T B},$$

$$\text{Cbar} = \begin{bmatrix} 0 & C_o \end{bmatrix} = \text{C T}', \quad \text{T}' = \text{T}^{-1}.$$

This decomposition places the unobservable modes *on top of* the observable ones, opposite to what happens in (16.2). Moreover, the nonsingular matrix T is chosen to be orthogonal.

The command [Abar,Bbar,Cbar,T] = obsvf(A,B,C,tol) further specifies the tolerance tol used for the selection of the unobservable modes. □

16.9 EXERCISES

16.1 (Diagonal Systems). Consider the system

$$\dot{x} = \begin{bmatrix} 1 & 0 & 0 \\ 0 & 0 & 0 \\ 0 & 0 & -1 \end{bmatrix} x, \qquad y = \begin{bmatrix} c_1 & c_2 & c_3 \end{bmatrix} u,$$

where c_1, c_2, and c_3 are unknown scalars.

(a) Provide a necessary and sufficient condition on the c_i so that the system is detectable.

(b) Generalize the previous result for an arbitrary system with a single output and diagonal matrix A.

Hint: Take a look at Exercise 15.1 in Lecture 15. □

LECTURE 17

Minimal Realizations

CONTENTS

This lecture addresses the issue of constructing state-space realizations of the smallest possible order for SISO transfer functions.

1. Minimal Realizations
2. Markov Parameters
3. Similarity of Minimal Realizations
4. Order of a Minimal SISO Realization
5. MATLAB® Commands
6. Exercises

17.1 MINIMAL REALIZATIONS

We recall from Lecture 4 that, given a transfer function $\hat{G}(s)$, we say that

$$\dot{x} = Ax + Bu, \qquad y = Cx + Du, \qquad x \in \mathbb{R}^n, \ u \in \mathbb{R}^k, \ y \in \mathbb{R}^m \qquad \text{(CLTI)}$$

is a realization of $\hat{G}(s)$ if

$$\hat{G}(s) = C(sI - A)^{-1}B + D.$$

The size n of the state-space vector x is called the *order* of the realization. We saw in Exercise 4.5 that a transfer function can have realizations of different orders, which justifies the following definition.

MATLAB® Hint 36.
`minreal(sys)`
computes a minimal realization of the system `sys`. ▶ p. 163

Definition 17.1 (Minimal realization). A realization of $\hat{G}(s)$ is called *minimal* or *irreducible* if there is no realization of $\hat{G}(s)$ of smaller order. ☐

The minimality of a realization is intimately related to controllability and observability, as expressed by the following theorem.

Theorem 17.1. *Every minimal realization must be both controllable and observable.* ☐

Proof of Theorem 17.1. This theorem can be easily proved by contradiction. Assuming that a realization is either not controllable or not observable, by the Kalman decomposition theorem one could find another realization of smaller order that realizes the same transfer function, which would contradict minimality. ☐

17.2 MARKOV PARAMETERS

It turns out that controllability and observability are not only necessary but also sufficient for minimality. To prove this, one needs to introduce the so-called Markov parameters. We saw in Lecture 6 that

$$(sI - A)^{-1} = \mathcal{L}[e^{At}] =: \mathcal{L}\left[\sum_{i=0}^{\infty} \frac{t^i}{i!} A^i\right].$$

Since

$$\mathcal{L}\left[\frac{t^i}{i!}\right] = s^{-(i+1)},$$

we conclude that

$$(sI - A)^{-1} = \sum_{i=0}^{\infty} s^{-(i+1)} A^i.$$

Therefore

$$\hat{G}(s) = C(sI - A)^{-1}B + D = D + \sum_{i=0}^{\infty} s^{-(i+1)} C A^i B. \tag{17.1}$$

The matrices D, $C A^i B$, $i \geq 0$ are called the *Markov parameters*, which are also related to the system's impulse response. To see how, we recall that

$$G(t) = \mathcal{L}^{-1}[\hat{G}(s)] = \mathcal{L}^{-1}[C(sI - A)^{-1}B + D] = Ce^{At}B + D\,\delta(t).$$

Taking derivatives of the right-hand side, we conclude that

$$\frac{d^i G(t)}{dt^i} = C A^i e^{At} B, \qquad \forall i \geq 1,\ t > 0$$

from which we obtain the following relationship between the impulse response and the Markov parameters:

$$\lim_{t \downarrow 0^+} \frac{d^i G(t)}{dt^i} = C A^i B, \qquad \forall i \geq 1. \tag{17.2}$$

The following result basically follows from the formulas derived above.

Theorem 17.2. *Two realizations*

$$\begin{cases} \dot{x} = Ax + Bu \\ y = Cx + Du \end{cases} \qquad\qquad \begin{cases} \dot{\bar{x}} = \bar{A}\bar{x} + \bar{B}u \\ y = \bar{C}\bar{x} + \bar{D}u \end{cases}$$

are zero-state equivalent if and only if they have the same Markov parameters; i.e.,

$$D = \bar{D}, \qquad\qquad C A^i B = \bar{C} \bar{A}^i \bar{B}, \quad \forall i \geq 0. \qquad\qquad \square$$

Proof of Theorem 17.2. From (17.1) we conclude that if the Markov parameters are the same then the two realization have the same transfer function.

Conversely, if two realizations have the same transfer function, then they must have the same D matrix, since this matrix is equal to the limit of the transfer function as $s \to \infty$. In addition, they must have the same impulse response, and we conclude from (17.2) that the remaining Markov parameters $C A^i B$, $i \geq 0$ must also be the same. ■

We are now ready to prove one of the key results regarding minimal realizations, which completely characterizes minimality in terms of controllability and observability.

Theorem 17.3. *A realization is minimal if and only if it is both controllable and observable.* □

Proof of Theorem 17.3. We have already shown in Theorem 17.1 that if a realization is minimal, then it must be both controllable and observable. We prove the converse by contradiction. Assume that

$$\dot{x} = Ax + Bu, \qquad y = Cx + Du, \qquad x \in \mathbb{R}^n \qquad (17.3)$$

is a controllable and observable realization of $\hat{G}(s)$, but this realization is not minimal; i.e., there exists another realization

$$\dot{\bar{x}} = \bar{A}\bar{x} + \bar{B}u, \qquad y = \bar{C}\bar{x} + Du, \qquad \bar{x} \in \mathbb{R}^{\bar{n}} \qquad (17.4)$$

for $\hat{G}(s)$ with $\bar{n} < n$. Denoting by $\mathcal{C}$ and $\mathcal{O}$ the controllability and observability matrices of (17.3), we have

$$\mathcal{OC} = \begin{bmatrix} C \\ CA \\ \vdots \\ CA^{n-1} \end{bmatrix} \begin{bmatrix} B & AB & \cdots & A^{n-1}B \end{bmatrix}$$

$$= \underbrace{\begin{bmatrix} CB & CAB & \cdots & CA^{n-1}B \\ CAB & CA^2B & \cdots & CA^nB \\ \vdots & \vdots & & \vdots \\ CA^{n-1}B & CA^nB & \cdots & CA^{2n-2}B \end{bmatrix}}_{\text{Markov parameters}}. \qquad (17.5)$$

Note. The matrix in (17.5) has rank n because the n columns of $\mathcal{O}$ are linearly independent and $\mathcal{C}$ also has n columns that are linearly independent. As these columns of $\mathcal{C}$ multiply $\mathcal{O}$, they give rise to n columns of $\mathcal{OC}$ that are linearly independent. On the other hand, the rank of $\mathcal{OC}$ cannot be larger than n, since it is given by the product of two rank n matrices.

Moreover, since (17.3) is controllable and observable, both $\mathcal{C}$ and $\mathcal{O}$ have rank n, and therefore the matrix in the right-hand side of (17.5) also has rank n.

Suppose now that we compute

$$\bar{\mathcal{O}}\bar{\mathcal{C}} = \begin{bmatrix} \bar{C} \\ \bar{C}\bar{A} \\ \vdots \\ \bar{C}\bar{A}^{n-1} \end{bmatrix} \begin{bmatrix} \bar{B} & \bar{A}\bar{B} & \cdots & \bar{A}^{n-1}\bar{B} \end{bmatrix}$$

Note. The matrices $\bar{C}$ and $\bar{\mathcal{O}}$ defined in (17.6) are *not* the controllability and observability matrices of (17.4) because the powers of $\bar{A}$ go up to $n > \bar{n}$.

$$= \begin{bmatrix} \bar{C}\bar{B} & \bar{C}\bar{A}\bar{B} & \cdots & \bar{C}\bar{A}^{n-1}\bar{B} \\ \bar{C}\bar{A}\bar{B} & \bar{C}\bar{A}^2\bar{B} & \cdots & \bar{C}\bar{A}^n\bar{B} \\ \vdots & \vdots & & \vdots \\ \bar{C}\bar{A}^{n-1}\bar{B} & \bar{C}\bar{A}^n\bar{B} & \cdots & \bar{C}\bar{A}^{2n-2}\bar{B} \end{bmatrix}. \qquad (17.6)$$

$$\underbrace{}_{\text{Markov parameters}}$$

Note. Why? because rank $AB \le$ min{rank A, rank B}.

Since (17.3) and (17.4) realize the same transfer function, they must have the same Markov parameter (cf. Theorem 17.2), and therefore $\mathcal{O}\mathcal{C} = \bar{\mathcal{O}}\bar{\mathcal{C}}$. But since $\bar{\mathcal{C}}$ has only $\bar{n} < n$ columns, its rank must be lower than n and therefore

$$\text{rank } \bar{\mathcal{O}}\bar{\mathcal{C}} \le \text{rank } \bar{\mathcal{C}} \le \bar{n} < n,$$

which contradicts that fact that rank $\bar{\mathcal{O}}\bar{\mathcal{C}} = \text{rank } \mathcal{O}\mathcal{C} = n$. ∎

17.3 SIMILARITY OF MINIMAL REALIZATIONS

The definition of minimal realization automatically guarantees that all minimal realizations have the same order, but minimal realizations are even more closely related.

Theorem 17.4. *All minimal realizations of a transfer function are algebraically equivalent.* □

The following concept is needed to prove Theorem 17.4.

Notation 3 Left and right inverses are often called *pseudoinverses*.

Definition 17.2 (Pseudoinverse). Given a full column rank matrix M, $M'M$ is nonsingular and $M^\ell := (M'M)^{-1}M'$ is called the *left inverse* of M. This matrix has the property that $M^\ell M = I$. Given a full row rank matrix N, NN' is nonsingular and $N^{\mathrm{r}} := N'(NN')^{-1}$ is called the *right inverse* of N. This matrix has the property that $NN^{\mathrm{r}} = I$. □

MATLAB® Hint 37. pinv(M) computes the pseudoinverse of M. ▶ p. 163

Proof of Theorem 17.4. To prove this theorem, let

$$\begin{cases} \dot{x} = Ax + Bu \\ y = Cx + Du \end{cases} \qquad \begin{cases} \dot{\bar{x}} = \bar{A}\bar{x} + \bar{B}u \\ y = \bar{C}\bar{x} + \bar{D}u \end{cases} \qquad x, \bar{x} \in \mathbb{R}^n$$

be two minimal realizations of the same transfer function. From Theorem 17.3 we know that these two realizations are controllable and observable. Moreover, since the two realizations must have the same Markov parameters, we must have

$$\mathcal{O}\mathcal{C} = \bar{\mathcal{O}}\bar{\mathcal{C}}, \qquad (17.7)$$

where $\mathcal{C}$ and $\mathcal{O}$ are the controllability and observability matrices of the first system, whereas $\bar{\mathcal{C}}$ and $\bar{\mathcal{O}}$ are the controllability and observability matrices of the second one (cf. proof of Theorem 17.3).

From the controllability of the first system, we conclude that $\mathcal{C}$ is full row rank, and therefore we can define

$$T := \bar{\mathcal{C}}\mathcal{C}^{\mathrm{r}} = \bar{\mathcal{C}}\mathcal{C}'(\mathcal{C}\mathcal{C}')^{-1}.$$

To verify that this matrix is nonsingular, we compute

$$\mathcal{O}^\ell \bar{\mathcal{O}} T = (\mathcal{O}'\mathcal{O})^{-1}\mathcal{O}'\bar{\mathcal{O}}\bar{\mathcal{C}}\mathcal{C}'(\mathcal{C}\mathcal{C}')^{-1} = (\mathcal{O}'\mathcal{O})^{-1}\mathcal{O}'\mathcal{O}\mathcal{C}\mathcal{C}'(\mathcal{C}\mathcal{C}')^{-1} = I,$$

where we used (17.7). This shows that $T^{-1} = \mathcal{O}^\ell \bar{\mathcal{O}}$, and therefore T is invertible. We show next that this matrix T provides a similarity transformation between the two realizations.

Right-multiplying (17.7) by $\mathcal{C}^r := \mathcal{C}'(\mathcal{C}\mathcal{C}')^{-1}$, we conclude that

$$\mathcal{O} = \bar{\mathcal{O}} T \quad \Leftrightarrow \quad \begin{bmatrix} C \\ CA \\ \vdots \end{bmatrix} = \begin{bmatrix} \bar{C} \\ \bar{C}\bar{A} \\ \vdots \end{bmatrix} T, \tag{17.8}$$

and therefore

$$C = \bar{C}T. \tag{17.9}$$

From the observability of the second system, we conclude that $\bar{\mathcal{O}}$ is full column rank, and therefore we can left-multiply (17.8) by $\bar{\mathcal{O}}^\ell := (\bar{\mathcal{O}}'\bar{\mathcal{O}})^{-1}\bar{\mathcal{O}}'$, which yields

$$T = \bar{\mathcal{O}}^\ell \mathcal{O} = (\bar{\mathcal{O}}'\bar{\mathcal{O}})^{-1}\bar{\mathcal{O}}'\mathcal{O}.$$

Left-multiplying (17.7) by $\bar{\mathcal{O}}^\ell$, we conclude that

$$T\mathcal{C} = \bar{\mathcal{C}} \quad \Leftrightarrow \quad T \begin{bmatrix} B & AB & \cdots \end{bmatrix} = \begin{bmatrix} \bar{B} & \bar{A}\bar{B} & \cdots \end{bmatrix},$$

and therefore

$$TB = \bar{B}. \tag{17.10}$$

On the other hand, since the two realizations have the same Markov parameters, we also have

$$\mathcal{O}A\mathcal{C} = \bar{\mathcal{O}}\bar{A}\bar{\mathcal{C}}.$$

Left-multiplying this equation by $\bar{\mathcal{O}}^\ell$ and right-multiplying by $\mathcal{C}^r$, we conclude that

$$TA = \bar{A}T \quad \Leftrightarrow \quad TAT^{-1} = \bar{A}. \tag{17.11}$$

This concludes the proof, since (17.9), (17.10), and (17.11) confirm that T provides a similarity transformation between the two realizations. ∎

17.4 ORDER OF A MINIMAL SISO REALIZATION

Notation. A polynomial is *monic* if its highest order coefficient is equal to 1.

Notation. Two polynomial are *coprime* if they have no common roots.

Any proper SISO rational function $\hat{g}(s)$ can be written as

$$\hat{g}(s) = \frac{n(s)}{d(s)}, \tag{17.12}$$

where $d(s)$ is a monic polynomial and $n(s)$ and $d(s)$ are coprime. In this case, the right-hand side of (17.12) is called a *coprime fraction*, $d(s)$ is called the *pole (or characteristic polynomial)* of $\hat{g}(s)$, and the degree of $d(s)$ is called the *degree of the transfer function $\hat{g}(s)$*. The roots of $d(s)$ are called the *poles of the transfer function*, and the roots of $n(s)$ are called the *zeros of the transfer function*.

Theorem 17.5. *A SISO realization*

$$\dot{x}/x^+ = Ax + Bu, \qquad y = Cx + Du, \qquad x \in \mathbb{R}^n,\ u, y \in \mathbb{R}, \qquad \text{(LTI)}$$

of $\hat{g}(s)$ is minimal if and only if its order n is equal to the degree of $\hat{g}(s)$. In this case, the pole polynomial $d(s)$ of $\hat{g}(s)$ is equal to the characteristic polynomial of A; i.e., $d(s) = \det(sI - A)$. $\qquad\square$

Proof of Theorem 17.5. The direct gain D of a realization does not affect its minimality, so we may ignore it in this proof. We thus take $\hat{g}(s)$ to be strictly proper.

To prove this theorem, it suffices to show that if $\hat{g}(s)$ can be written as the following coprime fraction

$$\hat{g}(s) = \frac{n(s)}{d(s)} = \frac{\beta_1 s^{n-1} + \beta_2 s^{n-2} + \cdots + \beta_{n-1} s + \beta_n}{s^n + \alpha_1 s^{n-1} + \alpha_2 s^{n-2} + \cdots + \alpha_{n-1} s + \alpha_n},$$

then it has a realization of order n that is both controllable and observable. Minimality then results from Theorem 17.3. To proceed, consider then the realization in the controllable canonical form derived in Exercise 4.4:

$$A = \begin{bmatrix} -\alpha_1 & -\alpha_2 & \cdots & -\alpha_{n-1} & -\alpha_n \\ 1 & 0 & \cdots & 0 & 0 \\ 0 & 1 & \cdots & 0 & 0 \\ \vdots & \vdots & \ddots & \vdots & \vdots \\ 0 & 0 & \cdots & 1 & 0 \end{bmatrix}, \qquad B = \begin{bmatrix} 1 \\ 0 \\ \vdots \\ 0 \\ 0 \end{bmatrix},$$

$$C = \begin{bmatrix} \beta_1 & \beta_2 & \cdots & \beta_{n-1} & \beta_n \end{bmatrix}. \qquad (17.13)$$

We have seen in Exercise 12.3 that this realization is controllable. We show that it is observable using the eigenvector test. Let $x \neq 0$ be an eigenvector A; i.e.,

$$Ax = \lambda x \quad \Leftrightarrow \quad \begin{cases} -\sum_{i=1}^{n} \alpha_i x_i = \lambda x_1 \\ x_i = \lambda x_{i+1}, & i \in \{1, 2, \ldots, n\}. \end{cases}$$

Therefore

$$\begin{cases} -\sum_{i=1}^{n} \alpha_i \lambda^{n-i} x_n = \lambda^n x_n \\ x_i = \lambda^{n-i} x_n, & \forall i \in \{1, \ldots, n\} \end{cases} \quad \Leftrightarrow \quad \begin{cases} d(\lambda) x_n = 0 \\ x_i = \lambda^{n-i} x_n, & \forall i \in \{1, \ldots, n\} \end{cases}.$$

Since $x \neq 0$, we must have $x_n \neq 0$, and therefore λ is a root of $d(s)$. On the other hand,

$$Cx = \sum_{i=1}^{n} \beta_i x_i = \sum_{i=1}^{n} \beta_i \lambda^{n-i} x_n = n(\lambda) x_n.$$

Since $d(s)$ and $n(s)$ are coprime and λ is a root of $d(s)$, it cannot be a root of $n(s)$. From this and the fact that $x_n \neq 0$, we conclude that $Cx \neq 0$, and therefore the pair (A, C) must be observable.

To conclude the proof, we need to show that the pole polynomial $d(s)$ is equal to the characteristic polynomial of the A matrix of any minimal realization. It is straightforward to show that this is so for the matrix A in the minimal realization (17.13). Moreover, since by Theorem 17.4 all minimal realizations must be algebraically equivalent, it then follows that all minimal realizations have matrices A with the same characteristic polynomial. ∎

This result has the following immediate consequence.

Note. We shall see in Lecture 19 that this result also holds for MIMO systems. (See Corollary 19.1)

Corollary 17.1. *Assuming that the SISO realization* (LTI) *of $\hat{g}(s)$ is minimal, the transfer function $\hat{g}(s)$ is BIBO stable if and only if the realization* (LTI) *is (internally) asymptotically stable.* □

17.5 MATLAB® COMMANDS

MATLAB® Hint 36 (`minreal`). The command `msys=minreal(sys)` computes a minimal realization of the system `sys`, which can either be in state-space or transfer function form. When `sys` is in state-space form, `msys` is a state-space system from which all uncontrollable and unobservable modes were removed. When `sys` is in transfer function form, `msys` is a transfer function from which all common poles and zeros have been canceled.

The command `msys=minreal(sys,tol)` further specifies the tolerance `tol` used for zero-pole cancellation and for uncontrollable and unobservable mode elimination.

The command `[msys,T]=minreal(sys)` also returns an orthogonal matrix `T` (i.e., $T^{-1} = T'$) such that (TAT', TB, CT') is a Kalman decomposition of (A, B, C). □

MATLAB® Hint 37 (`pinv`). The function `pinv(M)` computes the pseudoinverse of the matrix `M`. □

17.6 EXERCISES

17.1 (Minimal Realization). Consider the LTI system with realization

$$A = \begin{bmatrix} -1 & 0 \\ 0 & -1 \end{bmatrix}, \qquad B = \begin{bmatrix} 1 & 0 \\ 0 & 1 \end{bmatrix}, \qquad C = \begin{bmatrix} -1 & 1 \end{bmatrix}, \qquad D = \begin{bmatrix} 2 \\ 1 \end{bmatrix}.$$

Is this realization minimal? If not, find a minimal realization with the same transfer function. □

17.2 (Repeated eigenvalues). Consider the SISO LTI system

$$\dot{x}/x^+ = Ax + Bu, \qquad y = Cx + Du, \qquad x \in \mathbb{R}^n,\ u, y \in \mathbb{R}.$$

(a) Assume that A is a diagonal matrix and B, C are column/row vectors with entries b_i and c_i, respectively. Write the controllability and observability matrices for this system.

(b) Show that if A is a diagonal matrix with repeated eigenvalues, then the pair (A, B) cannot be controllable and the pair (A, C) cannot be observable.

(c) Can you find a SISO minimal realization for which the matrix A is diagonalizable with repeated eigenvalues? Justify your answer.

(d) Can you find a SISO minimal realization for which the matrix A is **not** diagonalizable with repeated eigenvalues? Justify your answer.

Hint: An example suffices to justify the answer "yes" in (c) or (d).   □

Linear Systems II—Advanced Material

PART V
Poles and Zeros of MIMO Systems

LECTURE 18

Smith-McMillan Form

Contents

This lecture introduces the concepts of poles and zeros for MIMO transfer functions.

18.1 INFORMAL DEFINITION OF POLES AND ZEROS

Note. For discrete-time transfer functions, the Laplace transform variable s should be replaced by the $\mathcal{Z}$-transform variable z.

Consider a SISO transfer function

$$g(s) = \frac{n(s)}{d(s)}. \tag{18.1}$$

The following concepts were introduced in Lecture 17.

1. The *poles* of $g(s)$ are the values of $s \in \mathbb{C}$ for which $g(s)$ becomes unbounded (technically not defined due to a division by zero).

2. The *zeros* of $g(s)$ are the values of $s \in \mathbb{C}$ for which $g(s) = 0$.

Notation. Two polynomial are *coprime* if they have no common roots.

When the polynomials $n(s)$ and $d(s)$ in (18.1) are coprime, then the zeros are simply the roots of $n(s)$ and the poles are the roots of $d(s)$. We recall that the number of roots of $n(s)$ (i.e., the number of poles) is equal to the dimension of a minimal realization for $g(s)$.

The most useful generalizations of these concepts for a MIMO transfer function $G(s)$ turn out to be as follows.

1. The *poles* of $G(s)$ are the values of $s \in \mathbb{C}$ for which *at least one* of the entries of $G(s)$ becomes unbounded.

2. The rank of $G(s)$ takes the same value for almost all values of $s \in \mathbb{C}$, but for some $s \in \mathbb{C}$, the rank of $G(s)$ drops. These values are called the *transmission zeros* of $G(s)$.

The above is is just an informal definition, because it is not clear how to compute the rank of $G(s)$ at the locations of the poles of $G(s)$. Therefore this definition does not permit us to determine whether a pole is also a transmission zero. It turns out that in MIMO systems one can have transmission zeros "on top" of poles. We shall return to this later.

The above definition for the poles also does not provide us a way to determine the multiplicity of poles or zeros. It turns out that also in the MIMO case the total number of poles (with the appropriate multiplicities) gives us the dimension of a minimal realization. In the remainder of this lecture, we introduce the formal definitions of poles and zero, which use the so-called Smith-McMillan form of a transfer function.

18.2 POLYNOMIAL MATRICES: SMITH FORM

A *real polynomial matrix* is a matrix-valued function whose entries are polynomials with real coefficients. We denote by $\mathbb{R}[s]^{m \times k}$ the set of $m \times k$ real polynomial matrices on the variable s.

Given a real polynomial matrix

$$P(s) = \begin{bmatrix} p_{11}(s) & p_{12}(s) & \cdots & p_{1k}(s) \\ p_{21}(s) & p_{22}(s) & \cdots & p_{2k}(s) \\ \vdots & \vdots & \ddots & \vdots \\ p_{m1}(s) & p_{m2}(s) & \cdots & p_{mk}(s) \end{bmatrix} \in \mathbb{R}[s]^{m \times k},$$

the *minors of $P(s)$ of order i* are the determinants of all square $i \times i$ submatrices of $P(s)$.

The *determinantal divisors of $P(s)$* are polynomials

$$\{D_i(s) : 0 \leq i \leq r\},$$

where $D_0(s) = 1$ and $D_i(s)$ is the monic greatest common divisor of all nonzero minors of $P(s)$ of order i. The integer r is called the *rank of $P(s)$* and is the maximum order of a nonzero minor of $P(s)$. All minors of order larger than r are equal to zero, and therefore there are no divisors of order larger than r.

The determinantal divisors provide insight into the linear independence of the rows and columns of $P(s)$, as s ranges over $\mathbb{C}$.

Lemma 18.1 (Determinantal divisors). *For every $s_0 \in \mathbb{C}$,*

$$\text{rank } P(s_0) \quad \begin{cases} = r & s_0 \text{ is not a root of } D_r(s) \\ < r & s_0 \text{ is a root of } D_r(s). \end{cases} \qquad \square$$

Proof of Lemma 18.1. First note that the rank of $P(s_0)$ can never exceed r, because if it had $r + 1$ independent rows or columns for some $s_0 \in \mathbb{C}$, one could use these to construct a nonsingular $(r + 1) \times (r + 1)$ submatrix. In this case, $P(s)$ would have a determinantal divisor of order $r + 1$, which would be nonzero at s_0.

Notation. The *monic greatest common divisor (gcd)* of a family of polynomial is the monic polynomial of greatest order that divides all the polynomials in the family. Informally, the gcd set of roots is the intersection of the sets of roots of all polynomials (sets taken with repetitions).

Note. In words: the rank of $P(s)$ drops precisely at the roots of $D_r(s)$.

At every $s_0 \in \mathbb{C}$ that is not a root of $D_r(s)$, one can find an $r \times r$ submatrix of $P(s_0)$ that is nonsingular and therefore has linearly independent rows and columns. However, at the roots of $D_r(s)$, all $r \times r$ submatrices are singular and therefore have linearly independent rows and columns. $\qquad\square$

Example 18.1 (Smith form). The real polynomial matrix

$$P(s) := \begin{bmatrix} s(s+2) & 0 \\ 0 & (s+1)^2 \\ (s+1)(s+2) & s+1 \\ 0 & s(s+1) \end{bmatrix} \qquad (18.2)$$

has the following minors, determinantal divisors, and invariant factors:

Order	Minors	Determinantal divisor	Invariant factors
$i=0$	None	$D_0(s) = 1$	
$i=1$	$s(s+2)$, $(s+1)^2$	$D_1(s) = 1$	$\epsilon_1(s) = 1$
	$(s+1)(s+2)$, $s+1$,		
	$s(s+1)$		
$i=2$	$s(s+2)(s+1)^2_{\text{rows 1,2}}$,	$D_2(s) = (s+1)(s+2)$	$\epsilon_2(s) = (s+1)(s+2)$
$(r=2)$	$s(s+2)(s+1)_{\text{rows 1,3}}$,		
	$s^2(s+2)(s+1)_{\text{rows 1,4}}$,		
	$-(s+1)^3(s+2)_{\text{rows 2,3}}$,		
	$0_{\text{rows 2,4}}$,		
	$s(s+1)^2(s+2)_{\text{rows 3,4}}$		

$\qquad\square$

The determinantal divisors allow us to define the so-called Smith form of a polynomial matrix.

Note. For scalar polynomials ($m=k=1$), the Smith form is the polynomial itself, scaled to be monic.

Note. Each $D_{i-1}(s)$ divides $D_i(s)$, and therefore all the $\epsilon_i(s)$ are actually polynomials (not rational functions). Moreover, each $\epsilon_{i-1}(s)$ divides $\epsilon_i(s)$.

Note. The product of the invariant factors equals the highest-order determinantal divisor: $D_r(s) = \epsilon_1(s)\epsilon_2(s)\cdots\epsilon_r(s)$.

Definition 18.1 (Smith form). The *Smith form* of a real polynomial matrix $P(s) \in \mathbb{R}[s]^{m \times k}$ is the diagonal real polynomial matrix defined by

$$S_P(s) := \begin{bmatrix} \epsilon_1(s) & 0 & \cdots & 0 & 0 & \cdots & 0 \\ 0 & \epsilon_2(s) & \cdots & 0 & 0 & \cdots & 0 \\ \vdots & \vdots & \ddots & \vdots & \vdots & & \vdots \\ 0 & 0 & \cdots & \epsilon_r(s) & 0 & \cdots & 0 \\ 0 & 0 & \cdots & 0 & 0 & \cdots & 0 \\ \vdots & \vdots & \ddots & \vdots & \vdots & & \vdots \\ 0 & 0 & \cdots & 0 & 0 & \cdots & 0 \end{bmatrix} \in \mathbb{R}[s]^{m \times k},$$

where $r := \operatorname{rank} P(s)$ and

$$\epsilon_i(s) := \frac{D_i(s)}{D_{i-1}(s)}, \quad i \in \{1, 2, \ldots, r\},$$

which are called the *invariant factors of $P(s)$*. $\qquad\square$

Example 18.2 (Smith form, continued). The Smith form of the real polynomial matrix (18.2) in Example 18.1 is given by

$$S_P(s) = \begin{bmatrix} 1 & 0 \\ 0 & (s+1)(s+2) \\ 0 & 0 \\ 0 & 0 \end{bmatrix}. \qquad \Box$$

The importance of the Smith form stems from the fact that a matrix can always be transformed into its Smith form by left- and right-multiplication by very special polynomial matrices. A square real polynomial matrix $U(s)$ is called *unimodular* if its inverse is also a polynomial matrix. A matrix is unimodular if and only if its determinant is a nonzero constant (independent of s).

Note. Why? Recall the adjoint formula for matrix inversion used in Section 4.3.1.

Example 18.3 (Unimodular matrices). The first matrix is unimodular, but not the second

$$U(s) = \begin{bmatrix} 1 & s \\ 0 & 2 \end{bmatrix}, \qquad U^{-1}(s) = \frac{1}{\det U(s)} \begin{bmatrix} 2 & -s \\ 0 & 1 \end{bmatrix} = \begin{bmatrix} 1 & -\frac{s}{2} \\ 0 & \frac{1}{2} \end{bmatrix}$$

$$P(s) = \begin{bmatrix} 1 & 1 \\ 0 & s \end{bmatrix}, \qquad P^{-1}(s) = \frac{1}{\det P(s)} \begin{bmatrix} s & -1 \\ 0 & 1 \end{bmatrix} = \begin{bmatrix} 1 & -\frac{1}{s} \\ 0 & \frac{1}{s} \end{bmatrix} \qquad \Box$$

Note. The matrices $L(s)$ and $R(s)$ can be found using a procedure similar to Gauss elimination [11, Section 2.2].

Theorem 18.1 (Smith form factorization). *For every real polynomial matrix $P(s) \in \mathbb{R}[s]^{m \times k}$ with Smith form $S_P(s)$, there exist unimodular real polynomial matrices $L(s) \in \mathbb{R}[s]^{m \times m}$, $R(s) \in \mathbb{R}[s]^{k \times k}$ such that*

$$P(s) = L(s) S_P(s) R(s). \qquad \Box$$

18.3 RATIONAL MATRICES: SMITH-MCMILLAN FORM

Attention! Note the curved brackets (s), as opposed to square brackets $[s]$.

A *real rational matrix* is a matrix-valued function whose entries are ratios of polynomials with real coefficients. We denote by $\mathbb{R}(s)^{m \times k}$ the set of $m \times k$ real rational matrices on the variable s.

Any real rational matrix $G(s) \in \mathbb{R}(s)^{m \times k}$ can be written as

$$G(s) = \frac{1}{d(s)} N(s), \tag{18.3}$$

Notation. The *monic least common denominator (lcd)* of a family of polynomials is the monic polynomial of smallest order that is divided by all the polynomials in the family.

where $d(s)$ is the monic least common denominator of all entries of $G(s)$ and $N(s) \in \mathbb{R}[s]^{m \times k}$ is a polynomial matrix.

Note. Each $\epsilon_{i-1}(s)$ divides $\epsilon_i(s)$, and therefore each $\psi_i(s)$ divides $\psi_{i-1}(s)$.

Attention! The fractions $\frac{\eta_i(s)}{\psi_i(s)}$ in the diagonal of the Smith-McMillan form are not necessarily proper.

Definition 18.2 (Smith-McMillan form). The *Smith-McMillan form* of the real rational matrix $G(s) \in \mathbb{R}(s)^{m \times k}$ in (18.3) is the diagonal real rational matrix defined by

$$SM_G(s) := \frac{1}{d(s)} S_N(s) = \begin{bmatrix} \frac{\eta_1(s)}{\psi_1(s)} & 0 & \cdots & 0 & 0 & \cdots & 0 \\ 0 & \frac{\eta_2(s)}{\psi_2(s)} & \cdots & 0 & 0 & \cdots & 0 \\ \vdots & \vdots & \ddots & \vdots & \vdots & & \vdots \\ 0 & 0 & \cdots & \frac{\eta_r(s)}{\psi_r(s)} & 0 & \cdots & 0 \\ 0 & 0 & \cdots & 0 & 0 & \cdots & 0 \\ \vdots & \vdots & \ddots & \vdots & \vdots & & \vdots \\ 0 & 0 & \cdots & 0 & 0 & \cdots & 0 \end{bmatrix} \in \mathbb{R}(s)^{m \times k},$$

$$(18.4)$$

Note. For scalar rational matrices ($m = k = 1$), the Smith-McMillan form is the rational matrix itself, with common factors canceled and scaled to have monic numerator and denominator.

where $S_N(s) \in \mathbb{R}[s]^{m \times k}$ denotes the Smith form of $N(s)$. All the common factors in the entries of (18.4) should be canceled, which means that the pairs of (monic) polynomials $\{\eta_i(s), \psi_i(s)\}$ are all coprime. □

From the Smith form factorization in Theorem 18.1, we know that there exists unimodular real polynomial matrices $L(s) \in \mathbb{R}[s]^{m \times m}$, $R(s) \in \mathbb{R}[s]^{k \times k}$ such that

$$N(s) = L(s) S_N(s) R(s),$$

from which we conclude the following.

Theorem 18.2 (Smith-McMillan factorization). *For every real rational matrix $G(s) \in \mathbb{R}(s)^{m \times k}$ with Smith-McMillan form $SM_G(s)$, there exist unimodular real polynomial matrices $L(s) \in \mathbb{R}[s]^{m \times m}$, $R(s) \in \mathbb{R}[s]^{k \times k}$ such that*

$$G(s) = \frac{1}{d(s)} N(s) = L(s) SM_G(s) R(s).$$ □

18.4 MCMILLAN DEGREE, POLES, AND ZEROS

The Smith-McMillan form is especially useful to define poles and zeros for rational matrices.

Definition 18.3 (Poles and zeros). For a real rational matrix $G(s) \in \mathbb{R}(s)^{m \times k}$ with Smith-McMillan form (18.4), the polynomial

$$p_G(s) := \psi_1(s) \psi_2(s) \cdots \psi_r(s)$$

MATLAB® Hint 38. `eig` and `tzero` *do not* necessarily compute the poles and transmission zeros of a real rational matrix. ▶ p. 176 (See also MATLAB® Hint 40, p. 186)

is called the *pole (or characteristic) polynomial of $G(s)$*, its degree is called the *McMillan degree of $G(s)$*, and its roots are called the *poles of $G(s)$*. The polynomial

$$z_G(s) := \eta_1(s) \eta_2(s) \cdots \eta_r(s)$$

is called the *zero polynomial of $G(s)$*, and its roots are called the *transmission zeros of $G(s)$*. □

Example 18.4 (Smith-McMillan form). Consider the real rational matrix

$$G(s) := \begin{bmatrix} \frac{s+2}{s+1} & 0 \\ 0 & \frac{s+1}{s} \\ \frac{s+2}{s} & \frac{1}{s} \\ 0 & 1 \end{bmatrix} = \frac{1}{s(s+1)} N(s), \quad N(s) := \begin{bmatrix} s(s+2) & 0 \\ 0 & (s+1)^2 \\ (s+1)(s+2) & s+1 \\ 0 & s(s+1) \end{bmatrix}.$$

Since we saw in Example 18.2 that the Smith form of $N(s)$ is given by

$$S_N(s) = \begin{bmatrix} 1 & 0 \\ 0 & (s+1)(s+2) \\ 0 & 0 \\ 0 & 0 \end{bmatrix},$$

we conclude that the Smith-McMillan form of $G(s)$ is given by

$$SM_G(s) = \frac{1}{d(s)} S_N(s) = \begin{bmatrix} \frac{1}{s(s+1)} & 0 \\ 0 & \frac{s+2}{s} \\ 0 & 0 \\ 0 & 0 \end{bmatrix}.$$

This rational matrix has zero and pole polynomials

$$z_G(s) = s+2, \qquad\qquad p_G(s) = s^2(s+1),$$

which means that its McMillan degree is 3, and it has a single transmission zero $\{-2\}$ and three poles $\{0, 0, -1\}$. $\qquad\square$

Attention! For scalar rational functions

$$h(s) = k\,\frac{n(s)}{d(s)},$$

Note. This emphasizes a rather trivial fact: a scalar rational function cannot have a zero and a pole at the same location. However, we shall see shortly that this is not true for matrix rational functions.

where $n(s)$ and $d(s)$ are coprime monic polynomials, the Smith-McMillan form is simply $\frac{n(s)}{d(s)}$, and the zero and characteristic polynomials are $z_h(s) := n(s)$ and $p_h(s) := d(s)$, respectively. $\qquad\square$

Attention! For square matrices $G(s) \in \mathbb{R}(s)^{m \times m}$ with $r = m$, one can get some insight into the zero and pole polynomials without computing the Smith-McMillan form because, due to the Smith-McMillan factorization in Theorem 18.2, we have

$$\det G(s) = \det L(s)\, SM_G(s)\, R(s) = k \det SM_G(s) = k\,\frac{z_G(s)}{p_G(s)}, \qquad (18.5)$$

where $k := \det L(s) \det R(s)$.

Unfortunately, when the zero and characteristic polynomials have common roots, the corresponding poles and zeros do not appear in $\det G(s)$. E.g., the rational function

Attention! This example shows that matrix rational functions can have a pole and a zero at the same location ("without cancellation").

$$G(s) = \begin{bmatrix} \frac{1}{s(s-2)} & 0 \\ 0 & \frac{s-2}{s} \end{bmatrix}$$

has a zero at $\{2\}$ and poles at $\{0, 0, 2\}$ and McMillan degree 3, but

$$\det G(s) = \frac{1}{s^2},$$

which indicates a double pole at the origin, but "hides" the (unstable) pole and zero at 2.

However, even in this case we may use (18.5) to compute the transmission zeros if we have a way to directly compute the pole polynomial $p_G(s)$. We shall see in Lecture 19 that this is possible if $G(s)$ is a transfer matrix for which we have a minimal realization. □

18.5 TRANSMISSION-BLOCKING PROPERTY OF TRANSMISSION ZEROS

Consider a continuous-time real rational transfer matrix

$$\hat{G}(s) := C(sI - A)^{-1}B + D \in \mathbb{R}(s)^{m \times k},$$

with minimal realization

$$\dot{x} = Ax + Bu, \qquad y = Cx + Du, \qquad x \in \mathbb{R}^n, \ u \in \mathbb{R}^k, \ y \in \mathbb{R}^m, \qquad \text{(CLTI)}$$

Note. Recall that unimodal matrices are nonsingular for every $s \in \mathbb{C}$.

and let z_0 be a transmission zero of $\hat{G}(s)$ that is not simultaneously a pole. The rank of its Smith-McMillan form $SM_{\hat{G}}(s)$ drops for $s = z_0$, and since there are unimodal polynomial matrices $L(s)$ and $R(s)$ for which

$$\hat{G}(z_0) = L(z_0)SM_{\hat{G}}(z_0)R(z_0)$$

(cf. Smith-McMillan factorization in Theorem 18.2), we conclude that the rank of $\hat{G}(s)$ also drops for $s = z_0$. This indicates that the columns of $\hat{G}(z_0)$ must be linearly dependent. Therefore there exists a nonzero vector $u_0 \in \mathbb{C}^k$ for which

$$\hat{G}(z_0)u_0 = 0.$$

Consider the following input and initial condition for the system (CLTI) :

Note. We shall see in Lecture 19 (Theorem 19.3) that, for minimal realizations, if z_0 is not a pole, then it cannot be an eigenvalue of A, and $z_0 I - A$ is invertible. This guarantees that the initial condition proposed in (18.6) is well defined.

$$u(t) = e^{z_0 t}u_0, \quad \forall t \geq 0, \qquad x(0) = x_0 := (z_0 I - A)^{-1}Bu_0. \qquad (18.6)$$

By direct substitution, we conclude that

$$x(t) = e^{z_0 t}x_0 \qquad (18.7)$$

is a solution to (CLTI) (cf. Exercise 18.2). Moreover, using (18.6) and (18.7), we obtain

$$y(t) = Cx(t) + Du(t) = e^{z_0 t}\left(C(z_0 I - A)^{-1}Bu_0 + Du_0 \right)$$

$$= e^{z_0 t}\hat{G}(z_0)u_0 = 0, \quad \forall t \geq 0.$$

This reasoning allows us to state the following property of transmission zeros.

Note. The
discrete-time version
of this property will
be discussed in
Exercise 18.3.

Property P18.1 (Transmission blocking). For every transmission zero z_0 of the continuous-time transfer matrix $\hat{G}(s)$ that is not a pole, there exists a nonzero input of the form $u(t) = e^{z_0 t} u_0$, $\forall t \geq 0$ and appropriate initial conditions for which the output is identically zero.   □

Attention! For "fat" transfer matrices corresponding to systems with more inputs that outputs (overactuated), *for every* $s_0 \in \mathbb{C}$ it is always possible to find a vector $u_0 \in \mathbb{C}^k$ for which $\hat{G}(s_0) u_0 = 0$. For such systems, the blocking property of transmission zeros is somewhat trivial.   □

18.6 MATLAB® COMMANDS

MATLAB® Hint 38 (`eig` and `tzero`). Contrary to what is advertised, the functions `eig(tf)` and `tzero(tf)` *do not* necessarily compute the poles and transmission zeros of the transfer function `tf`. This occurs when the MATLAB® representation of the transfer function contains "uncanceled" poles and zeros (which may be quite difficult to spot by inspection in the MIMO case). We shall see in MATLAB® Hint 40 (p. 186) how to make sure that these functions return the correct values.   □

18.7 EXERCISES

18.1 (Smith-McMillan form). Compute the Smith-McMillan form of

$$G(s) := \begin{bmatrix} \frac{s}{s+1} & 0 \\ \frac{1}{s+1} & \frac{s+1}{s^2} \\ 0 & \frac{1}{s} \end{bmatrix}.$$ □

18.2 (Solution to exponential input). Show that $x(t) = e^{z_0 t} x_0$ is a solution to (CLTI) for the following input and initial condition:

$$u(t) = e^{z_0 t} u_0, \quad \forall t \geq 0, \qquad x(0) = x_0 := (z_0 I - A)^{-1} B u_0.$$ □

18.3 (Transmission-blocking property in discrete time). Derive the transmission-blocking property (P18.1) for a discrete-time LTI system, which can be stated as follows. Consider a discrete-time real rational transfer matrix

$$\hat{G}(z) := C(zI - A)^{-1} B + D \in \mathbb{R}(z)^{m \times k},$$

with minimal realization

$$x^+ = Ax + Bu, \qquad y = Cx + Du, \qquad x \in \mathbb{R}^n, \ u \in \mathbb{R}^k, \ y \in \mathbb{R}^m, \qquad \text{(DLTI)}$$

and let z_0 be a transmission zero of $\hat{G}(z)$ that is not simultaneously a pole. There exists a nonzero input of the form $u(t) = z_0^t u_0$, $\forall t \geq 0$ and appropriate initial conditions for which the output is identically zero.   □

LECTURE 19

State-Space Zeros, Minimality, and System Inverses

CONTENTS

This lecture explores the connection between the poles and zeros of a transfer function (as defined in Lecture 18) and properties of its state-space realization. It also introduces the notion of system inverse and its connection to poles and zeros.

19.1 POLES OF TRANSFER FUNCTIONS VERSUS EIGENVALUES OF STATE-SPACE REALIZATIONS

Notation. In this lecture, we jointly present the results for continuous and discrete time and use a slash to separate the two cases. In discrete time, the Laplace

Consider the LTI system

$$\dot{x}/x^+ = Ax + Bu, \qquad y = Cx + Du, \qquad x \in \mathbb{R}^n,\ u \in \mathbb{R}^k,\ y \in \mathbb{R}^m, \qquad \text{(LTI)}$$

with a real rational transfer matrix

$$\hat{G}(s) := C(sI - A)^{-1}B + D \in \mathbb{R}(s)^{m \times k}.$$

Note. We recall that

$$M^{-1} = \frac{1}{\det M}(\text{adj}\, M)',$$

$$\text{adj}\, M := [\text{cof}_{ij}\, M],$$

where $\text{cof}_{ij}\, M$ denotes the ijth cofactor of M, i.e, the determinant of the M submatrix obtained by removing row i and column j multiplied by $(-1)^{i+j}$.

$$\hat{G}(s) = \frac{1}{\Delta(s)}\Big(C[\text{adj}(sI - A)]'B + \Delta(s)D\Big) = \frac{1}{d(s)}N(s),$$

where $\Delta(s) := \det(sI - A)$ is the characteristic polynomial of A, $d(s)$ is the monic least common denominator of all entries of $\hat{G}(s)$, and $N(s) \in \mathbb{R}[s]^{m \times k}$ is a polynomial matrix. Since $C[\text{adj}(sI - A)]'B + \Delta(s)D$ is a polynomial matrix, $d(s)$ must be equal to $\Delta(s)$, aside from possible cancellations. Therefore

$$\{\text{roots of } d(s)\} \subset \{\text{roots of } \Delta(s)\} = \{\text{eigenvalues of } A\}.$$

The Smith-McMillan form of $\hat{G}(s)$ is given by

$$SM_G(s) := \frac{1}{d(s)} S_N(s) = \begin{bmatrix} \frac{\eta_1(s)}{\psi_1(s)} & 0 & \cdots & 0 & 0 & \cdots & 0 \\ 0 & \frac{\eta_2(s)}{\psi_2(s)} & \cdots & 0 & 0 & \cdots & 0 \\ \vdots & \vdots & \ddots & \vdots & \vdots & & \vdots \\ 0 & 0 & \cdots & \frac{\eta_r(s)}{\psi_r(s)} & 0 & \cdots & 0 \\ 0 & 0 & \cdots & 0 & 0 & \cdots & 0 \\ \vdots & \vdots & \ddots & \vdots & \vdots & & \vdots \\ 0 & 0 & \cdots & 0 & 0 & \cdots & 0 \end{bmatrix} \in \mathbb{R}(s)^{m \times k},$$

where $S_N(s) \in \mathbb{R}[s]^{m \times k}$ denotes the Smith form of $N(s)$, and the $\psi_i(s)$ are obtained by dividing the invariant factors $\epsilon_i(s)$ of $N(s)$ by $d(s)$. Therefore all the roots of the $\psi_i(s)$ must also be roots of $d(s)$. We thus conclude that

$$\{\text{poles of } \hat{G}(s)\} = \{\text{roots of } p_{\hat{G}}(s)\} = \{\text{all roots of the } \psi_i(s)\}$$
$$\subset \{\text{roots of } d(s)\} \subset \{\text{eigenvalues of } A\}. \quad \square$$

19.2 TRANSMISSION ZEROS OF TRANSFER FUNCTIONS VERSUS INVARIANT ZEROS OF STATE-SPACE REALIZATIONS

Consider the continuous-time LTI system

$$\dot{x} = Ax + Bu, \qquad y = Cx + Du, \qquad x \in \mathbb{R}^n,\ u \in \mathbb{R}^k,\ y \in \mathbb{R}^m. \quad \text{(CLTI)}$$

Taking the Laplace transform of each side of the two equations in (CLTI), we obtain

$$s\hat{x}(s) - x(0) = A\hat{x}(s) + B\hat{u}(s), \qquad \hat{y}(s) = C\hat{x}(s) + D\hat{u}(s),$$

Note. We recall that for unilateral Laplace transforms $\mathcal{L}[\dot{x}(t)] = s\hat{x}(s) - x(0)$ (cf. Section 3.4).

which can be rewritten as

$$P(s)\begin{bmatrix} -\hat{x}(s) \\ \hat{u}(s) \end{bmatrix} = \begin{bmatrix} -x(0) \\ \hat{y}(s) \end{bmatrix},$$

where the real polynomial matrix

$$P(s) := \begin{bmatrix} sI - A & B \\ -C & D \end{bmatrix} \in \mathbb{R}[s]^{(n+m) \times (n+k)}$$

is called the *Rosenbrock's system matrix*. This matrix is used to introduce a notion of zeros for state-space realizations.

Note. The invariant zero polynomial is precisely the rth-order determinantal divisor $D_r(s)$ of $P(s)$.

MATLAB® Hint 39. `tzero(sys)` computes the invariant zeros of the state-space system `sys`, which are *not* necessarily the transmission zeros.   ▶ p. 186

Definition 19.1 (Invariant zeros). The *invariant zero polynomial of the state-space system (CLTI)* is the monic greatest common divisor $z_P(s)$ of all nonzero minors of order $r := \operatorname{rank} P(s)$. The roots of $z_P(s)$ are called the *invariant zeros of the state-space system (CLTI)*.   □

Example 19.1 (Invariant zeros). Consider the continuous-time LTI system

$$\dot{x} = \begin{bmatrix} 0 & -1 & 1 \\ 1 & -2 & 1 \\ 0 & 1 & -1 \end{bmatrix} x + \begin{bmatrix} 1 & 0 \\ 1 & 1 \\ 1 & 2 \end{bmatrix} u, \qquad y = \begin{bmatrix} 0 & 1 & 0 \end{bmatrix} x.$$

Its Rosenbrock's system matrix is given by

$$P(s) = \begin{bmatrix} s & 1 & -1 & 1 & 0 \\ -1 & s+2 & -1 & 1 & 1 \\ 0 & -1 & s+1 & 1 & 2 \\ 0 & -1 & 0 & 0 & 0 \end{bmatrix},$$

whose 4th-order minors are

$$\underbrace{(s+1)(s+2)}_{\text{``minus'' column 5}}, \underbrace{(s+1)(s+2)}_{\text{``minus'' column 4}}, \underbrace{-(s+2)}_{\text{``minus'' column 3}}, \underbrace{0}_{\text{``minus'' column 2}}, \underbrace{(s+2)}_{\text{``minus'' column 1}};$$

therefore, the invariant zero polynomial is $z_P(s) = s + 2$, which means that the state-space system has a single invariant zero at $s = -2$.   □

Notation. The vector $\begin{bmatrix} x_0' & -u_0' \end{bmatrix}' \in \mathbb{C}^{n+k}$ is called the *invariant zero direction*.

Note. If z_0 is not an eigenvalue of A, then u_0 must be nonzero. Why?

Invariant zeros also have a blocking property. If z_0 is an invariant zero, then the rank of $P(s)$ drops for $s = z_0$ (cf. Lemma 18.1), which indicates that the columns of $P(z_0)$ must be linearly dependent. Therefore there exists a nonzero vector $\begin{bmatrix} x_0' & -u_0' \end{bmatrix}' \in \mathbb{C}^{n+k}$ for which

$$P(z_0) \begin{bmatrix} x_0 \\ -u_0 \end{bmatrix} = 0.$$

Consider the input and initial condition for the system (CLTI)

$$u(t) = e^{z_0 t} u_0, \quad \forall t \geq 0, \qquad\qquad x(0) = x_0.$$

By direct substitution, we conclude that

$$x(t) = e^{z_0 t} x_0$$

is a solution to (CLTI) (cf. Exercise 18.2). Moreover,

$$y(t) = Cx(t) + Du(t) = e^{z_0 t} (Cx_0 + Du_0) = 0, \quad \forall t \geq 0.$$

This reasoning allows us to state the following property of invariant zeros.

Property P19.1 (Transmission blocking). For every invariant zero z_0 of the state-space system (CLTI) that is not an eigenvalue of A, there exists a nonzero input of the form $u(t) = e^{z_0 t} u_0, \forall t \geq 0$ and appropriate initial conditions for which the output is identically zero.

Attention!
Transmission zeros
are defined in the
frequency domain for
transfer matrices,
whereas invariant
zeros are defined in
the time domain for
state-space
realizations.

In view of this property and the previously observed blocking property of the transmission zeros of a transfer function, it is not surprising that there is a strong connection between transmission and invariant zeros. Consider the LTI system

$$\dot{x}/x^+ = Ax + Bu, \qquad y = Cx + Du, \qquad x \in \mathbb{R}^n,\ u \in \mathbb{R}^k,\ y \in \mathbb{R}^m, \qquad \text{(LTI)}$$

with a real rational transfer matrix

$$\hat{G}(s) := C(sI - A)^{-1}B + D \in \mathbb{R}(s)^{m \times k}.$$

Attention! The
system (LTI) may have
more invariant zeros
than the transmission
zeros of $\hat{G}(s)$.

Theorem 19.2. *The following inclusion holds:*

$$\left\{ \text{transmission zeros of } \hat{G}(s) \right\} \subset \left\{ \text{invariant zeros of (LTI)} \right\}. \qquad \square$$

Proof of Theorem 19.2. The Rosenbrock's system matrix can be factored as

$$
\begin{aligned}
P(s) &:= \begin{bmatrix} sI - A & B \\ -C & D \end{bmatrix}_{(n+m) \times (n+k)} \\
&= \begin{bmatrix} sI - A & 0_{n \times m} \\ -C & I_{m \times m} \end{bmatrix}_{(n+m) \times (n+m)} \begin{bmatrix} I_{n \times n} & (sI - A)^{-1}B \\ 0_{m \times n} & \hat{G}(s) \end{bmatrix}_{(n+m) \times (n+k)}.
\end{aligned}
$$

Therefore, if z_0 is a transmission zero of $\hat{G}(s)$, the rank of $\hat{G}(s)$ drops at $s = z_0$. In this case, the rank of $P(s)$ also drops, making z_0 an invariant zero of (LTI). $\square$

19.3 ORDER OF MINIMAL REALIZATIONS

Consider the LTI system

$$\dot{x}/x^+ = Ax + Bu, \qquad y = Cx + Du, \qquad x \in \mathbb{R}^n,\ u \in \mathbb{R}^k,\ y \in \mathbb{R}^m \qquad \text{(LTI)}$$

with a real rational transfer matrix

$$\hat{G}(s) := C(sI - A)^{-1}B + D \in \mathbb{R}(s)^{m \times k}.$$

We just saw that

$$\left\{ \text{poles of } \hat{G}(s) \right\} \subset \left\{ \text{eigenvalues of } A \right\} \qquad (19.1\text{a})$$

$$\left\{ \text{transmission zeros of } \hat{G}(s) \right\} \subset \left\{ \text{invariant zeros of (LTI)} \right\}. \qquad (19.1\text{b})$$

It turns out that these inclusions actually hold with equality for minimal realizations.

Note. Recall that the
McMillan degree is
the degree of the pole
polynomial and
therefore the number
of poles of $\hat{G}(s)$.

Theorem 19.3. *The realization (LTI) is minimal if and only if n is equal to the McMillan degree of $\hat{G}(s)$. In this case,*

$$p_{\hat{G}}(s) = \det(sI - A), \qquad z_{\hat{G}}(s) = z_P(s), \qquad (19.2)$$

where $p_{\hat{G}}(s)$ and $z_{\hat{G}}(s)$ denote the pole and zero polynomial of $G(s)$, respectively, and $z_P(s)$ denotes the invariant polynomial of (LTI). Therefore the inclusions in (19.1) hold with equality. $\square$

The proof of this result can be found in [1, pp. 301–305, 397–398] and can be constructed (rather tediously) by computing a minimal realization directly from the Smith-McMillan form.

MATLAB® Hint 40.
Theorem 19.3 justifies
computing the
transmission zeros
and the poles of the
transfer function of
the system sys using
`tzero(minreal(sys))`
and
`eig(minreal(sys))`,
respectively. ▶ p. 186

This result has the following immediate consequence.

Corollary 19.1. *Assuming that the realization* (LTI) *of* $\hat{G}(s)$ *is minimal, the transfer matrix* $\hat{G}(s)$ *is BIBO stable if and only if the realization* (LTI) *is (internally) asymptotically stable.* $\square$

Proof of Corollary 19.1. We saw in Lecture 8 that asymptotic stability of (LTI) is equivalent to all eigenvalues of A having strictly negative real parts. In addition, we saw in Lecture 9 that BIBO stability of $\hat{G}(s)$ is equivalent to the poles of $\hat{G}(s)$ having strictly negative real parts. Since for minimal realization the set of eigenvalues of A is the same as the set of poles of $\hat{G}(s)$, we conclude that these two notions are equivalent.

$\square$

Example 19.2. Consider the continuous-time LTI system

$$\dot{x} = \begin{bmatrix} 0 & -1 & 1 \\ 1 & -2 & 1 \\ 0 & 1 & -1 \end{bmatrix} x + \begin{bmatrix} 1 & 0 \\ 1 & 1 \\ 1 & 2 \end{bmatrix} u, \qquad y = \begin{bmatrix} 0 & 1 & 0 \end{bmatrix} x. \tag{19.3}$$

For this state-space realization, the eigenvalues of A are at $\{0, -1, -2\}$, and we have a single invariant zero at $s = -2$ (cf. Example 19.1).

The transfer function of (19.3) is given by

$$\hat{G}(s) = \begin{bmatrix} 0 & 1 & 0 \end{bmatrix} \begin{bmatrix} s & 1 & -1 \\ -1 & s+2 & -1 \\ 0 & -1 & s+1 \end{bmatrix}^{-1} \begin{bmatrix} 1 & 0 \\ 1 & 1 \\ 1 & 2 \end{bmatrix} = \begin{bmatrix} \frac{1}{s} & \frac{1}{s} \end{bmatrix} = \frac{1}{s} \begin{bmatrix} 1 & 1 \end{bmatrix}.$$

Since the Smith form of $\begin{bmatrix} 1 & 1 \end{bmatrix}$ is simply $\begin{bmatrix} 1 & 0 \end{bmatrix}$, the Smith-McMillan form of $\hat{G}(s)$ is given by

$$SM_{\hat{G}}(s) = \begin{bmatrix} \frac{1}{s} & 0 \end{bmatrix}.$$

Attention!
Cancellations are not easy to find in MIMO systems. Note that the eigenvalue of -1 does not appear in a minimal realization, but does not appear to "cancel" with any invariant zero.

Therefore this system has no transmission zeros and a single pole at $s = 0$. For this system, we observe strict inclusions in (19.1), which can be explained when we perform a Kalman decomposition of (19.3). This decomposition results in

$$\dot{\bar{x}} = \begin{bmatrix} 0 & 0 & -\sqrt{2} \\ 0 & -1 & -\sqrt{3} \\ 0 & 0 & -2 \end{bmatrix} \bar{x} + \begin{bmatrix} \sqrt{3} & \sqrt{3} \\ 0 & -\sqrt{2} \\ 0 & 0 \end{bmatrix} u, \qquad y = \begin{bmatrix} \frac{\sqrt{3}}{3} & 0 & \frac{\sqrt{6}}{3} \end{bmatrix} x,$$

which shows an uncontrollable (but observable) mode at $s = -2$ and an unobservable (but controllable) mode at $s = -1$. $\square$

Attention! We saw in Section 18.4 that for *square transfer matrices* $\hat{G}(s) \in \mathbb{R}(s)^{m \times m}$,

$$\det \hat{G}(s) = k \frac{z_{\hat{G}}(s)}{p_{\hat{G}}(s)}$$

for some constant k. If we have available a minimal realization of $\hat{G}(s)$, then it is straightforward to compute $p_{\hat{G}}(s)$ using (19.2). This allows us to compute the zero polynomial of $\hat{G}(s)$ as

$$z_{\hat{G}}(s) = \frac{1}{k} \det(sI - A) \det \hat{G}(s).$$

The constant k is completely specified by the requirement that $z_{\hat{G}}(s)$ be monic. $\qquad\square$

19.4 SYSTEM INVERSE

Consider the LTI system

$$\dot{x}/x^+ = Ax + Bu, \qquad y = Cx + Du, \qquad x \in \mathbb{R}^n,\ u \in \mathbb{R}^k,\ y \in \mathbb{R}^m. \qquad \text{(LTI)}$$

Attention! Note the "flipped" sizes of the input and output signals.

We say that the system

$$\dot{\bar{x}}/\bar{x}^+ = \bar{A}\bar{x} + \bar{B}\bar{u}, \qquad \bar{y} = \bar{C}\bar{x} + \bar{D}\bar{u}, \qquad \bar{x} \in \mathbb{R}^{\bar{n}},\ \bar{u} \in \mathbb{R}^m,\ \bar{y} \in \mathbb{R}^k \qquad (19.4)$$

is an *inverse* for (LTI) if in both cascade connections in Figure 19.1 the cascade's output exactly replicates the cascade's input for *zero initial conditions of all systems*.

Since the definition of the system inverse reflects only responses to zero initial conditions, it can be completely defined in terms of transfer functions. The system (19.4) is an inverse for (LTI) if and only if

Note. The left equality in (19.5) corresponds to Figure 19.1(a) and the right to Figure 19.1(b).

$$\hat{\bar{G}}(s)\hat{G}(s) = I, \qquad \hat{G}(s)\hat{\bar{G}}(s) = I, \qquad (19.5)$$

where $\hat{G}(s)$ and $\hat{\bar{G}}(s)$ are the transfer matrices of (LTI) and (19.4), respectively. From the properties of matrix inverses, we know that (19.5) is possible only if the system is square $(m = k)$ and, in this case, either one of the equalities in (19.5) is equivalent to

Note. When $m \neq k$, it may still be possible for one of the equalities in (19.5) to hold. In this case, we say that the system has a *left* or a *right inverse*. These names are inspired by the equations in (19.5) and *not* by the connections in Figure 19.1, where the inverses appear on the other side.

$$\hat{\bar{G}}(s) = G^{-1}(s).$$

Attention! The definition of inverse guarantees only that the inputs and outputs of the cascade match for zero initial conditions. When the initial conditions are nonzero, these are generally not equal. However, if both the system and its inverse are asymptotically stable, then the effect of initial conditions disappears and we still have asymptotic matching between inputs and outputs; i.e.,

$$\lim_{t \to \infty} \|u(t) - \bar{y}(t)\| = 0$$

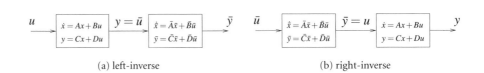

(a) left-inverse   (b) right-inverse

Figure 19.1. Cascade interconnections.

Note. The system (19.4) in Figure 19.1(b) can be viewed as an *open-loop* controller that achieves perfect (asymptotic) tracking of the reference signal $\bar{u}$.

in Figure 19.1(a) and

$$\lim_{t \to \infty} \|\bar{u}(t) - y(t)\| = 0$$

in Figure 19.1(b). □

19.5   EXISTENCE OF AN INVERSE

Consider the LTI system

$$\dot{x}/x^+ = Ax + Bu, \qquad y = Cx + Du, \qquad x \in \mathbb{R}^n, \ u, y \in \mathbb{R}^m. \qquad \text{(LTI)}$$

Assuming that the matrix D is nonsingular, from the output equation, we conclude that

$$y = Cx + Du \quad \Leftrightarrow \quad u = -D^{-1}Cx + D^{-1}y$$

and therefore

$$\dot{x}/x^+ = (A - BD^{-1}C)x + BD^{-1}y.$$

This shows that the state-space system

Note. Alternatively, for every input y to (19.6), when one feeds its output u to (LTI), one recovers y. In fact, one not only recovers y from u (or vice versa), but the state of the system and its inverse follow exactly the same trajectory.

$$\dot{x}/x^+ = \bar{A}x + \bar{B}y, \qquad u = \bar{C}x + \bar{D}y, \qquad x \in \mathbb{R}^n, \ u, y \in \mathbb{R}^m \qquad (19.6)$$

with

$$\bar{A} := A - BD^{-1}C, \qquad \bar{B} := BD^{-1}, \qquad \bar{C} := -D^{-1}C, \qquad \bar{D} := D^{-1}$$

takes y as an input and exactly recovers u at its output (keeping the states of both systems equal at all times), which means that (19.6) is an inverse of (LTI) and vice versa. This leads to the following result.

MATLAB® Hint 41. `inv(sys)` computes the inverse of the system `sys`. ▶ p. 186

Theorem 19.4 (Invertible system). *The system (LTI) has an inverse if and only if D is a nonsingular matrix. Moreover,*

$$\hat{G}(s)^{-1} = \hat{\bar{G}}(s),$$

where

Note. We shall see in Exercise 19.2 that if (LTI) is a minimal realization, then (19.6) is also minimal.

$$\hat{G}(s) := C(sI - A)^{-1}B + D, \qquad \hat{\bar{G}}(s) := \bar{C}(sI - \bar{A})^{-1}\bar{B} + \bar{D}. \qquad \square$$

Proof of Theorem 19.4. The fact that nonsingularity of D is sufficient for the existence of an inverse was proved by explicitly constructing the inverse.

Note. We recall that the transfer function of an LTI system is always proper.

We use a contradiction argument to prove necessity. Assume that the system has an inverse with (proper) transfer function $\hat{G}(s)^{-1}$, but D is singular. In this case, there exists a nonzero vector v for which $Dv = 0$. Therefore all entries of

$$h(s) := \hat{G}(s)v = C(sI - A)^{-1}Bv$$

are strictly proper, because of the absence of a direct feed-through term in $h(s)$. But then, since $\hat{G}(s)^{-1}$ is proper, $\hat{G}(s)^{-1}h(s)$ is also strictly proper. This contradicts the fact that

$$\hat{G}(s)^{-1}h(s) = \hat{G}(s)^{-1}\hat{G}(s)v = v. \qquad \square$$

19.6 POLES AND ZEROS OF AN INVERSE

Given a real rational matrix $\hat{G}(s) \in \mathbb{R}(s)^{m \times m}$ with Smith-McMillan form

$$SM_G(s) = \begin{bmatrix} \frac{\eta_1(s)}{\psi_1(s)} & 0 & \cdots & 0 & 0 & \cdots & 0 \\ 0 & \frac{\eta_2(s)}{\psi_2(s)} & \cdots & 0 & 0 & \cdots & 0 \\ \vdots & \vdots & \ddots & \vdots & \vdots & & \vdots \\ 0 & 0 & \cdots & \frac{\eta_r(s)}{\psi_r(s)} & 0 & \cdots & 0 \\ 0 & 0 & \cdots & 0 & 0 & \cdots & 0 \\ \vdots & \vdots & \ddots & \vdots & \vdots & & \vdots \\ 0 & 0 & \cdots & 0 & 0 & \cdots & 0 \end{bmatrix} \in \mathbb{R}(s)^{m \times m},$$

we saw in Theorem 18.2 that there exist unimodular real polynomial matrices $L(s)$, $R(s) \in \mathbb{R}[s]^{m \times m}$ such that

$$\hat{G}(s) = L(s) SM_{\hat{G}}(s) R(s).$$

Since $L(s)$ and $R(s)$ are unimodal, $\hat{G}(s)$ has an inverse if and only if $SM_{\hat{G}}(s)$ is invertible, which happens only when $m = r$. In this case,

$$\hat{G}(s)^{-1} = R(s)^{-1} SM_{\hat{G}}(s)^{-1} L(s)^{-1},$$

where

$$SM_{\hat{G}}(s)^{-1} = \begin{bmatrix} \frac{\psi_1(s)}{\eta_1(s)} & 0 & \cdots & 0 \\ 0 & \frac{\psi_2(s)}{\eta_2(s)} & \cdots & 0 \\ \vdots & \vdots & \ddots & \vdots \\ 0 & 0 & \cdots & \frac{\psi_r(s)}{\eta_r(s)} \end{bmatrix} \in \mathbb{R}(s)^{m \times m},$$

is the Smith-McMillan form of $\hat{G}(s)^{-1}$ (up to a change in order of the columns and rows). From this, we obtain the following generalization to MIMO systems of a fact that is trivial for the SISO case.

Properties (Inverse). Assume that the system with transfer matrix $\hat{G}(s)$ is invertible.

P19.2 The poles of the transfer matrix $\hat{G}(s)^{-1}$ are the transmission zeros of its inverse $\hat{G}(s)$ and vice versa.

P19.3 $\hat{G}(s)^{-1}$ is BIBO stable if and only if every transmission zero of $\hat{G}(s)$ has a strictly negative real part.

Note 10. We saw in Section 19.5 that when D is nonsingular, the system

$$\dot{x}/x^+ = Ax + Bu, \qquad y = Cx + Du, \qquad x \in \mathbb{R}^n, \, u, y \in \mathbb{R}^m \qquad \text{(LTI)}$$

is invertible, and its inverse $\hat{G}(s)^{-1}$ has the realization

$$\dot{x}/x^+ = \bar{A}x + \bar{B}y, \qquad u = \bar{C}x + \bar{D}y, \qquad x \in \mathbb{R}^n, \, u, y \in \mathbb{R}^m,$$

Note 10. P19.2 allows us to compute the transmission zeros of $\hat{G}(s)$ by finding the poles of its inverse $\hat{G}(s)^{-1}$. ▶ p. 184

Notation. A system whose transmission zeros have strictly negative real parts is called *(strictly) minimum phase.*

with

$$\bar{A} := A - BD^{-1}C, \qquad \bar{B} := BD^{-1}, \qquad \bar{C} := -D^{-1}C, \qquad \bar{D} := D^{-1}.$$

Note. Why is the realization for the inverse minimal? See Exercise 19.2.

Note. This also shows that invertible systems with McMillan degree n have exactly n transmission zeros.

Assuming that (LTI) is a minimal realization, then this realization for the inverse is also minimal and, as we saw in Theorem 19.3, the poles of $\hat{G}(s)^{-1}$ are simply the eigenvalues of $\bar{A} := A - BD^{-1}C$. In view of P19.2, this makes it very simple to compute the transmission zeros of (square) invertible systems. We just have to compute the eigenvalues of $\bar{A} := A - BD^{-1}C$. $\qquad\qquad\square$

19.7 FEEDBACK CONTROL OF STABLE SYSTEMS WITH STABLE INVERSES

Consider a system with transfer function $\hat{G}(s)$ that is a BIBO stable system with a BIBO stable inverse. Given an arbitrary desired transfer function $\hat{Q}(s)$ from a reference input r to the output y, one can always design an open-loop controller that results in this transfer function. Such a control law is shown in Figure 19.2.

Note. To verify that the closed-loop transfer function from r to y in Figure 19.3 is indeed equal to $\hat{Q}(s)$, note that the signals labeled a and b cancel each other, since $a = y$ and $b = \bar{u} = y$.

Note. We shall confirm in Exercise 19.3 that the transfer function of the controller in the dashed box in Figure 19.3 is indeed given by (19.7).

The same desired transfer function can also be achieved with a feedback controller with the transfer function

$$\hat{C}(s) = \hat{G}(s)^{-1}\big(I - \hat{Q}(s)\big)^{-1}\hat{Q}(s). \tag{19.7}$$

This control law is shown in Figure 19.3.

One may ask why one would choose the feedback control in Figure 19.3 instead of an open-loop control as in Figure 19.2. It turns out that, in general, the closed-loop controller is more robust with respect to modeling uncertainty. In particular, suppose we have a SISO process with transfer function $\hat{g}(s)$ and that both $\hat{g}(s)$ and $\hat{g}(s)^{-1}$ are stable. To achieve a desired transfer function

$$\hat{q}(s) = \frac{k}{s+k}, \quad k \geq 0$$

from a reference r to y, we could either use the open-loop controller

$$\hat{\bar{g}}(s) = \hat{g}(s)^{-1}\hat{q}(s) = \frac{k}{s+k}\hat{g}(s)^{-1}$$

in Figure 19.2 or the closed-loop controller

$$\hat{c}(s) = \hat{g}(s)^{-1}\frac{q(s)}{1-q(s)} = \frac{k}{s}\hat{g}(s)^{-1}$$

in Figure 19.3. Since both controllers guarantee the same transfer function $\hat{q}(s)$ from r to y and $\hat{q}(0) = 1$, we conclude that both result in zero steady-state error for step responses.

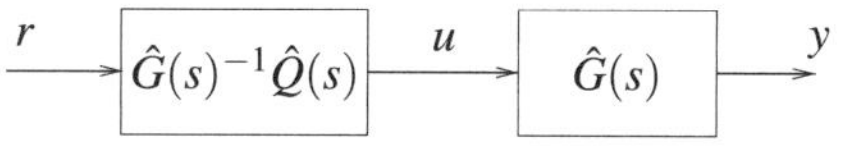

Figure 19.2. Open-loop control.

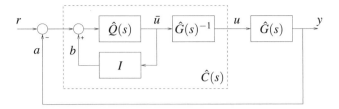

Figure 19.3. Feedback controller that achieves a closed-loop transfer function $\hat{Q}(s)$ from r to y.

Suppose now that the transfer function $\hat{g}_{\text{true}}(s)$ of the true plant is not exactly the inverse of the transfer function $\hat{g}_{\text{cont}}(s)$ used to construct the controller. For the open-loop design, we obtain the following transfer functions from r to y,

$$\hat{g}_{\text{true}}(s)\hat{\bar{g}}(s) = \hat{g}_{\text{true}}(s)\hat{g}_{\text{cont}}(s)^{-1}\hat{q}(s),$$

and for the closed-loop design, we obtain

$$\frac{\hat{g}_{\text{true}}(s)\hat{c}(s)}{1 + \hat{g}_{\text{true}}(s)\hat{c}(s)} = \frac{k\hat{g}_{\text{true}}(s)\hat{g}_{\text{cont}}(s)^{-1}}{s + k\hat{g}_{\text{true}}(s)\hat{g}_{\text{cont}}(s)^{-1}}. \tag{19.8}$$

The latter transfer function has the interesting property that *no matter what is the mismatch between $\hat{g}_{\text{true}}(s)$ and $\hat{g}_{\text{cont}}(s)$, the transfer function from r to y is still equal to 1 at $s = 0$*, and therefore the closed-loop design preserves zero steady-state error for step responses.

Attention! Note, however, that the open-loop controller never results in an unstable system, as long as the process remains stable, whereas the closed-loop controller may become unstable if the poles of (19.8) leave the left-hand side complex half-plane.

19.8 MATLAB® COMMANDS

MATLAB® Hint 39 (`tzero`). The function `tzero(sys)` computes the invariant zeros of the state-space system `sys`.

Contrary to what is advertised, it *does not* necessarily compute the transmission zeros of the system (cf. MATLAB® Hint 40, p. 186). You may verify this by trying this function on the system in Examples 19.1 and 19.2. □

MATLAB® Hint 40 (`eig` and `tzero`). The functions `tzero(minreal(sys))` and `eig(minreal(sys))` return the transmission zeros and poles of the transfer function of the system `sys`, which can either be in state-space or transfer function form.

Note that one needs to "insert" the function `minreal` to make sure that one does indeed get only the poles and transmission zeros, because otherwise one may get extra poles and transmission zeros. This is especially important for MIMO systems, for which one cannot find these extra poles and zeros simply by inspection (cf. Example 19.2). □

MATLAB® Hint 41 (`inv`). The function `inv(sys)` computes the inverse of the system `sys`. When `sys` is a state-space model, `inv` returns a state-space model, and when `sys` is a transfer function, it returns a transfer function. □

19.9 EXERCISES

19.1 (Transmission zeros). Verify that all transmission zeros of the following transfer matrix have a strictly negative real part and therefore that both $\hat{G}(s)$ and $\hat{G}(s)^{-1}$ are BIBO stable:

$$\hat{G}(s) = \begin{bmatrix} \frac{s+2}{s+1} & \frac{s-1}{s+2} \\ 0 & \frac{s+2}{s+3} \end{bmatrix}. \qquad \square$$

19.2 (Controllability, observability, and minimality of the inverse). Consider the LTI system (LTI) and its inverse (19.6).

(a) Show that (A, B) is controllable if and only if $(\bar{A}, \bar{B})$ is controllable.

(b) Show that (A, C) is observable if and only if $(\bar{A}, \bar{C})$ is observable.

(c) Show that (A, B, C, D) is minimal if and only if $(\bar{A}, \bar{B}, \bar{C}, \bar{D})$ is minimal.

Hint: Use the eigenvector tests. $\qquad \square$

19.3. Verify that the transfer function of the controller in the dashed box in Figure 19.3 is indeed given by (19.7). $\qquad \square$

PART VI
LQR/LQG Optimal Control

LECTURE 20

Linear Quadratic Regulation (LQR)

CONTENTS

This lecture introduces the most general form of the linear quadratic regulation problem and solves it using an appropriate feedback invariant.

20.1 DETERMINISTIC LINEAR QUADRATIC REGULATION (LQR)

Attention! Note the *negative feedback* and the *absence of a reference signal* in Figure 20.1.

Figure 20.1 shows the feedback configuration for the *linear quadratic regulation (LQR) problem*. The process is assumed to be a continuous-time LTI system of the form

$$\dot{x} = Ax + Bu, \qquad\qquad x \in \mathbb{R}^n, \ u \in \mathbb{R}^k,$$
$$y = Cx, \qquad\qquad y \in \mathbb{R}^m,$$
$$z = Gx + Hu, \qquad\qquad z \in \mathbb{R}^\ell,$$

and has two distinct outputs.

1. The *measured output* $y(t)$ corresponds to the signal(s) that can be measured

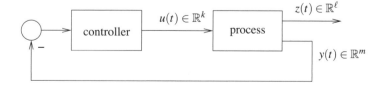

Figure 20.1. Linear quadratic regulation (LQR) feedback configuration

Note. Measured
outputs are typically
determined by the
available sensors.

Note. Controlled
outputs are selected
by the controller
designer and should
be viewed as design
parameters.

and are therefore available for control.

2. The *controlled output* $z(t)$ corresponds to the signal(s) that one would like to
   make as small as possible in the shortest possible time.

   Sometimes $z(t) = y(t)$, which means that our control objective is simply to
   make the measured output very small. At other times one may have

   $$z(t) = \begin{bmatrix} y(t) \\ \dot{y}(t) \end{bmatrix},$$

   which means that we want to make both the measured output $y(t)$ and its
   derivative $\dot{y}(t)$ very small. Many other options are possible.

20.2   OPTIMAL REGULATION

The LQR problem is defined as follows. Find the control input $u(t), t \in [0, \infty)$ that
makes the following criterion as small as possible:

$$J_{\text{LQR}} := \int_0^\infty \|z(t)\|^2 + \rho \|u(t)\|^2 dt, \tag{20.1}$$

where ρ is a positive constant. The term

$$\int_0^\infty \|z(t)\|^2 dt$$

corresponds to the *energy of the controlled output*, and the term

$$\int_0^\infty \|u(t)\|^2 dt$$

corresponds to the *energy of the control signal*. In LQR one seeks a controller that
minimizes both energies. However, decreasing the energy of the controlled output
will require a large control signal, and a small control signal will lead to large controlled outputs. The role of the constant ρ is to establish a trade-off between these
conflicting goals.

1. When we chose ρ very large, the most effective way to decrease J_{LQR} is to
   employ a small control input, at the expense of a large controlled output.

2. When we chose ρ very small, the most effective way to decrease J_{LQR} is to
   obtain a very small controlled output, even if this is achieved at the expense of
   employing a large control input.

Note 11. A simple
choice for the
matrices $\bar{Q}$ and $\bar{R}$ is
given by *Bryson's
rule.*   ▶ p. 196

Often the optimal LQR problem is defined more generally and consists of finding
the control input that minimizes

$$J_{\text{LQR}} := \int_0^\infty z(t)' \bar{Q} z(t) + \rho \, u(t)' \bar{R} u(t) \, dt, \tag{20.2}$$

where $\bar{Q} \in \mathbb{R}^{\ell \times \ell}$ and $\bar{R} \in \mathbb{R}^{m \times m}$ are symmetric positive-definite matrices and ρ is a positive constant.

We shall consider the most general form for a quadratic criterion, which is

$$J_{\text{LQR}} := \int_0^\infty x(t)'Qx(t) + u(t)'Ru(t) + 2x(t)'Nu(t)dt. \qquad \text{(J-LQR)}$$

Since $z = Gx + Hu$, the criterion in (20.1) is a special form of the criterion (J-LQR) with

$$Q = G'G, \qquad\qquad R = H'H + \rho I, \qquad\qquad N = G'H$$

and (20.2) is a special form of the criterion (J-LQR) with

$$Q = G'\bar{Q}G, \qquad\qquad R = H'\bar{Q}H + \rho\bar{R}, \qquad\qquad N = G'\bar{Q}H.$$

20.3   FEEDBACK INVARIANTS

Given a continuous-time LTI system

$$\dot{x} = Ax + Bu, \qquad\qquad x \in \mathbb{R}^n,\ u \in \mathbb{R}^k, \qquad\qquad \text{(AB-CLTI)}$$

Note. A *functional* maps functions (in this case signals, i.e. functions of time) to scalar values (in this case real numbers).

we say that a functional

$$H\big(x(\cdot); u(\cdot)\big)$$

that involves the system's input and state is a *feedback invariant* for the system (AB-CLTI) if, when computed along a solution to the system, its value depends only on the initial condition $x(0)$ and not on the specific input signal $u(\cdot)$.

Note. This concept was already introduced in Lecture 10, where Proposition 20.1 was proved.   ▶p.88

Proposition 20.1 (Feedback invariant). *For every symmetric matrix P, the functional*

$$H\big(x(\cdot); u(\cdot)\big) := -\int_0^\infty \big(Ax(t) + Bu(t)\big)'Px(t) + x(t)'P\big(Ax(t) + Bu(t)\big)\,dt$$

is a feedback invariant for the system (AB-CLTI), *as long as* $\lim_{t \to \infty} x(t) = 0$.   □

20.4   FEEDBACK INVARIANTS IN OPTIMAL CONTROL

Suppose that we are able to express a criterion J to be minimized by an appropriate choice of the input $u(\cdot)$ in the form

$$J = H\big(x(\cdot); u(\cdot)\big) + \int_0^\infty \Lambda\big(x(t), u(t)\big)dt, \qquad\qquad (20.3)$$

where H is a feedback invariant and the function $\Lambda(x, u)$ has the property that for every $x \in \mathbb{R}^n$

$$\min_{u \in \mathbb{R}^k} \Lambda(x, u) = 0.$$

Note. If one wants to restrict the optimization to solutions that lead to an asymptotically stable closed-loop system, then H needs to be a feedback invariant only for inputs that lead to $x(t) \to 0$ (as in Proposition 20.1). However, in this case one must check that (20.4) does indeed lead to $x(t) \to 0$.

In this case, the control

$$u(t) = \arg \min_{u \in \mathbb{R}^k} \Lambda(x, u), \qquad (20.4)$$

minimizes the criterion J, and the optimal value of J is equal to the feedback invariant

$$J = H\big(x(\cdot); u(\cdot)\big).$$

Note that it is not possible to get a lower value for J, since (1) the feedback invariant $H\big(x(\cdot); u(\cdot)\big)$ will never be affected by u, and (2) a smaller value for J would require the integral in the right-hand side of (20.3) to be negative, which is not possible, since $\Lambda\big(x(t), u(t)\big)$ can at best be as low as zero.

20.5 OPTIMAL STATE FEEDBACK

It turns out that the LQR criterion

$$J_{\text{LQR}} := \int_0^\infty x(t)'Qx(t) + u(t)'Ru(t) + 2x(t)'Nu(t)dt \qquad \text{(J-LQR)}$$

Attention! To keep the formulas short, in the remainder of this section we drop the time dependence (t) when the state x and the input u appear in time integrals.

can be expressed as in (20.3) for an appropriate choice of feedback invariant. In fact, the feedback invariant in Proposition 20.1 will work, provided that we choose the matrix P appropriately. To check that this is so, we add and subtract this feedback invariant to the LQR criterion and conclude that

$$
\begin{aligned}
J_{\text{LQR}} &:= \int_0^\infty x'Qx + u'Ru + 2x'Nu \, dt \\
&= H\big(x(\cdot); u(\cdot)\big) \\
&\quad + \int_0^\infty x'Qx + u'Ru + 2x'Nu + (Ax + Bu)'Px + x'P(Ax + Bu) \, dt \\
&= H\big(x(\cdot); u(\cdot)\big) + \int_0^\infty x'(A'P + PA + Q)x + u'Ru + 2u'(B'P + N')x \, dt.
\end{aligned}
$$

By completing the square, we can group the quadratic term in u with the cross-term in u times x:

$$
\begin{aligned}
(u' + x'K')R(u + Kx) \\
= u'Ru + x'(PB + N)R^{-1}(B'P + N')x + 2u'(B'P + N')x,
\end{aligned}
$$

where

$$K := R^{-1}(B'P + N'),$$

from which we conclude that

$$
\begin{aligned}
J_{\text{LQR}} = H\big(x(\cdot); u(\cdot)\big) + \int_0^\infty x'\big(A'P + PA + Q - (PB + N)R^{-1}(B'P + N')\big)x \\
+ (u' + x'K')R(u + Kx) \, dt.
\end{aligned}
$$

Notation.
Equation (20.5) is
called an *algebraic
Riccati equation
(ARE)*.

If we are able to select the matrix P so that

$$A'P + PA + Q - (PB + N)R^{-1}(B'P + N') = 0, \qquad (20.5)$$

we obtain precisely an expression such as (20.3) with

$$\Lambda(x, u) := (u' + x'K')R(u + Kx),$$

which has a minimum equal to zero for

$$u = -Kx, \qquad\qquad K := R^{-1}(B'P + N'),$$

Notation. Recall that
a matrix is *Hurwitz* or
a *stability matrix* if all
its eigenvalues have a
negative real part.

Note. *Asymptotic
stability of the closed
loop is needed, because
we need to make sure
that the proposed input
$u(t)$ leads to the
assumed fact that
$\lim_{t\to\infty} x(t)Px(t) = 0$.*

leading to the closed-loop system

$$\dot{x} = \big(A - BR^{-1}(B'P + N')\big)x.$$

The following has been proved.

Theorem 20.1. *Assume that there exists a symmetric solution P to the algebraic Riccati equation (20.5) for which $A - BR^{-1}(B'P + N')$ is a stability matrix. Then the feedback law*

$$u(t) := -Kx(t), \quad \forall t \geq 0, \qquad K := R^{-1}(B'P + N') \qquad (20.6)$$

minimizes the LQR criterion (J-LQR) *and leads to*

$$J_{\mathrm{LQR}} := \int_0^\infty x'Qx + u'Ru + 2x'Nu\ dt = x'(0)Px(0). \qquad\qquad \square$$

MATLAB® Hint 42.
lqr solves the ARE
(20.5) and computes
the optimal state
feedback
(20.6).   ▶ p. 195

20.6   LQR IN MATLAB®

Example. See
Example 22.1.

MATLAB® Hint 42 (lqr). The command [K,P,E]=lqr(A,B,Q,R,N) solves the algebraic Riccati equation

$$\mathrm{A}'\mathrm{P} + \mathrm{PA} + \mathrm{Q} - (\mathrm{PB} + \mathrm{N})R^{-1}(\mathrm{B}'\mathrm{P} + \mathrm{N}') = 0$$

and computes the (negative feedback) optimal state feedback matrix gain

$$\mathrm{K} = \mathrm{R}^{-1}(\mathrm{B}'\mathrm{P} + \mathrm{N}')$$

that minimizes the LQR criteria

$$J := \int_0^\infty x'\mathrm{Q}x + u'\mathrm{R}u + 2x'\mathrm{N}u\ dt$$

for the continuous-time process

$$\dot{x} = \mathrm{A}x + \mathrm{B}u.$$

This command also returns the poles E of the closed-loop system

$$\dot{x} = (\mathrm{A} - \mathrm{BK})x. \qquad\qquad \square$$

20.7 ADDITIONAL NOTES

Note 11 (Bryson's rule). A simple and reasonable choice for the matrices $\bar{Q}$ and $\bar{R}$ in (20.2) is given by *Bryson's rule* [6, p. 537]. Select $\bar{Q}$ and $\bar{R}$ diagonal, with

$$\bar{Q}_{ii} = \frac{1}{\text{maximum acceptable value of } z_i^2}, \qquad i \in \{1, 2, \ldots, \ell\},$$

$$\bar{R}_{jj} = \frac{1}{\text{maximum acceptable value of } u_j^2}, \qquad j \in \{1, 2, \ldots, k\},$$

which corresponds to the following criterion

$$J_{\text{LQR}} := \int_0^\infty \Big(\sum_{i=1}^\ell \bar{Q}_{ii}\, z_i(t)^2 + \rho \sum_{j=1}^m \bar{R}_{jj}\, u(t)^2 \Big) dt.$$

In essence, Bryson's rule scales the variables that appear in J_{LQR} so that the *maximum acceptable value for each term is 1*. This is especially important when the units used for the different components of u and z make the values for these variables numerically very different from each other.

Although Bryson's rule usually gives good results, often it is just the starting point for a trial-and-error iterative design procedure aimed at obtaining desirable properties for the closed-loop system. We shall pursue this further in Section 22.3. □

20.8 EXERCISES

20.1 (Feedback invariant). Consider the nonlinear system

$$\dot{x} = f(x, u), \qquad\qquad x \in \mathbb{R}^n,\ u \in \mathbb{R}^k$$

and a continuously differentiable function $V : \mathbb{R}^n \to \mathbb{R}$, with $V(0) = 0$. Verify that the functional

$$H\big(x(\cdot); u(\cdot)\big) := -\int_0^\infty \frac{\partial V}{\partial x}\big(x(t)\big) f\big(x(t), u(t)\big)\, dt$$

is a feedback invariant as long as $\lim_{t \to \infty} x(t) = 0$. □

20.2 (Nonlinear optimal control). Consider the nonlinear system

$$\dot{x} = f(x, u), \qquad\qquad x \in \mathbb{R}^n,\ u \in \mathbb{R}^k.$$

Use the feedback invariant in Exercise 20.1 to construct a result parallel to Theorem 20.1 for the minimization of the criterion

$$J := \int_0^\infty Q(x) + u' R(x) u\, dt,$$

where $R(x)$ is a state-dependent positive-definite matrix. □

LECTURE 21

The Algebraic Riccati Equation (ARE)

CONTENTS

This lecture addresses the existence of solutions to the algebraic Riccati equation.

21.1 THE HAMILTONIAN MATRIX

The construction of the optimal LQR feedback law in Theorem 20.1 required the existence of a symmetric solution P to the ARE,

$$A'P + PA + Q - (PB + N)R^{-1}(B'P + N') = 0, \qquad (21.1)$$

for which $A - BR^{-1}(B'P + N')$ is a stability matrix. To study the solutions of this equation, it is convenient to expand the last term in the left-hand side of (21.1), which leads to

$$(A - BR^{-1}N')'P + P(A - BR^{-1}N') + Q - NR^{-1}N' - PBR^{-1}B'P = 0. \qquad (21.2)$$

This equation can be compactly rewritten as

$$\begin{bmatrix} P & -I \end{bmatrix} \mathbf{H} \begin{bmatrix} I \\ P \end{bmatrix} = 0, \qquad (21.3)$$

where

$$\mathbf{H} := \begin{bmatrix} A - BR^{-1}N' & -BR^{-1}B' \\ -Q + NR^{-1}N' & -(A - BR^{-1}N')' \end{bmatrix} \in \mathbb{R}^{2n \times 2n}$$

is called the *Hamiltonian matrix associated with* (21.1).

21.2 DOMAIN OF THE RICCATI OPERATOR

Notation. We write
$\mathbf{H} \in$ Ric when $\mathbf{H}$ is in
the domain of the
Riccati operator.

A Hamiltonian matrix $\mathbf{H}$ is said to be in the *domain of the Riccati operator* if there exist square matrices $\mathbf{H}_-, P \in \mathbb{R}^{n \times n}$ such that

$$\mathbf{H} M = M \mathbf{H}_-, \qquad\qquad M := \begin{bmatrix} I \\ P \end{bmatrix}, \qquad\qquad (21.4)$$

where $\mathbf{H}_-$ is a stability matrix and I is the $n \times n$ identity matrix.

Theorem 21.1. *Suppose that $\mathbf{H}$ is in the domain of the Riccati operator and let $P, \mathbf{H}_- \in \mathbb{R}^{n \times n}$ be as in (21.4). Then the following properties hold.*

1. *P satisfies the ARE (21.1),*

Notation. In general
the ARE has multiple
solutions, but only the
one in (21.4) makes
the closed-loop
system asymptotically
stable. This solution is
called the *stabilizing
solution*.

2. *$A - B R^{-1}(B'P + N') = \mathbf{H}_-$ is a stability matrix, and*

3. *P is a symmetric matrix.* $\qquad\qquad\qquad\qquad\qquad\qquad\qquad\quad \square$

Proof of Theorem 21.1. To prove statement 1, we left-multiply (21.4) by the matrix $\begin{bmatrix} P & -I \end{bmatrix}$ and obtain (21.3).

To prove statement 2, we just look at the top n rows of the matrix equation (21.4):

$$\begin{bmatrix} A - B R^{-1} N' & -B R^{-1} B' \\ \cdot & \cdot \end{bmatrix} \begin{bmatrix} I \\ P \end{bmatrix} = \begin{bmatrix} I \\ \cdot \end{bmatrix} \mathbf{H}_-,$$

from which $A - B R^{-1}(B'P + N') = \mathbf{H}_-$ follows.

To prove statement 3, we left-multiply (21.4) by $\begin{bmatrix} -P' & I \end{bmatrix}$ and obtain

$$\begin{bmatrix} -P' & I \end{bmatrix} \mathbf{H} \begin{bmatrix} I \\ P \end{bmatrix} = (P - P')\mathbf{H}_-. \qquad\qquad (21.5)$$

Note. We shall
confirm in
Exercise 21.1 that the
matrix in the
left-hand side of
(21.5) is indeed
symmetric.

Moreover, using the definition of $\mathbf{H}$, we can conclude that the matrix in the left-hand side of (21.5) is symmetric. Therefore

$$(P - P')\mathbf{H}_- = \mathbf{H}_-'(P' - P) = -\mathbf{H}_-'(P - P'). \qquad\qquad (21.6)$$

Note. This same
argument was used in
the proof of the
Lyapunov stability
theorem (Theorem
8.2).

Multiplying this equation on the left and right by $e^{\mathbf{H}_-'t}$ and $e^{\mathbf{H}_-t}$, respectively, we conclude that

$$e^{\mathbf{H}_-'t}(P - P')\mathbf{H}_- e^{\mathbf{H}_-t} + e^{\mathbf{H}_-'t}\mathbf{H}_-'(P - P')e^{\mathbf{H}_-t} = 0$$

$$\Leftrightarrow \quad \frac{d}{dt} e^{\mathbf{H}_-'t}(P - P')e^{\mathbf{H}_-t} = 0,$$

$\forall t$, which means that $e^{\mathbf{H}_-'t}(P - P')e^{\mathbf{H}_-t}$ is constant. However, since $\mathbf{H}_-$ is a stability matrix, this quantity must also converge to zero as $t \to \infty$. Therefore it must actually be identically zero. Since $e^{\mathbf{H}_-t}$ is nonsingular, we conclude that we must have $P = P'$. $\qquad\qquad\qquad\qquad\qquad\qquad\qquad\qquad\qquad\qquad\qquad\qquad\quad \blacksquare$

21.3 STABLE SUBSPACES

Given a square matrix M, suppose that we factor its characteristic polynomial as a product of polynomials

$$\Delta(s) = \det(sI - M) = \Delta_-(s)\Delta_+(s),$$

where all the roots of $\Delta_-(s)$ have a negative real part and all roots of $\Delta_+(s)$ have positive or zero real parts. The *stable subspace* of M is defined by

$$\mathcal{V}_- := \ker \Delta_-(M)$$

and has a few important properties, as listed below.

Note. See Exercise 21.3.

Note. From P21.1, we can see that the dimension of $\mathcal{V}_-$ is equal to the number of eigenvalues of M with a negative real part (with repetitions).

Properties (Stable subspaces). Let $\mathcal{V}_-$ be the stable subspace of M. Then

P21.1 $\dim \mathcal{V}_- = \deg \Delta_-(s)$, and

P21.2 for every matrix V whose columns form a basis for $\mathcal{V}_-$, there exists a stability matrix M_- whose characteristic polynomial is $\Delta_-(s)$ such that

$$MV = VM_-. \tag{21.7}$$

21.4 STABLE SUBSPACE OF THE HAMILTONIAN MATRIX

Our goal now is to find the conditions under which the Hamiltonian matrix $\mathbf{H} \in \mathbb{R}^{2n \times 2n}$ belongs to the domain of the Riccati operator, i.e., those for which there exist symmetric matrices $\mathbf{H}_-, P \in \mathbb{R}^{n \times n}$ such that

$$\mathbf{H} M = M \mathbf{H}_-, \qquad\qquad M := \begin{bmatrix} I \\ P \end{bmatrix},$$

where $\mathbf{H}_-$ is a stability matrix and I is the $n \times n$ identity matrix. From the properties of stable subspaces, we conclude that such a matrix $\mathbf{H}_-$ exists if we can find a basis for the stable subspace $\mathcal{V}_-$ of $\mathbf{H}$ of the appropriate form $M = \begin{bmatrix} I & P' \end{bmatrix}'$. For this to be possible, the stable subspace has to have dimension precisely equal to n, which is the key issue of concern. We shall see shortly that the structure $\begin{bmatrix} I & P' \end{bmatrix}'$ for M is relatively simple to produce.

21.4.1 DIMENSION OF THE STABLE SUBSPACE OF $\mathbf{H}$

To investigate the dimension of $\mathcal{V}_-$, we need to compute the characteristic polynomial of $\mathbf{H}$. To do this, note that

$$\mathbf{H} \begin{bmatrix} 0 & I \\ -I & 0 \end{bmatrix} = \begin{bmatrix} BR^{-1}B' & A - BR^{-1}N' \\ (A - BR^{-1}N')' & -Q + NR^{-1}N' \end{bmatrix} = \begin{bmatrix} 0 & -I \\ I & 0 \end{bmatrix} \mathbf{H}'.$$

Therefore, defining $J := \begin{bmatrix} 0 & I \\ -I & 0 \end{bmatrix}$,

$$\mathbf{H} = -J\mathbf{H}'J^{-1}.$$

Since the characteristic polynomial is invariant with respect to similarity transformations and matrix transposition, we conclude that

$$\Delta(s) := \det(sI - \mathbf{H}) = \det(sI + J\mathbf{H}'J^{-1}) = \det(sI + \mathbf{H}')$$
$$= \det(sI + \mathbf{H}) = (-1)^{2n}\det((-s)I - \mathbf{H}) = \Delta(-s),$$

which shows that if λ is an eigenvalue of $\mathbf{H}$, then $-\lambda$ is also an eigenvalue of $\mathbf{H}$ with the same multiplicity. We thus conclude that the $2n$ eigenvalues of $\mathbf{H}$ are distributed symmetrically with respect to the imaginary axis. To check that we actually have n eigenvalues with a negative real part and another n with a positive real part, we need to make sure that $\mathbf{H}$ has no eigenvalues over the imaginary axis. This point is addressed by the following result.

Lemma 21.1. *Assume that $Q - NR^{-1}N' \geq 0$. When the pair (A, B) is stabilizable and the pair $(A - BR^{-1}N', Q - NR^{-1}N')$ is detectable, then*

1. *the Hamiltonian matrix $\mathbf{H}$ has no eigenvalues on the imaginary axis, and*

2. *its stable subspace $\mathcal{V}_-$ has dimension n.*   □

Attention! The best LQR controllers are obtained for choices of the controlled output z for which $N = G'H = 0$ (cf. Lecture 22). In this case, Lemma 21.1 simply requires stabilizability of (A, B) and detectability of (A, G) (cf. Exercise 21.4).   □

Proof of Lemma 21.1. To prove this result by contradiction, let $x := \begin{bmatrix} x_1' & x_2' \end{bmatrix}'$, $x_1, x_2 \in \mathbb{C}^n$ be an eigenvector of $\mathbf{H}$ associated with an eigenvalue $\lambda := j\omega$, $\omega \in \mathbb{R}$. This means that

Notation. The symbol $(\cdot)^*$ denotes complex conjugate transpose.

Attention! The notation used here differs from that of MATLAB®. Here $(\cdot)'$ denotes transpose and $(\cdot)^*$ denotes complex conjugate transpose, whereas in MATLAB®, $(\cdot).'$ denotes transpose and $(\cdot)'$ denotes complex conjugate transpose.

$$\begin{bmatrix} j\omega I - A + BR^{-1}N' & BR^{-1}B' \\ Q - NR^{-1}N' & j\omega + (A - BR^{-1}N')' \end{bmatrix} \begin{bmatrix} x_1 \\ x_2 \end{bmatrix} = 0. \qquad (21.8)$$

Using the facts that (λ, x) is an eigenvalue/eigenvector pair of $\mathbf{H}$ and that this matrix is real-valued, one concludes that

$$\begin{bmatrix} x_2^* & x_1^* \end{bmatrix} \mathbf{H} \begin{bmatrix} x_1 \\ x_2 \end{bmatrix} + \begin{bmatrix} x_1^* & x_2^* \end{bmatrix} \mathbf{H}' \begin{bmatrix} x_2 \\ x_1 \end{bmatrix}$$

$$= \begin{bmatrix} x_2^* & x_1^* \end{bmatrix} (\mathbf{H}x) + (\mathbf{H}x)^* \begin{bmatrix} x_2 \\ x_1 \end{bmatrix}$$

$$= \begin{bmatrix} x_2^* & x_1^* \end{bmatrix} j\omega \begin{bmatrix} x_1 \\ x_2 \end{bmatrix} + \left(j\omega \begin{bmatrix} x_1 \\ x_2 \end{bmatrix} \right)^* \begin{bmatrix} x_2 \\ x_1 \end{bmatrix}$$

$$= j\omega(x_2^*x_1 + x_1^*x_2) - j\omega(x_1^*x_2 + x_2^*x_1) = 0. \qquad (21.9)$$

On the other hand, using the definition of $\mathbf{H}$, one concludes that the left-hand side of (21.9) is given by

$$\begin{bmatrix} x_2^* & x_1^* \end{bmatrix} \begin{bmatrix} A - BR^{-1}N' & -BR^{-1}B' \\ -Q + NR^{-1}N' & -(A - BR^{-1}N')' \end{bmatrix} \begin{bmatrix} x_1 \\ x_2 \end{bmatrix}$$

$$+ \begin{bmatrix} x_1^* & x_2^* \end{bmatrix} \begin{bmatrix} (A - BR^{-1}N')' & -Q + NR^{-1}N' \\ -BR^{-1}B' & -(A - BR^{-1}N') \end{bmatrix} \begin{bmatrix} x_2 \\ x_1 \end{bmatrix}$$
$$= -2x_1^*(Q - NR^{-1}N')x_1 - 2x_2^*(BR^{-1}B')x_2.$$

Since this expression must equal zero and $R^{-1} > 0$, we conclude that

$$(Q - NR^{-1}N')x_1 = 0, \qquad\qquad B'x_2 = 0.$$

From this and (21.8) we also conclude that

$$(j\omega I - A + BR^{-1}N')x_1 = 0, \qquad\qquad (j\omega + A')x_2 = 0.$$

But then we have an eigenvector x_2 of A' in the kernel of B' and an eigenvector x_1 of $A - BR^{-1}N'$ in the kernel of $Q - NR^{-1}N'$. Since the corresponding eigenvalues do not have negative real parts, this contradicts the stabilizability and detectability assumptions.

The fact that $\mathcal{V}_-$ has dimension n follows from the discussion preceding the statement of the lemma.   ∎

21.4.2 BASIS FOR THE STABLE SUBSPACE OF **H**

Suppose that the assumptions of Lemma 21.1 hold and let

$$V := \begin{bmatrix} V_1 \\ V_2 \end{bmatrix} \in \mathbb{R}^{2n \times n}$$

be a matrix whose n columns form a basis for the stable subspace $\mathcal{V}_-$ of **H**. Assuming that $V_1 \in \mathbb{R}^{n \times n}$ is nonsingular, then

Note. Under the assumptions of Lemma 21.1, V_1 is always nonsingular, as shown in [5, Theorem 6.5, p. 202].

$$V V_1^{-1} = \begin{bmatrix} I \\ P \end{bmatrix}, \qquad P := V_2 V_1^{-1}$$

is also a basis for $\mathcal{V}_-$. Therefore, we conclude from property P21.2 that there exists a stability matrix **H**$_-$ such that

$$\mathbf{H} \begin{bmatrix} I \\ P \end{bmatrix} = \begin{bmatrix} I \\ P \end{bmatrix} \mathbf{H}_-, \tag{21.10}$$

and therefore **H** belongs to the domain of the Riccati operator. Combining Lemma 21.1 with Theorem 21.1, we obtain the following main result regarding the solution to the ARE.

Theorem 21.2. *Assume that $Q - NR^{-1}N' \geq 0$. When the pair (A, B) is stabilizable and the pair $(A - BR^{-1}N', Q - NR^{-1}N')$ is detectable,*

1. **H** *is in the domain of the Riccati operator,*

2. *P satisfies the ARE* (21.1),

3. $A - BR^{-1}(B'P + N') = \mathbf{H}_-$ is a stability matrix, and

4. P is symmetric,

Note 12. When the pair $(A - BR^{-1}N', Q - NR^{-1}N')$ is observable, one can show that P is also positive-definite.   ▶ p. 202

where P, $\mathbf{H}_- \in \mathbb{R}^{n \times n}$ are as in (21.10). Moreover, the eigenvalues of $\mathbf{H}_-$ are the eigenvalues of $\mathbf{H}$ with a negative real part.   □

Attention! It is insightful to interpret the results of Theorem 21.2, when applied to the minimization of

$$J_{\mathrm{LQR}} := \int_0^\infty z'\bar{Q}z + \rho\, u'\bar{R}u \; dt, \quad z := Gx + Hu, \quad \rho, \bar{Q}, \bar{R} > 0,$$

which corresponds to

$$Q = G'\bar{Q}G, \qquad\qquad R = H'\bar{Q}H + \rho\bar{R}, \qquad\qquad N = G'\bar{Q}H.$$

When $N = 0$, we conclude that Theorem 21.2 requires the detectability of the pair $(A, Q) = (A, G'\bar{Q}G)$. Since $\bar{Q} > 0$, it is straightforward to verify (e.g., using the eigenvector test) that this is equivalent to the detectability of the pair (A, G), which means that the system must be detectable through the controlled output z.

The need for (A, B) to be stabilizable is quite reasonable, because otherwise it is not possible to make $x \to 0$ for every initial condition. The need for (A, G) to be detectable can be intuitively understood by the fact that if the system had unstable modes that did not appear in z, it could be possible to make J_{LQR} very small, even though the state x might be exploding.   □

Note 12. To prove that P is positive-definite, we rewrite the ARE

$$(A - BR^{-1}N')'P + P(A - BR^{-1}N') + Q - NR^{-1}N' - PBR^{-1}B'P = 0$$

in (21.2) as

$$\mathbf{H}_-'P + P\mathbf{H}_- = -S, \quad S := (Q - NR^{-1}N') + PBR^{-1}B'P.$$

The positive definiteness of P then follows from the Lyapunov observability test as long as we are able to establish the observability of the pair $(\mathbf{H}_-, S)$.

To show that the pair $(\mathbf{H}_-, S)$ is observable, we use the eigenvector test. To prove this by contradiction, assume that x is an eigenvector of $\mathbf{H}_-$ that lies in the kernel of S; i.e.,

$$\big(A - BR^{-1}(B'P + N')\big)x = \lambda x, \quad Sx = \big((Q - NR^{-1}N') + PBR^{-1}B'P\big)x = 0.$$

Since $Q - NR^{-1}N'$ and $PBR^{-1}B'P$ are both symmetric positive-semidefinite matrices, the equation $Sx = 0$ implies that

$$x'\big((Q - NR^{-1}N') + PBR^{-1}B'P\big)x = 0 \;\Rightarrow\; (Q - NR^{-1}N')x = 0, \quad B'Px = 0.$$

We thus conclude that

$$(A - BR^{-1}N')x = \lambda x, \qquad\qquad (Q - NR^{-1}N')x = 0,$$

which contradicts the fact that the pair $(A - BR^{-1}N', Q - NR^{-1}N')$ is observable.

                                                                                    □

21.1. Verify that for every matrix P, the following matrix is symmetric:

$$\begin{bmatrix} -P' & I \end{bmatrix} \mathbf{H} \begin{bmatrix} I \\ P \end{bmatrix},$$

where $\mathbf{H}$ is the Hamiltonian matrix.   □

21.2 (Invariance of stable subspaces). Show that the stable subspace $\mathcal{V}_-$ of a matrix M is always M-invariant.   □

21.3 (Properties of stable subspaces). Prove Properties P21.1 and P21.2.

Hint: Transform M into its Jordan normal form.   □

21.4. Show that detectability of (A, G) is equivalent to detectability of (A, Q) with $Q := G'G$.

Hint: Use the eigenvector test and note that the kernels of G and G'G are exactly the same.   □

LECTURE 22

Frequency Domain and Asymptotic Properties of LQR

CONTENTS

This lecture discusses several important properties of LQR controllers.

22.1 KALMAN'S EQUALITY

Consider the continuous-time LTI process

$$\dot{x} = Ax + Bu, \qquad z = Gx + Hu, \qquad x \in \mathbb{R}^n, \; u \in \mathbb{R}^k, \; z \in \mathbb{R}^\ell,$$

for which one wants to minimize the LQR criterion

$$J_{\text{LQR}} := \int_0^\infty \|z(t)\|^2 + \rho \|u(t)\|^2 \, dt, \tag{22.1}$$

where ρ is a positive constant. Throughout this whole lecture we assume that

$$N := G'H = 0, \tag{22.2}$$

Attention! This condition is *not* being added for simplicity. We shall see in Example 22.1 that, without it, the results in this section are not valid.

for which the optimal control is given by

$$u = -Kx, \qquad K := R^{-1}B'P, \qquad R := H'H + \rho I,$$

where P is the stabilizing solution to the ARE

$$A'P + PA + G'G - PBR^{-1}B'P = 0.$$

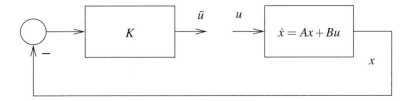

Figure 22.1. State feedback open-loop gain.

We saw in the Lecture 20 that under appropriate stabilizability and detectability assumptions, the LQR control results in a closed-loop system that is asymptotically stable.

LQR controllers also have desirable properties in the frequency domain. To understand why, consider the open-loop transfer matrix from the process input u to the controller output $\bar{u}$ (Figure 22.1). The state-space model from u to $\bar{u}$ is given by

$$\dot{x} = Ax + Bu, \qquad\qquad \bar{u} = -Kx,$$

which corresponds to the following open-loop *negative-feedback* $k \times k$ transfer matrix

$$\hat{L}(s) = K(sI - A)^{-1}B.$$

Another important open-loop transfer matrix is that from the control signal u to the controlled output z,

$$\hat{T}(s) = G(sI - A)^{-1}B + H.$$

These transfer matrices are related by the so-called *Kalman's equality*:

Kalman's equality. *For the LQR criterion in (22.1) with (22.2), we have*

$$\left(I + \hat{L}(-s)'\right)R\left(I + \hat{L}(s)\right) = R + H'H + \hat{T}(-s)'\hat{T}(s). \tag{22.3}$$

Kalman's equality has many important consequences. One of them is *Kalman's inequality*, which is obtained by setting $s = j\omega$ in (22.3) and using the fact that for real-rational transfer matrices

$$\hat{L}(-j\omega)' = \hat{L}(j\omega)^*, \quad \hat{T}(-j\omega)' = \hat{T}(j\omega)^*, \quad H'H + \hat{T}(j\omega)^*\hat{T}(j\omega) \geq 0.$$

Kalman's inequality. *For the LQR criterion in (22.1) with (22.2), we have*

$$\left(I + \hat{L}(j\omega)\right)^*R\left(I + \hat{L}(j\omega)\right) \geq R, \quad \forall\omega \in \mathbb{R}. \tag{22.4}$$

22.2 FREQUENCY DOMAIN PROPERTIES: SINGLE-INPUT CASE

We focus our attention in single-input processes ($k = 1$), for which $\hat{L}(s)$ is a scalar transfer function. Dividing both sides of Kalman's inequality (22.4) by the scalar R, we obtain

$$|1 + \hat{L}(j\omega)| \geq 1, \quad \forall\omega \in \mathbb{R},$$

which expresses the fact that *the Nyquist plot of $\hat{L}(j\omega)$ does not enter a circle of radius 1 around the point -1 of the complex plane.* This is represented graphically in Figure 22.2 and has several significant implications, which are discussed next.

Note. Kalman's equality follows directly from simple algebraic manipulations of the ARE (cf. Exercise 22.1).

Note 13. For multiple input systems, similar conclusions could be drawn, based on the *multivariable Nyquist criterion.* ▶ p. 216

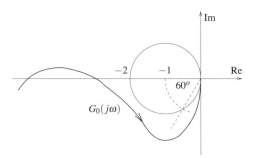

Figure 22.2. Nyquist plot for a LQR state feedback controller.

POSITIVE GAIN MARGIN.   If the process gain is multiplied by a constant $k > 1$, its Nyquist plot simply expands radially, and therefore the number of encirclements does not change. This corresponds to a *positive gain margin of* $+\infty$.

NEGATIVE GAIN MARGIN.   If the process gain is multiplied by a constant $0.5 < k < 1$, its Nyquist plot contracts radially, but the number of encirclements still does not change. This corresponds to a *negative gain margin of* $20 \log_{10}(.5) = -6$ dB.

PHASE MARGIN.   If the process phase increases by $\theta \in [-60°, 60°]$, its Nyquist plot rotates by θ, but the number of encirclements still does not change. This corresponds to a *phase margin of* $\pm 60°$.

SENSITIVITY AND COMPLEMENTARY SENSITIVITY FUNCTIONS.   The sensitivity and the complementary sensitivity functions are given by

$$\hat{S}(s) := \frac{1}{1 + \hat{L}(s)}, \qquad \hat{T}(s) := 1 - \hat{S}(s) = \frac{\hat{L}(s)}{1 + \hat{L}(s)},$$

respectively. Kalman's inequality guarantees that

$$|\hat{S}(j\omega)| \leq 1, \quad |\hat{T}(j\omega) - 1| \leq 1, \quad |\hat{T}(j\omega)| \leq 2, \quad \Re[\hat{T}(j\omega)] \geq 0, \quad \forall \omega \in \mathbb{R}.$$

$$(22.5)$$

Note. The first inequality results directly from the fact that $|1 + \hat{L}(j\omega)| \geq 1$, the second from the fact that $\hat{T}(s) = 1 - \hat{S}(s)$, and the last two from the fact that the second inequality shows that $\hat{T}(j\omega)$ must belong to a circle of radius 1 around the point $+1$.

We recall the following facts about the sensitivity function.

1. A small sensitivity function is desirable for good disturbance rejection. Generally, this is especially important at *low frequencies*.

2. A complementary sensitivity function close to 1 is desirable for good reference tracking. Generally, this is especially important at *low frequencies*.

3. A small complementary sensitivity function is desirable for good noise rejection. Generally, this is especially important at *high frequencies*.

Attention! Kalman's inequality is valid only when $N = G'H = 0$. When this is not the case, LQR controllers can exhibit significantly worse properties in terms of gain and phase margins. To some extent, this limits the controlled outputs that should be placed in z. For example, consider the process $\dot{x} = Ax + Bu$, $y = Cx$ and suppose that we want to regulate

$$z = y = Cx.$$

This leads to $G = C$ and $H = 0$. Therefore $G'H = 0$, for which Kalman's inequality holds. However, choosing

$$z = \begin{bmatrix} y \\ \dot{y} \end{bmatrix} = \begin{bmatrix} Cx \\ C\dot{x} \end{bmatrix} = \begin{bmatrix} Cx \\ CAx + CBu \end{bmatrix} = \begin{bmatrix} C \\ CA \end{bmatrix} x + \begin{bmatrix} 0 \\ CB \end{bmatrix} u,$$

leads to

<table>
<tr><td>

Note. If the transfer function from u to y has two more poles than zeros, then one can show that $CB = 0$ and $H = 0$. In this case, Kalman's inequality holds also for this choice of z.

</td><td>

$$G = \begin{bmatrix} C \\ CA \end{bmatrix}, \qquad\qquad H = \begin{bmatrix} 0 \\ CB \end{bmatrix},$$

and therefore

$$G'H = A'C'CB,$$

which may not be equal to zero. $\square$

</td></tr>
</table>

22.3 LOOP SHAPING USING LQR: SINGLE-INPUT CASE

Note. *Loop shaping* consists of designing the controller to meet specifications on the open-loop gain $\hat{L}(s)$. A brief review of this control design method can be found in Section 22.8.

Using Kalman's inequality, we saw that any LQR controller automatically provides some upper bounds on the magnitude of the sensitivity function and its complementary. However, these bounds are frequency-independent and may not result in appropriate loop shaping.

We discuss next a few rules that allow us to perform loop shaping using LQR. We continue to restrict our attention to the single-input case ($k = 1$).

LOW-FREQUENCY OPEN-LOOP GAIN.   Dividing both sides of Kalman's equality (22.3) by the scalar $R := H'H + \rho$, we obtain

$$|1 + \hat{L}(j\omega)|^2 = 1 + \frac{H'H}{H'H + \rho} + \frac{\|\hat{T}(j\omega)\|^2}{H'H + \rho}.$$

Therefore, for the range of frequencies for which $|\hat{L}(j\omega)| \gg 1$ (typically low frequencies),

$$|\hat{L}(j\omega)| \approx |1 + \hat{L}(j\omega)| \approx \frac{\|\hat{T}(j\omega)\|}{\sqrt{H'H + \rho}},$$

MATLAB® Hint 43.
`sigma(sys)` draws the norm-Bode plot of the system `sys`.   ▶ p. 216

which means that the open-loop gain for the optimal feedback $\hat{L}(s)$ follows the shape of the Bode plot from u to the controlled output z. To understand the implications of this formula, it is instructive to consider two fairly typical cases.

1. When $z = y$, with $y := Cx$ scalar, we have

$$|\hat{L}(j\omega)| \approx \frac{|\hat{T}(j\omega)|}{\sqrt{H'H + \rho}},$$

where

$$\hat{T}(s) := C(sI - A)^{-1}B$$

is the transfer function from the control input u to the measured output y. In this case,

Note. Although the magnitude of $\hat{L}(j\omega)$ mimics the magnitude of $\hat{T}(j\omega)$, the phase of the open-loop gain $\hat{L}(j\omega)$ always leads to a stable closed loop with an appropriate phase margin.

(a) the shape of the magnitude of the open-loop gain $\hat{L}(j\omega)$ is determined by the magnitude of the transfer function from the control input u to the measured output y, and

(b) the parameter ρ moves the magnitude Bode plot up and down (more precisely $H'H + \rho$).

2. When $z = \begin{bmatrix} y & \gamma\dot{y} \end{bmatrix}'$, with $y := Cx$ scalar, i.e.,

$$z = \begin{bmatrix} y \\ \gamma\dot{y} \end{bmatrix} = \begin{bmatrix} Cx \\ \gamma CAx + \gamma CBu \end{bmatrix} \quad \Rightarrow \quad G = \begin{bmatrix} C \\ \gamma CA \end{bmatrix}, \quad H = \begin{bmatrix} 0 \\ \gamma CB \end{bmatrix},$$

we conclude that

$$\hat{T}(s) = \begin{bmatrix} \hat{P}(s) \\ \gamma s \hat{P}(s) \end{bmatrix} = \begin{bmatrix} 1 \\ \gamma s \end{bmatrix} \hat{P}(s), \qquad \hat{P}(s) := C(sI - A)^{-1}B,$$

and therefore

$$|\hat{L}(j\omega)| \approx \frac{\sqrt{1 + \gamma^2\omega^2}\,|\hat{P}(j\omega)|}{\sqrt{H'H + \rho}} = \frac{|1 + j\gamma\omega|\,|\hat{P}(j\omega)|}{\sqrt{H'H + \rho}}. \tag{22.6}$$

In this case, the low-frequency open-loop gain mimics the process transfer function from u to y, with an extra zero at $1/\gamma$ and scaled by $\frac{1}{\sqrt{H'H+\rho}}$. Thus

(a) ρ moves the magnitude Bode plot up and down (more precisely $H'H + \rho$), and

(b) large values for γ lead to a low-frequency zero and generally result in a larger phase margin (above the minimum of $60°$) and a smaller overshoot in the step response. However, this is often achieved at the expense of a slower response.

Attention! It sometimes happens that the above two choices for z still do not provide a sufficiently good low-frequency open-loop response. In such cases, one may actually add dynamics to more accurately shape $\hat{L}(s)$. For example, suppose that one wants a very large magnitude for $\hat{L}(s)$ at a particular frequency ω_0 to reject a specific periodic disturbance. This could be achieved by including in z a filtered version of the output y obtained from a transfer function with a resonance close to ω_0 to increase the gain at this frequency. In this case, one could define

$$z = \begin{bmatrix} y \\ \gamma_1 \dot{y} \\ \gamma_2 \bar{y} \end{bmatrix},$$

where $\bar{y}$ is obtained from y through a system with transfer function equal to

$$\frac{1}{(s + \epsilon)^2 + \omega_0^2}$$

for some small $\epsilon > 0$. Many other options are possible, allowing one to *precisely shape $\hat{L}(s)$ over the range of frequencies for which this transfer function has a large magnitude.* □

HIGH-FREQUENCY OPEN-LOOP GAIN. Figure 22.2 shows that the open-loop gain $\hat{L}(j\omega)$ can have at most $-90°$ phase for high-frequencies, and therefore the roll-off rate is at most -20 dB/decade. In practice, this means that for $\omega \gg 1$,

$$|\hat{L}(j\omega)| \approx \frac{c}{\omega \sqrt{H'H + \rho}},$$

for some constant c. Therefore the cross-over frequency is approximately given by

$$\frac{c}{\omega_{\text{cross}} \sqrt{H'H + \rho}} \approx 1 \quad \Leftrightarrow \quad \omega_{\text{cross}} \approx \frac{c}{\sqrt{H'H + \rho}}.$$

Thus

1. LQR controllers always exhibit a high-frequency magnitude decay of -20 dB/decade, and

2. the cross-over frequency is proportional to $1/\sqrt{H'H + \rho}$, and generally small values for $H'H + \rho$ result in faster step responses.

Attention! The (slow) -20 dB/decade magnitude decrease is the main shortcoming of state feedback LQR controllers, because it may not be sufficient to clear high-frequency upper bounds on the open-loop gain needed to reject disturbances and/or for robustness with respect to process uncertainty. We will see in Section 23.5 that this can actually be improved with output feedback controllers. □

22.4 LQR DESIGN EXAMPLE

Example 22.1 (Aircraft roll dynamics). Figure 22.3 shows the roll angle dynamics of an aircraft [15, p. 381]. Defining $x := \begin{bmatrix} \theta & \omega & \tau \end{bmatrix}'$, we can write the aircraft dynamics as

$$\dot{x} = Ax + Bu,$$

where

$$A := \begin{bmatrix} 0 & 1 & 0 \\ 0 & -0.875 & -20 \\ 0 & 0 & -50 \end{bmatrix}, \qquad B := \begin{bmatrix} 0 \\ 0 \\ 50 \end{bmatrix}.$$

OPEN-LOOP GAINS. Figure 22.4 shows Bode plots of the open-loop gain $\hat{L}(s) = K(sI - A)^{-1}B$ for several LQR controllers obtained for this system. The controlled output was chosen to be $z := \begin{bmatrix} \theta & \gamma\dot{\theta} \end{bmatrix}'$, which corresponds to

$$G := \begin{bmatrix} 1 & 0 & 0 \\ 0 & \gamma & 0 \end{bmatrix}, \qquad H := \begin{bmatrix} 0 \\ 0 \end{bmatrix}.$$

The controllers minimize the criterion (22.1) for several values of ρ and γ. The matrix gains K and the Bode plots of the open-loop gains can be computed using the following sequence of MATLAB® commands:

MATLAB® Hint 44.
See MATLAB® Hint
42.   ▶ p. 195

```
A = [0,1,0;0,-.875,-20;0,0,-50];   B = [0;0;50];
                                   % process dynamics
G = [1,0,0;0,gamma*1,0];           H = [0;0];
                                   % controlled output z
Q = G'*G;          R = H'*H+rho;   N = G'*H;
                                   % weight matrices
K=lqr(A,B,Q,R,N);                  % compute LQR gain
                                   % open-loop gain
G0=ss(A,B,K,0);

bode(G0);
```

for the different values of `gamma` and `rho`.

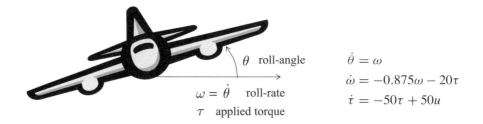

$$\dot{\theta} = \omega$$
$$\dot{\omega} = -0.875\omega - 20\tau$$
$$\dot{\tau} = -50\tau + 50u$$

θ roll-angle

$\omega = \dot{\theta}$ roll-rate

τ applied torque

Figure 22.3. Aircraft roll angle dynamics.

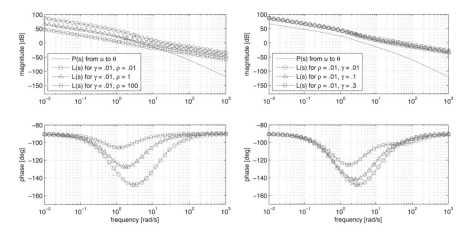

(a) Open-loop gain for several values of ρ. This parameter allows us to move the whole magnitude Bode plot up and down.

(b) Open-loop gain for several values of γ. Larger values for this parameter result in a larger phase margin.

Figure 22.4. Bode plots for the open-loop gain of the LQR controllers in Example 22.1. As expected, for low frequencies the open-loop gain magnitude matches that of the process transfer function from u to θ (but with significantly lower/better phase), and at high-frequencies the gain magnitude falls at -20 dB/decade.

Figure 22.4(a) shows the open-loop gain for several values of ρ, where we can see that ρ allows us to move the whole magnitude Bode plot up and down. Figure 22.4(b) shows the open-loop gain for several values of γ, where we can see that a larger γ results in a larger phase margin. As expected, for low frequencies the open-loop gain magnitude matches that of the process transfer function from u to θ (but with significantly lower/better phase), and at high frequencies the gain magnitude falls at -20 dB/decade.

Note. The use of LQR controllers to drive an output variable to a set point will be studied in detail later in Section 23.6.

STEP RESPONSES. Figure 22.5 shows step responses for the state feedback LQR controllers whose Bode plots for the open-loop gain are shown in Figure 22.4. Figure 22.5(a) shows that smaller values of ρ lead to faster responses, and Figure 22.5(b) shows that larger values for γ lead to smaller overshoots (but slower responses).

NYQUIST PLOTS. Figure 22.6 shows Nyquist plots of the open-loop gain $\hat{L}(s) = K(sI - A)^{-1}B$ for $\rho = 0.01$, but different choices of the controlled output z. In Figure 22.6(a) $z := \begin{bmatrix} \theta & \dot{\theta} \end{bmatrix}'$, which corresponds to

$$G := \begin{bmatrix} 1 & 0 & 0 \\ 0 & 1 & 0 \end{bmatrix}, \qquad\qquad H := \begin{bmatrix} 0 \\ 0 \end{bmatrix}.$$

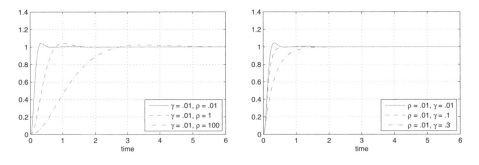

(a) Step response for several values of ρ. This parameter allows us to control the speed of the response.

(b) Step response for several values of γ. This parameter allows us to control the overshoot.

Figure 22.5. Closed-loop step responses for the LQR controllers in Example 22.1.

In this case, $H'G = [\,0\,0\,0\,]$, and Kalman's inequality holds, as can be seen in the Nyquist plot. In Figure 22.6(b), the controlled output was chosen to be $z := \begin{bmatrix} \theta & \dot\tau \end{bmatrix}'$, which corresponds to

$$G := \begin{bmatrix} 1 & 0 & 0 \\ 0 & 0 & -50 \end{bmatrix}, \qquad\qquad H := \begin{bmatrix} 0 \\ 50 \end{bmatrix}.$$

In this case, we have $H'G = [\,0\,0\,-2500\,]$, and Kalman's inequality does not hold. We can see from the Nyquist plot that the phase and gain margins are very small and there is little robustness with respect to unmodeled dynamics, since a small perturbation in the process can lead to an encirclement of the point -1. $\qquad\square$

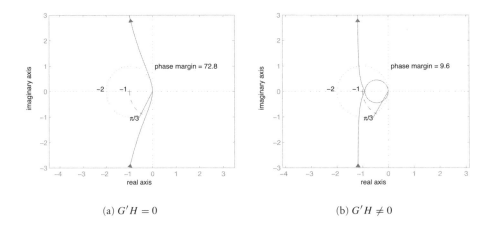

(a) $G'H = 0$ (b) $G'H \neq 0$

Figure 22.6. Nyquist plots for the open-loop gain of the LQR controllers in Example 22.1.

22.5 CHEAP CONTROL CASE

In view of the LQR criterion

$$J_{\text{LQR}} := \int_0^\infty \|z(t)\|^2 + \rho\|u(t)\|^2 \, dt,$$

by making ρ very small one does not penalize the energy used by the control signal. Based on this, one could expect that, as $\rho \to 0$,

1. the system's response becomes arbitrarily fast, and

2. the optimal value of the criterion converges to zero.

This limiting case is called *cheap control* and it turns out that whether or not the above conjectures are true depends on the transmission zeros of the system.

22.5.1 CLOSED-LOOP POLES

We saw in Lecture 21 (cf. Theorem 21.2) that the poles of the closed loop correspond to the stable eigenvalues of the Hamiltonian matrix

$$\mathbf{H} := \begin{bmatrix} A & -BR^{-1}B' \\ -G'G & -A' \end{bmatrix} \in \mathbb{R}^{2n \times 2n}, \qquad R := H'H + \rho I \in \mathbb{R}^{k \times k}.$$

Note.
Cf. Exercise 22.2.

To determine the eigenvalues of $\mathbf{H}$, we use the fact that

$$\det(sI - \mathbf{H}) = c\Delta(s)\Delta(-s) \det\left(R - H'H + \hat{T}(-s)'\hat{T}(s)\right), \tag{22.7}$$

where $c := (-1)^n \det R^{-1}$ and

Note. The transfer matrix $\hat{T}(s)$ that appears in (22.7) can be viewed as the transfer function from the control input u to the controlled output z.

$$\Delta(s) := \det(sI - A), \qquad \hat{T}(s) := G(sI - A)^{-1}B + H.$$

As $\rho \to 0$, $H'H \to R$, and therefore

$$\det(sI - \mathbf{H}) \to c\Delta(s)\Delta(-s) \det \hat{T}(-s)'\hat{T}(s). \tag{22.8}$$

We saw in Theorem 18.2 that there exist unimodular real polynomial matrices $L(s) \in \mathbb{R}[s]^{\ell \times \ell}$, $R(s) \in \mathbb{R}[s]^{k \times k}$ such that

$$\hat{T}(s) = L(s)SM_T(s)R(s), \tag{22.9}$$

where

$$SM_T(s) := \begin{bmatrix} \frac{\eta_1(s)}{\psi_1(s)} & \cdots & 0 & 0 \\ \vdots & \ddots & \vdots & \vdots \\ 0 & \cdots & \frac{\eta_r(s)}{\psi_r(s)} & 0 \\ 0 & \cdots & 0 & 0 \end{bmatrix} \in \mathbb{R}(s)^{\ell \times k}$$

is the Smith-McMillan form of $\hat{T}(s)$. To proceed, we should consider the square and nonsquare cases separately.

SQUARE TRANSFER MATRIX. When $\hat{T}(s)$ is square and full rank (i.e., $\ell = k = r$),

$$\det \hat{T}(-s)'\hat{T}(s) = \bar{c}\,\frac{\eta_1(-s)\cdots\eta_k(-s)\,\eta_1(s)\cdots\eta_k(s)}{\psi_1(-s)\cdots\psi_k(-s)\,\psi_1(s)\cdots\psi_k(s)} = \bar{c}\,\frac{z_T(-s)z_T(s)}{p_T(-s)p_T(s)},$$

where $z_T(s)$ and $p_T(s)$ are the zero and pole polynomials of $\hat{G}(s)$, respectively, and $\bar{c}$ is the (constant) product of the determinants of all the unimodular matrices. When the realization is minimal, $p_T(s) = \Delta(s)$ (cf. Theorem 19.3) and (22.9) simplifies to

$$\det(sI - \mathbf{H}) \to c\,\bar{c}\,z_T(s)z_T(-s).$$

Two conclusions can be drawn.

1. When $\hat{T}(s)$ has q transmission zeros

$$a_i + jb_i, \quad i \in \{1, 2, \ldots, q\},$$

then $2q$ of the eigenvalues of $\mathbf{H}$ converge to

$$\pm a_i \pm jb_i, \quad i \in \{1, 2, \ldots, q\}.$$

Therefore q closed-loop poles converge to

$$-|a_i| + jb_i, \quad i \in \{1, 2, \ldots, q\}.$$

Note. Recall that the poles of the closed-loop system are only the stable eigenvalues of $\mathbf{H}$, which converge to either $a_i + jb_i$ and $-a_i - jb_i$, depending on which of them has negative real part.

2. When $\hat{T}(s)$ does not have any transmission zero, $\mathbf{H}$ has no finite eigenvalues as $\rho \to 0$. Therefore all closed-loop poles must converge to infinity.

NONSQUARE TRANSFER MATRIX. When $\hat{T}(s)$ is not square and/or not full rank, by substituting (22.9) into (22.8), we obtain

$$\det(sI - \mathbf{H}) \to \bar{c}\Delta(s)\Delta(-s)\det\begin{bmatrix} \frac{\eta_1(-s)}{\psi_1(-s)} & \cdots & 0 \\ \vdots & \ddots & \vdots \\ 0 & \cdots & \frac{\eta_r(-s)}{\psi_r(-s)} \end{bmatrix}$$

$$\times L_r(-s)'L_r(s)\begin{bmatrix} \frac{\eta_1(s)}{\psi_1(s)} & \cdots & 0 \\ \vdots & \ddots & \vdots \\ 0 & \cdots & \frac{\eta_r(s)}{\psi_r(s)} \end{bmatrix},$$

where $L_r(s) \in \mathbb{R}[s]^{\ell \times r}$ contains the leftmost r columns of $L(s)$. In this case, when the realization is minimal, we obtain

$$\det(sI - \mathbf{H}) \to \bar{c}\,z_T(s)z_T(-s)\det L_r(-s)'L_r(s),$$

which shows that for nonsquare matrices $\det(sI - \mathbf{H})$ generally has more roots than the transmission zeros of $\hat{T}(s)$. In this case, one needs to compute the stable roots of

$$\Delta(s)\Delta(-s)\det\hat{T}(-s)'\hat{T}(s)$$

to determine the asymptotic locations of the closed-loop poles.

Attention! This means that in general one wants to avoid transmission zeros from the control input u to the controlled output z, especially slow transmission zeros that will attract the poles of the closed loop. For nonsquare systems, one must pay attention to all the zeros of $\det \hat{T}(-s)' \hat{T}(s)$. □

22.5.2 COST

We saw in Lecture 20 that the minimum value of the LQR criterion is given by

$$J_{\mathrm{LQR}} := \int_0^\infty \|z(t)\|^2 + \rho \|u(t)\|^2 \, dt = x'(0) P_\rho x(0),$$

where ρ is a positive constant and P_ρ is the corresponding solution to the ARE

$$A' P_\rho + P_\rho A + G'G - P_\rho B R_\rho^{-1} B' P_\rho = 0, \qquad R_\rho := H'H + \rho I. \qquad (22.10)$$

The following result makes explicit the dependence of P_ρ on ρ, as this parameter converges to zero.

Theorem 22.1. *When $H = 0$, the solution to (22.10) satisfies*

$$\lim_{\rho \to 0} P_\rho \begin{cases} = 0 & \ell = k \text{ and all transmission zeros of } \hat{T}(s) \text{ have negative or zero real} \\ & \qquad \text{parts,} \\ \neq 0 & \ell = k \text{ and } \hat{T}(s) \text{ has transmission zeros with positive real parts,} \\ \neq 0 & \ell > k. \end{cases}$$

□

Attention! This result shows a fundamental limitation due to unstable transmission zeros. It shows that when there are transmission zeros from the input u to the controlled output z, it is not possible to reduce the energy of z arbitrarily, even if one is willing to spend much control energy. □

Attention! Suppose that $\ell = k$ and all transmission zeros of $\hat{T}(s)$ have negative or zero real parts. Taking limits on both sides of (22.10) and using the fact that $\lim_{\rho \to 0} P_\rho = 0$, we conclude that

$$\lim_{\rho \to 0} \frac{1}{\rho} P_\rho B B' P_\rho = \lim_{\rho \to 0} \rho K_\rho' K_\rho = G'G,$$

where $K_\rho := R_\rho^{-1} B' P_\rho$ is the state feedback gain. Assuming that G is full row rank, this implies that

$$\lim_{\rho \to 0} \sqrt{\rho} K_\rho = SG,$$

for some orthogonal matrix S (cf. Exercise 22.3). This shows that asymptotically we have

$$K_\rho = \frac{1}{\sqrt{\rho}} SG,$$

and therefore the optimal control is of the form

$$u = K_\rho x = \frac{1}{\sqrt{\rho}} SGx = \frac{1}{\sqrt{\rho}} Sz;$$

i.e., for these systems *the cheap control problem corresponds to high-gain static feedback of the controlled output.*   □

22.6 MATLAB® COMMANDS

MATLAB® Hint 43 (sigma).   The command `sigma(sys)` draws the norm-Bode plot of the system `sys`. For scalar transfer functions, this command plots the usual magnitude Bode plot, but for MIMO transfer matrices, it plots the norm of the transfer matrix versus the frequency.   □

MATLAB® Hint 45 (nyquist).   The command `nyquist(sys)` draws the Nyquist plot of the system `sys`.

Especially when there are poles very close to the imaginary axis (e.g., because they were actually on the axis and you moved them slightly to the left), the automatic scale may not be very good, because it may be hard to distinguish the point -1 from the origin. In this case, you can use the zoom features of MATLAB® to see what is going on near -1. Try clicking on the magnifying glass and selecting a region of interest, or try left-clicking with the mouse and selecting "zoom on $(-1, 0)$" (without the magnifying glass selected).   □

22.7 ADDITIONAL NOTES

Note 13 (Multivariable Nyquist criterion). The Nyquist criterion is used to investigate the stability of the *negative-feedback* connection in Figure 22.7. It allows one to compute the number of unstable (i.e., in the closed right-hand side plane) poles of the closed-loop transfer matrix $(I + \hat{L}(s))^{-1}$ as a function of the number of unstable poles of the open-loop transfer matrix $\hat{L}(s)$.

To apply the criterion, we start by drawing the *Nyquist plot of $\hat{L}(s)$*, which is done by evaluating $\det(I + \hat{L}(j\omega))$ from $\omega = -\infty$ to $\omega = +\infty$ and plotting it in the complex plane. This leads to a closed curve that is always symmetric with respect to

Note. The Nyquist plot should be viewed as the image of a clockwise contour that goes along the axis and closes with a right-hand side loop at ∞.

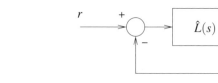

Figure 22.7. Negative feedback.

MATLAB® Hint 45.
nyquist(sys)
draws the Nyquist
plot of the system
sys.   ▶ p. 216

the real axis. This curve should be annotated with arrows indicating the direction corresponding to increasing ω.

Any poles of $\hat{L}(s)$ on the imaginary axis should be moved slightly to the left of the axis, because the criterion is valid only when $\hat{L}(s)$ is analytic on the imaginary axis. E.g.,

$$\hat{L}(s) = \frac{s+1}{s(s-3)} \quad\longrightarrow\quad \hat{L}_\epsilon(s) \approx \frac{s+1}{(s+\epsilon)(s-3)}$$

$$\hat{L}(s) = \frac{s}{s^2+4} = \frac{s}{(s+2j)(s-2j)} \quad\longrightarrow\quad \hat{L}_\epsilon(s) \approx \frac{s}{(s+\epsilon+2j)(s+\epsilon-2j)}$$

$$= \frac{s}{(s+\epsilon)^2+4}$$

for a small $\epsilon > 0$. The criterion should then be applied to the perturbed transfer matrix $\hat{L}_\epsilon(s)$. If we conclude that the closed loop is asymptotically stable for $\hat{L}_\epsilon(s)$ with very small $\epsilon > 0$, then the closed loop with $\hat{L}(s)$ is also asymptotically stable and vice versa.

Nyquist stability criterion. *The total number of* unstable *(closed-loop) poles of* $\big(I + \hat{L}(s)\big)^{-1}$ *(#CUP) is given by*

$$\#CUP = \#ENC + \#OUP,$$

where #OUP denotes the number of unstable (open-loop) poles of $\hat{L}(s)$ and #ENC is the number of clockwise encirclements by the multivariable Nyquist plot around the origin. To have a stable closed-loop system, one thus needs

$$\#ENC = -\#OUP. \qquad\qquad \square$$

Note. To compute
#ENC, we draw a ray
from the origin to ∞
in *any* direction and
add 1 each time the
Nyquist plot crosses
the ray in the
clockwise direction
(with respect to the
origin of the ray) and
subtract 1 each time it
crosses the ray in the
counterclockwise
direction. The final
count gives #ENC.

Attention! For the multivariable Nyquist criteria, we count encirclements *around the origin* and not around -1, because the multivariable Nyquist plot is shifted to the right by adding the I to in $\det\big(I + \hat{L}(j\omega)\big)$. $\qquad\qquad \square$

22.8   THE LOOP-SHAPING DESIGN METHOD (REVIEW)

Note. The
loop-shaping design
method is covered
extensively, e.g., in
[6].

The goal of this section is to briefly review the loop-shaping control design method for SISO systems. The basic idea behind loop shaping is to convert the desired specifications for the closed-loop system in Figure 22.8 into constraints on the open-loop gain

$$\hat{L}(s) := \hat{C}(s)\hat{P}(s).$$

The controller $\hat{C}(s)$ is then designed so that the open-loop gain $\hat{L}(s)$ satisfies these constraints. The shaping of $\hat{L}(s)$ can be done using the classical methods briefly mentioned in Section 22.8.2 and explained in much greater detail in [6, Chapter 6.7].

Attention! The review in this section is focused on the SISO case, so it does not address the state feedback case for systems with more than one state. However, we shall see in Lecture 23 that we can often recover the LQR open-loop gain just with output feedback.   ▶ p. 230

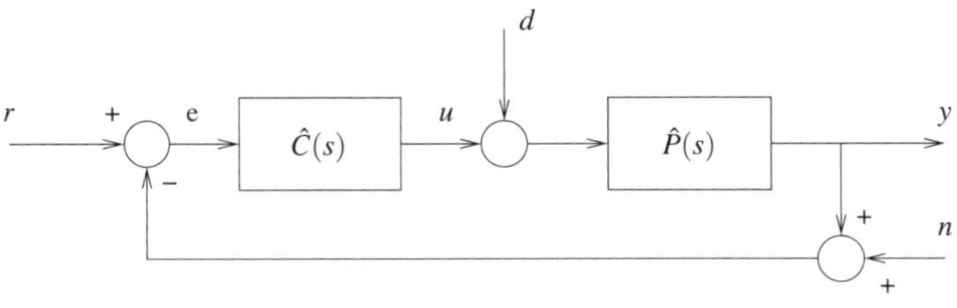

Figure 22.8. Closed-loop system.

However, it can also be done using LQR state feedback, as discussed in Section 22.3, or using LQG/LQR output feedback controllers, as we shall see in Section 23.5.

22.8.1   OPEN-LOOP VERSUS CLOSED-LOOP SPECIFICATIONS

We start by discussing how several closed-loop specifications can be converted into constraints on the open-loop gain $\hat{L}(s)$.

Notation. The distance between the phase of $\hat{L}(j\omega_c)$ and $-180°$ is called the *phase margin*.

STABILITY.   Assuming that the open-loop gain has no unstable poles, the stability of the closed-loop system is guaranteed as long as the phase of the open-loop gain is above $-180°$ at the cross-over frequency ω_c, i.e., at the frequency for which

$$|\hat{L}(j\omega_c)| = 1.$$

OVERSHOOT.   Larger phase margins generally correspond to a smaller overshoot for the step response of the closed-loop system. The following rules of thumb work well when the open-loop gain $\hat{L}(s)$ has a pole at the origin, an additional real pole, and no zeros.

Phase margin (deg)	Overshoot (%)
65	≤ 5
60	≤ 10
45	≤ 15

REFERENCE TRACKING.   Suppose that one wants the tracking error to be at least $k_T \ll 1$ times smaller than the reference, over the range of frequencies $[0, \omega_T]$. In the frequency domain, this can be expressed by

$$\frac{|\hat{E}(j\omega)|}{|\hat{R}(j\omega)|} \leq k_T, \quad \forall \omega \in [0, \omega_T], \tag{22.11}$$

where $\hat{E}(s)$ and $\hat{R}(s)$ denote the Laplace transforms of the tracking error $e := r - y$ and the reference signal r, respectively, in the absence of noise and disturbances. For the closed-loop system in Figure 22.8,

$$\hat{E}(s) = \frac{1}{1 + \hat{L}(s)} \hat{R}(s).$$

Therefore (22.11) is equivalent to

$$\frac{1}{|1 + \hat{L}(j\omega)|} \leq k_T, \quad \forall \omega \in [0, \omega_T] \quad \Leftrightarrow \quad |1 + \hat{L}(j\omega)| \geq \frac{1}{k_T}, \quad \forall \omega \in [0, \omega_T].$$

This condition is guaranteed to hold by requiring that

$$|\hat{L}(j\omega)| \geq \frac{1}{k_T} + 1, \quad \forall \omega \in [0, \omega_T]. \tag{22.12}$$

DISTURBANCE REJECTION.. Suppose that one wants input disturbances to appear in the output attenuated at least $k_D \ll 1$ times, over the range of frequencies $[0, \omega_D]$. In the frequency domain, this can be expressed by

$$\frac{|\hat{Y}(j\omega)|}{|\hat{D}(j\omega)|} \leq k_D, \quad \forall \omega \in [0, \omega_D], \tag{22.13}$$

where $\hat{Y}(s)$ and $\hat{D}(s)$ denote the Laplace transforms of the output y and the input disturbance d, respectively, in the absence of reference and measurement noise. For the closed-loop system in Figure 22.8,

$$\hat{Y}(s) = \frac{\hat{P}(s)}{1 + \hat{L}(s)} \hat{D}(s),$$

and therefore (22.13) is equivalent to

$$\frac{|\hat{P}(j\omega)|}{|1 + \hat{L}(j\omega)|} \leq k_D, \quad \forall \omega \in [0, \omega_D]$$

$$\Leftrightarrow \quad |1 + \hat{L}(j\omega)| \geq \frac{|\hat{P}(j\omega)|}{k_D}, \quad \forall \omega \in [0, \omega_D].$$

This condition is guaranteed to hold as long as one requires that

$$|\hat{L}(j\omega)| \geq \frac{|\hat{P}(j\omega)|}{k_D} + 1, \quad \forall \omega \in [0, \omega_D]. \tag{22.14}$$

NOISE REJECTION. Suppose that one wants measurement noise to appear in the output attenuated at least $k_N \ll 1$ times, over the range of frequencies $[\omega_N, \infty)$. In the frequency domain, this can be expressed by

$$\frac{|\hat{Y}(j\omega)|}{|\hat{N}(j\omega)|} \leq k_N, \quad \forall \omega \in [\omega_N, \infty), \tag{22.15}$$

where $\hat{Y}(s)$ and $\hat{N}(s)$ denote the Laplace transforms of the output y and the measurement noise n, respectively, in the absence of reference and disturbances. For the closed-loop system in Figure 22.8,

$$\hat{Y}(s) = -\frac{\hat{L}(s)}{1 + \hat{L}(s)} \hat{N}(s),$$

Table 22.1. Summary of the Relationship between Closed-loop Specifications and Open-loop Constraints for the Loop-shaping Design Method

Closed-loop specification	Open-loop constraint
Overshoot $\leq 10\%$ ($\leq 5\%$)	Phase margin $\geq 60°$ ($\geq 65°$)
$\dfrac{|\hat{E}(j\omega)|}{|\hat{R}(j\omega)|} \leq k_T, \quad \forall \omega \in [0, \omega_T]$	$|\hat{L}(j\omega)| \geq \dfrac{1}{k_T} + 1, \quad \forall \omega \in [0, \omega_T]$
$\dfrac{|\hat{Y}(j\omega)|}{|\hat{D}(j\omega)|} \leq k_D, \quad \forall \omega \in [0, \omega_D]$	$|\hat{L}(j\omega)| \geq \dfrac{|\hat{P}(j\omega)|}{k_D} + 1, \quad \forall \omega \in [0, \omega_D]$
$\dfrac{|\hat{Y}(j\omega)|}{|\hat{N}(j\omega)|} \leq k_N, \quad \forall \omega \in [\omega_N, \infty)$	$|\hat{L}(j\omega)| \leq \dfrac{k_N}{1 + k_N}, \quad \forall \omega \in [\omega_N, \infty)$

and therefore (22.15) is equivalent to

$$\frac{|\hat{L}(j\omega)|}{|1 + \hat{L}(j\omega)|} \leq k_N, \quad \forall \omega \in [\omega_N, \infty) \quad \Leftrightarrow \quad \left| 1 + \frac{1}{\hat{L}(j\omega)} \right| \geq \frac{1}{k_N}, \quad \forall \omega \in [\omega_N, \infty).$$

This condition is guaranteed to hold as long as one requires that

$$\left| \frac{1}{\hat{L}(j\omega)} \right| \geq \frac{1}{k_N} + 1, \quad \forall \omega \in [\omega_N, \infty) \quad \Leftrightarrow \quad |\hat{L}(j\omega)| \leq \frac{k_N}{1 + k_N}, \quad \forall \omega \in [\omega_N, \infty).$$

Table 22.1 and Figure 22.9 summarize the constraints on the open-loop gain $G_0(j\omega)$ discussed above.

Attention! The conditions derived above for the open-loop gain $\hat{L}(j\omega)$ are *sufficient* for the original closed-loop specifications to hold, but they are not *necessary*. When the open-loop gain "almost" verifies the conditions derived, it may be worth it to check directly whether it verifies the original closed-loop conditions. $\qquad\square$

22.8.2 OPEN-LOOP GAIN SHAPING

In classical lead/lag compensation, one starts with a basic unit-gain controller

$$\hat{C}(s) = 1$$

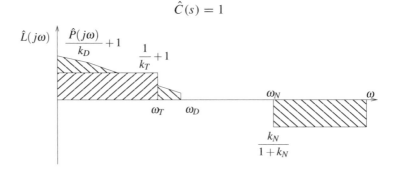

Figure 22.9. Typical open-loop specifications for the loop-shaping control design.

Note. One actually does not "add" to the controller. To be precise, one *multiplies* the controller by appropriate gain, lead, and lag blocks. However, this does correspond to additions in the magnitude (in dBs) and phase Bode plots.

and "adds" to it appropriate blocks to shape the desired open-loop gain

$$\hat{L}(s) := \hat{C}(s)\hat{P}(s),$$

so that it satisfies the appropriate open-loop constraints. This shaping can be achieved using three basic tools.

1. *Proportional gain.* Multiplying the controller by a constant k moves the magnitude Bode plot up and down, without changing its phase.

2. *Lead compensation.* Multiplying the controller by a lead block with transfer function

$$\hat{C}_{\text{lead}}(s) = \frac{Ts+1}{\alpha Ts+1}, \qquad \alpha < 1$$

increases the phase margin when placed at the cross-over frequency. Figure 22.10(a) shows the Bode plot of a lead compensator.

Note. A lead compensator also increases the cross-over frequency, so it may require some trial and error to get the peak of the phase right at the cross-over frequency.

3. *Lag compensation.* Multiplying the controller by a lag block with transfer function

$$\hat{C}_{\text{lag}}(s) = \frac{s/z+1}{s/p+1}, \qquad p < z$$

decreases the high-frequency gain. Figure 22.10(b) shows the Bode plot of a lag compensator.

Note. A lag compensator also increases the phase, so it can decrease the phase margin. To avoid this, one should only introduce lag compensation away from the cross-over frequency.

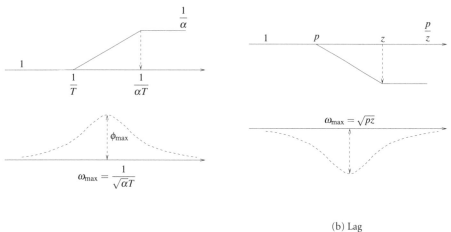

(a) Lead

(b) Lag

Figure 22.10. Bode plots of lead/lag compensators. The maximum lead phase angle is given by $\phi_{\text{max}} = \arcsin\frac{1-\alpha}{1+\alpha}$; therefore, to obtain a desired given lead angle ϕ_{max} one sets $\alpha = \frac{1-\sin\phi_{\text{max}}}{1+\sin\phi_{\text{max}}}$.

22.1 (Kalman equality). Prove Kalman's equality (22.3).

Hint: Add and subtract $(s\ P)$ to the ARE and then left- and right-multiply it by $-B'(sI+A')^{-1}$ and $(sI-A)^{-1}B$, respectively. □

22.2 (Eigenvalues of the Hamiltonian matrix). Show that (22.7) holds.

Hint: Use the following properties of the determinant:

$$\det \begin{bmatrix} M_1 & M_2 \\ M_3 & M_4 \end{bmatrix} = \det M_1 \det M_4 \det \left(I - M_3 M_1^{-1} M_2 M_4^{-1} \right), \tag{22.16a}$$

$$\det(I + XY) = \det(I + YX) \tag{22.16b}$$

Notation. A square matrix S is called *orthogonal* if its inverse exists and is equal to its transpose; i.e., $SS' = S'S = I$.

(cf., e.g., [9, equations 1-235 and 1-201]). □

22.3. Show that given two matrices $X, M \in \mathbb{R}^{n \times \ell}$ with M full row rank and $X'X = M'M$, there exists an orthogonal matrix $S \in \mathbb{R}^{\ell \times \ell}$ such that $M = SX$. □

LECTURE 23

Output Feedback

CONTENTS

This lecture addresses the feedback control problem when only the output (not the whole state) can be measured.

1. Certainty Equivalence
2. Deterministic Minimum-Energy Estimation (MEE)
3. Stochastic Linear Quadratic Gaussian (LQG) Estimation
4. LQR/LQG Output Feedback
5. Loop Transfer Recovery (LTR)
6. Optimal Set Point Control
7. LQR/LQG with MATLAB®
8. LTR Design Example
9. Exercises

23.1 CERTAINTY EQUIVALENCE

The state feedback LQR formulation considered in Lecture 20 suffered from the drawback that the optimal control law

$$u(t) = -Kx(t) \tag{23.1}$$

required the whole state x of the process to be measured. A possible approach to overcome this difficulty is to construct an estimate $\hat{x}$ of the state of the process based solely on the past values of the measured output y and control signal u, and then use

$$u(t) = -K\hat{x}(t)$$

instead of (23.1). This approach is usually known as *certainty equivalence* and leads to the architecture in Figure 23.1. In this lecture we consider the problem of constructing state estimates for use in certainty equivalence controllers.

23.2 DETERMINISTIC MINIMUM-ENERGY ESTIMATION (MEE)

Consider a continuous-time LTI system of the form

$$\dot{x} = Ax + Bu, \qquad y = Cx, \qquad x \in \mathbb{R}^n,\ u \in \mathbb{R}^k,\ y \in \mathbb{R}^m, \qquad \text{(CLTI)}$$

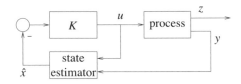

Figure 23.1. Certainty equivalence controller.

where u is the control signal and y is the measured output. Estimating the state x at some time t can be viewed as solving (CLTI) for the unknown $x(t)$, for given $u(\tau)$, $y(\tau)$, $\tau \leq t$.

Assuming that the model (CLTI) is exact and observable, we saw in Lecture 15 that $x(t)$ can be reconstructed exactly using the constructibility Gramian

$$x(t) = W_{Cn}(t_0, t)^{-1}$$
$$\times \left(\int_{t_0}^{t} e^{A'(\tau-t)} C' y(\tau) d\tau + \int_{t_0}^{t} \int_{\tau}^{t} e^{A'(\tau-t)} C' C e^{A(\tau-s)} B u(s) ds d\tau \right),$$

where

$$W_{Cn}(t_0, t) := \int_{t_0}^{t} e^{A'(\tau-t)} C' C e^{A(\tau-t)} d\tau$$

(cf. Theorem 15.2).

In practice, the model (CLTI) is never exact, and the measured output y is generated by a system of the form

$$\dot{x} = Ax + Bu + \bar{B}d, \quad y = Cx + n, \quad x \in \mathbb{R}^n, \ u \in \mathbb{R}^k, \ d \in \mathbb{R}^q, \ y \in \mathbb{R}^m,$$
$$(23.2)$$

where d represents a disturbance and n measurement noise. Since neither d nor n are known, solving (23.2) for x no longer yields a unique solution, since essentially any state value could explain the measured output for sufficiently large noise and disturbances.

Minimum-energy estimation (MEE) consists of finding a state trajectory

$$\dot{\bar{x}} = A\bar{x} + Bu + \bar{B}d, \quad y = C\bar{x} + n, \quad \bar{x} \in \mathbb{R}^n, \ u \in \mathbb{R}^k, \ d \in \mathbb{R}^q, \ y \in \mathbb{R}^m \quad (23.3)$$

that starts at rest as $t \to -\infty$ and is consistent with the past measured output y and control signal u for the *least amount of noise n and disturbance d*, measured by

Note 14. In particular, we assume that $\bar{x}(t) \to 0$ and $y(t) \to 0$, as $t \to -\infty$.

$$J_{\text{MEE}} := \int_{-\infty}^{t} n(\tau)' Q n(\tau) + d(\tau)' R d(\tau) d\tau, \quad (23.4)$$

where $Q \in \mathbb{R}^{m \times m}$ and $R \in \mathbb{R}^{q \times q}$ are symmetric positive-definite matrices. Once this trajectory has been found, based on the data collected on the interval $(-\infty, t]$, the minimum-energy state estimate is simply the most recent value of $\bar{x}$,

$$\hat{x}(t) = \bar{x}(t).$$

The role of the matrices Q and R can be understood as follows.

1. When we choose Q large, we are forcing the noise term to be small, which means that we "believe" in the measured output. This leads to state estimators that respond fast to changes in the output y.

2. When we choose R large, we are forcing the disturbance term to be small, which means that we "believe" in the past values of the state estimate. This leads to state estimators that respond cautiously (slowly) to unexpected changes in the measured output.

23.2.1   SOLUTION TO THE MEE PROBLEM

The MEE problem is solved by minimizing the quadratic cost

$$J_{\text{MEE}} = \int_{-\infty}^{t} \big(C\bar{x}(\tau) - y(\tau)\big)' Q\big(C\bar{x}(\tau) - y(\tau)\big) + d(\tau)' R d(\tau) d\tau$$

for the system (23.3) by appropriately choosing the disturbance $d(\cdot)$. We shall see in Section 23.2.4 that this minimization can be performed using arguments like the ones used to solve the LQR problem, leading to the following result.

Theorem 23.1 (Minimum-energy estimation). *Assume that there exists a symmetric positive-definite solution P to the following ARE*

$$(-A')P + P(-A) + C'QC - P\bar{B}R^{-1}\bar{B}'P = 0, \qquad (23.5)$$

for which $-A - \bar{B}R^{-1}\bar{B}'P$ is a stability matrix. Then the MEE estimator for (23.2) for the criteria (23.4) is given by

$$\dot{\hat{x}} = (A - LC)\hat{x} + Bu + Ly, \qquad\qquad L := P^{-1}C'Q. \qquad (23.6)$$

Note. The reader may recall that we had proposed a state estimator of this form in Lecture 16, but had not shown that it was optimal.

23.2.2   DUAL ALGEBRAIC RICCATI EQUATION

To determine conditions for the existence of an appropriate solution to the ARE (23.5), it is convenient to left- and right-multiply this equation by $S := P^{-1}$ and then multiply it by -1. This procedure yields an equivalent equation called the *dual algebraic Riccati equation,*

$$AS + SA' + \bar{B}R^{-1}\bar{B}' - SC'QCS = 0. \qquad (23.7)$$

The gain L can be written in terms of the solution S to the dual ARE as $L := SC'Q$.

To solve the MEE problem, one needs to find a symmetric positive-definite solution to the dual ARE for which $-A - \bar{B}R^{-1}\bar{B}'S^{-1}$ is a stability matrix. The results in Lecture 21 provide conditions for the existence of an appropriate solution to the dual ARE (23.5).

Note. See Exercise 23.1 for an alternative set of conditions that also guarantees a solution to the dual ARE.

Theorem 23.2 (Solution to the dual ARE). *Assume that the pair $(A, \bar{B})$ is controllable and that the pair (A, C) is detectable.*

1. *There exists a symmetric positive-definite solution S to the dual ARE (23.7), for which $A - LC$ is a stability matrix.*

2. *There exists a symmetric positive-definite solution $P := S^{-1}$ to the ARE (23.5), for which $-A - \bar{B}R^{-1}\bar{B}'P = -A - \bar{B}R^{-1}\bar{B}'S^{-1}$ is a stability matrix.* $\quad\square$

Proof of Theorem 23.2. Part 1 is a straightforward application of Theorem 21.2 for $N = 0$ and the following facts.

1. The stabilizability of (A', C') is equivalent to the detectability of (A, C),

2. the observability of $(A', \bar{B}')$ is equivalent to the controllability of $(A, \bar{B})$, and

3. $A' - C'L'$ is a stability matrix if and only if $A + LC$ is a stability matrix.

The fact that $P := S^{-1}$ satisfies (23.5) has already been established from the construction of the dual ARE (23.7). To prove part 2, it remains to show that $-A - \bar{B}R^{-1}\bar{B}'S^{-1}$ is a stability matrix. To do this, we rewrite (23.7) as

$$(-A - \bar{B}R^{-1}\bar{B}'S^{-1})S + S(-A' - S^{-1}\bar{B}R^{-1}\bar{B}') = -Y,$$

where $Y := SC'QCS + \bar{B}R^{-1}\bar{B}'$. The stability of $-A - \bar{B}R^{-1}\bar{B}'S^{-1}$ then follows from the Lyapunov stability theorem (Theorem 12.5), because the pair $(-A - \bar{B}R^{-1}\bar{B}'S^{-1}, Y)$ is controllable . $\quad\blacksquare$

Note.
Cf. Exercise 23.2.

23.2.3   CONVERGENCE OF THE ESTIMATES

The MEE estimator is often written as

$$\dot{\hat{x}} = A\hat{x} + Bu + L(y - \hat{y}), \qquad \hat{y} = C\hat{x}, \qquad L := SC'Q. \qquad (23.8)$$

Defining the state estimation error $e = x - \hat{x}$, we conclude from (23.8) and (23.2) that

$$\dot{e} = (A - LC)e + \bar{B}d - Ln.$$

Note. Why? Because the poles of the transfer matrices from d and n to e are the eigenvalues of $A - LC$.

Since $A - LC$ is a stability matrix, we conclude that, in the absence of measurement noise and disturbances, $e(t) \to 0$ as $t \to \infty$ and therefore $\|x(t) - \hat{x}(t)\| \to 0$ as $t \to \infty$.

In the presence of noise, we have BIBO stability from the inputs d and n to the "output" e, so $\hat{x}(t)$ may not converge to $x(t)$, but at least does not diverge from it.

23.2.4   PROOF OF THE MEE THEOREM

Due to the exogenous term $y(\tau)$ in the MEE criteria, we need a more sophisticated feedback invariant to solve this problem.

Note. Although H_0 must not depend on $d(\cdot)$ and $\bar{x}(\cdot)$, it may depend on $u(\cdot)$ and $y(\cdot)$, since these variables are given and are not being optimized.

Proposition 23.1 (Feedback invariant). *Suppose that the input $u(\cdot) \in \mathbb{R}^k$ and output $y(\cdot) \in \mathbb{R}^m$ to (23.3) are given up to some time $t > 0$. For every symmetric matrix P, differentiable signal $\beta : (-\infty, t] \to \mathbb{R}^n$, and scalar H_0 that does not depend on $d(\cdot)$ and $\bar{x}(\cdot)$, the functional*

$$H\big(\bar{x}(\cdot); d(\cdot)\big)$$

$$:= H_0 + \int_{-\infty}^{t} \bigg(\big(A\bar{x}(\tau) + Bu(\tau) + \bar{B}d(\tau) - \dot{\beta}(\tau)\big)' P\big(\bar{x}(\tau) - \beta(\tau)\big)$$

$$+ \big(\bar{x}(\tau) - \beta(\tau)\big)' P\big(A\bar{x}(\tau) + Bu(\tau) + \bar{B}d(\tau) - \dot{\beta}(\tau)\big)\bigg)d\tau$$

$$- \big(\bar{x}(t) - \beta(t)\big)' P\big(\bar{x}(t) - \beta(t)\big)$$

is a feedback invariant for (23.3), as long as $\lim_{\tau \to -\infty} \big(\bar{x}(\tau) - \beta(\tau)\big) = 0$. $\square$

Proof of Proposition 23.1. We can rewrite H as

Note. Here, by *feedback invariant* we mean that the value of $H\big(\bar{x}(\cdot); d(\cdot)\big)$ does not depend on the disturbance signal $d(\cdot)$ that needs to be optimized.

$$H\big(\bar{x}(\cdot); d(\cdot)\big)$$

$$= H_0 + \int_{-\infty}^{t} \bigg(\big(\dot{\bar{x}}(\tau) - \dot{\beta}(\tau)\big)' P\big(\bar{x}(\tau) - \beta(\tau)\big)$$

$$+ \big(\bar{x}(\tau) - \beta(\tau)\big)' P\big(\dot{\bar{x}}(\tau) - \dot{\beta}(\tau)\big)\bigg)d\tau - \big(\bar{x}(t) - \beta(t)\big)' P\big(\bar{x}(t) - \beta(t)\big)$$

$$= H_0 + \int_{-\infty}^{t} \frac{d\big(\bar{x}(\tau) - \beta(\tau)\big)' P\big(\bar{x}(\tau) - \beta(\tau)\big)}{d\tau} \, d\tau$$

$$- \big(\bar{x}(t) - \beta(t)\big)' P\big(\bar{x}(t) - \beta(t)\big)$$

$$= H_0 + \lim_{\tau \to -\infty} \big(\bar{x}(\tau) - \beta(\tau)\big)' P\big(\bar{x}(\tau) - \beta(\tau)\big) = H_0,$$

as long as $\lim_{\tau \to -\infty} \big(\bar{x}(\tau) - \beta(\tau)\big) = 0$. $\blacksquare$

If we now add and subtract this feedback invariant to our J_{MEE} criterion, we obtain

Note. To keep the formulas short, we do not explicitly include the dependency on τ for the signals inside the integral.

$$J_{\text{MEE}} = H\big(\bar{x}(\cdot); d(\cdot)\big) - H_0 + \big(\bar{x}(t) - \beta(t)\big)' P\big(\bar{x}(t) - \beta(t)\big)$$

$$+ \int_{-\infty}^{t} \bigg(\bar{x}'(-A'P - PA + C'QC)\bar{x} + y'Qy + 2\beta'P(Bu - \dot{\beta})$$

$$- 2\bar{x}'(-A'P\beta + PBu + C'Qy - P\dot{\beta}) + d'Rd$$

$$- 2d'\bar{B}'P(\bar{x} - \beta)\bigg)d\tau.$$

In preparation for a minimization with respect to d, we complete the square to combine all the terms that contain d into a single quadratic form, which, after tedious manipulation, eventually leads to

$$J_{\text{MEE}} = H\big(\bar{x}(\cdot); d(\cdot)\big) - H_0 + \big(\bar{x}(t) - \beta(t)\big)' P\big(\bar{x}(t) - \beta(t)\big)$$

$$+ \int_{-\infty}^{t} \bigg(\bar{x}'(-A'P - PA + C'QC - P\bar{B}R^{-1}\bar{B}'P)\bar{x}$$

$$-2\bar{x}'\Big((-A'P - PBR^{-1}\bar{B}'P)\beta + PBu + C'Qy - P\dot{\beta}\Big)$$
$$+ y'Qy + 2\beta'P(Bu - \dot{\beta}) - \beta'P\bar{B}R^{-1}\bar{B}'P\beta$$
$$+ \big(d - R^{-1}\bar{B}'P(\bar{x} - \beta)\big)'R\big(d - R^{-1}\bar{B}'P(\bar{x} - \beta)\big)\Big)d\tau. \quad (23.9)$$

Suppose now that we pick

1. the matrix P to be the solution to the ARE (23.5),

2. the signal β to satisfy

$$P\dot{\beta} = -(A'P + P\bar{B}R^{-1}\bar{B}'P)\beta + PBu + C'Qy = 0$$
$$\Leftrightarrow \quad \dot{\beta} = (A - P^{-1}C'QC)\beta + Bu + P^{-1}C'Qy, \quad (23.10)$$

   initialized so that $\lim_{\tau \to -\infty} \beta(\tau) = 0$, and

3. the scalar H_0 given by

Note. Since β depends only on $u(\cdot)$ and $y(\cdot)$, the scalar H_0 also depends only on these signals, as stated in Proposition 23.1.

$$H_0 := \int_{-\infty}^{t} \Big(y'Qy + 2\beta'P(Bu - \dot{\beta}) - \beta'P\bar{B}R^{-1}\bar{B}'P\beta\Big)d\tau.$$

In this case, (23.9) becomes simply

$$J_{\text{MEE}} = H\big(\bar{x}(\cdot); d(\cdot)\big) + \big(\bar{x}(t) - \beta(t)\big)'P\big(\bar{x}(t) - \beta(t)\big)$$
$$+ \int_{-\infty}^{t} \big(d - R^{-1}\bar{B}'P(\bar{x} - \beta)\big)'R\big(d - R^{-1}\bar{B}'P(\bar{x} - \beta)\big)d\tau,$$

Note. It is very convenient that equation (23.10), which generates $\beta(\cdot)$, does not depend on the final time t at which the estimate is being computed. Because of this, we can continuously obtain from this equation the current state estimate $\hat{x}(t) = \beta(t)$.

which, since $H\big(\bar{x}(\cdot); d(\cdot)\big)$ is a feedback invariant, shows that J_{MEE} can be minimized by selecting

$$\bar{x}(t) = \beta(t), \qquad d(\tau) = R^{-1}\bar{B}'P\big(\bar{x}(\tau) - \beta(\tau)\big), \quad \forall \tau < t.$$

These choices, together with the differential equation (23.3), completely define the optimal trajectory $\bar{x}(\tau), \tau \leq t$ that minimizes J_{MEE}. Moreover, (23.10) computes exactly the MEE $\hat{x}(t) = \bar{x}(t) = \beta(t)$ at the final time t. Note that under the choice of $d(\tau), \forall \tau < t$, we conclude from (23.3) and (23.10) that

$$(\dot{\bar{x}} - \dot{\beta})$$
$$= A\bar{x} + Bu + \bar{B}R^{-1}\bar{B}'P(\bar{x} - \beta) - (A - P^{-1}C'QC)\beta - Bu - P^{-1}C'Qy$$
$$= (A + \bar{B}R^{-1}\bar{B}'P)(\bar{x} - \beta) + P^{-1}C'Q(C\beta - y).$$

Note. Recall that $y \to 0$ (cf. Note 14, p. 224) and also that $\beta \to 0$ as $t \to -\infty$.

Therefore $\bar{x} - \beta \to 0$ as $t \to -\infty$, because $-A - \bar{B}R^{-1}\bar{B}'P$ is a stability matrix, as stated in Proposition 23.1. ∎

23.3 STOCHASTIC LINEAR QUADRATIC GAUSSIAN (LQG) ESTIMATION

The MEE introduced before also has a stochastic interpretation. To state it, we consider again the continuous-time LTI system

$$\dot{x} = Ax + Bu + \bar{B}d, \quad y = Cx + n, \quad x \in \mathbb{R}^n, \ u \in \mathbb{R}^k, \ d \in \mathbb{R}^q, \ y \in \mathbb{R}^m,$$

but now assume that the disturbance d and the measurement noise n are uncorrelated zero-mean Gaussian white-noise stochastic processes with covariance matrices

$$\mathrm{E}[d(t)d'(\tau)] = \delta(t - \tau)R^{-1}, \quad \mathrm{E}[n(t)n'(\tau)] = \delta(t - \tau)Q^{-1}, \quad R, Q > 0. \tag{23.11}$$

Note. In this context, the estimator (23.6) is usually called a *Kalman filter*.

MATLAB® Hint 46.
`kalman` computes the optimal MEE/LQG estimator gain L. ▶ p. 235

The MEE state estimate $\hat{x}(t)$ given by equation (23.6) in Section 23.2 also minimizes the asymptotic norm of the estimation error,

$$J_{\mathrm{LQG}} := \lim_{t \to \infty} \mathrm{E}[\|x(t) - \hat{x}(t)\|^2].$$

This is consistent with what we saw before regarding the roles of the matrices Q and R in MEE:

1. A large Q corresponds to little measurement noise and leads to state estimators that respond fast to changes in the measured output.

2. A large R corresponds to small disturbances and leads to state estimates that respond cautiously (slowly) to unexpected changes in the measured output.

23.4   LQR/LQG OUTPUT FEEDBACK

We now go back to the problem of designing an output feedback controller for the continuous-time LTI process

$$\dot{x} = Ax + Bu + \bar{B}d, \qquad x \in \mathbb{R}^n, \ u \in \mathbb{R}^k, \ d \in \mathbb{R}^q, \tag{23.12a}$$

$$y = Cx + n, \qquad y, n \in \mathbb{R}^m, \tag{23.12b}$$

$$z = Gx + Hu, \qquad z \in \mathbb{R}^\ell. \tag{23.12c}$$

Suppose that we designed a state feedback controller

$$u = -Kx \tag{23.13}$$

that solves an LQR problem and constructed an LQG/MEE state estimator

$$\dot{\hat{x}} = (A - LC)\hat{x} + Bu + Ly.$$

MATLAB® Hint 47.
`reg(sys,K,L)` computes the LQG/LQR *positive* output feedback controller for the process `sys` with regulator gain `K` and estimator gain `L`. ▶ p. 235

We can obtain an output feedback controller by using the estimated state $\hat{x}$ in (23.13), instead of the true state x. This leads to the output feedback controller

$$\dot{\hat{x}} = (A - LC)\hat{x} + Bu + Ly = (A - LC - BK)\hat{x} + Ly, \quad u = -K\hat{x}, \tag{23.14}$$

with *negative-feedback* transfer matrix given by

$$\hat{C}(s) = K(sI - A + LC + BK)^{-1}L.$$

This is usually known as an *LQG/LQR output feedback controller*. Since both $A - BK$ and $A - LC$ are stability matrices, the separation principle (cf. Theorem 16.10 and Exercise 23.3) guarantees that this controller makes the closed-loop system asymptotically stable.

23.5   LOOP TRANSFER RECOVERY (LTR)

We saw in Lecture 22 that a state feedback controller

$$u = -Kx$$

for the process (23.12) has desirable robustness properties and that we can even shape its open-loop gain

$$\hat{L}(s) = K(sI - A)^{-1}B$$

by appropriately choosing the LQR weighting parameter ρ and the controlled output z.

Suppose now that the state is not accessible and that we constructed an LQG/LQR output feedback controller with *negative-feedback* transfer matrix given by

$$\hat{C}(s) = K(sI - A + LC + BK)^{-1}L,$$

Note. This ARE would arise from the solution to an MEE problem with cost (23.4) or an LQG problem with disturbance and noise satisfying (23.11).

where $L := SC'Q$ and S is a solution to the dual ARE

$$AS + SA' + \bar{B}R^{-1}\bar{B}' - SC'QCS = 0$$

for which $A - LC$ is a stability matrix.

In general there is no guarantee that LQG/LQR controllers will inherit the open-loop gain of the original state feedback design. However, for processes that do not have transmission zeros in the closed right-hand side complex plane, one can recover the LQR open-loop gain by appropriately designing the state estimator.

Note. $\bar{B} = B$ corresponds to an *input disturbance,* since the process becomes $\dot{x} = Ax + B(u + d)$.

Theorem 23.3 (Loop transfer recovery). *Suppose that the transfer matrix*

$$\hat{P}(s) := C(sI - A)^{-1}B$$

from u to y is square $(k = m)$ *and has* no transmission zeros in the closed right half-plane. *Selecting*

$$\bar{B} := B, \qquad\qquad R := I, \qquad\qquad Q := \sigma I, \quad \sigma > 0,$$

Note. In general, the larger $\omega_{\max}$ is, the larger σ needs to be for the gains to match.

the open-loop gain for the output feedback LQG/LQR controller converges to the open-loop gain for the LQR state feedback controller over a range of frequencies $[0, \omega_{\max}]$ as we make $\sigma \to +\infty$, i.e.,

$$C(j\omega)P(j\omega) \xrightarrow{\quad \sigma \to +\infty \quad} \hat{L}(j\omega), \quad \forall\omega \in [0, \omega_{\max}]. \qquad \square$$

MATLAB® Hint 48. In terms of the input parameters to the kalman function, this corresponds to making QN = I and RN = $\bar{\sigma}I$, with $\bar{\sigma} :=$ $1/\sigma \to 0$.   ▶ p. 235

Attention! The following items should be kept in mind regarding Theorem 23.3.

1. To achieve loop-gain recovery, we need to chose $Q = \sigma I$, *regardless of the noise statistics.*

2. One should not make σ larger than necessary, because we do not want to recover the (slow) -20 dB/decade magnitude decrease at high frequencies. In practice we should make σ *just large enough to get loop recovery until just above or at cross-over.* For larger values of ω, the output feedback controller may actually behave much better than the state feedback one.

3. When the process has *zeros in the right half-plane,* loop-gain recovery will generally work only up to the frequencies of the nonminimum-phase zeros.

   When the zeros are in the *left half-plane but close to the axis,* the closed loop will not be very robust with respect to uncertainty in the position of the zeros. This is because the controller will attempt to cancel these zeros.   □

23.6   OPTIMAL SET POINT CONTROL

Consider again the continuous-time LTI process

$$\dot{x} = Ax + Bu, \qquad z = Gx + Hu, \qquad x \in \mathbb{R}^n,\ u \in \mathbb{R}^k,\ z \in \mathbb{R}^\ell, \qquad (23.15)$$

but suppose that now one wants the controlled output z to converge as fast as possible to a given nonzero constant *set point value* r, corresponding to an equilibrium point $(x_{\text{eq}}, u_{\text{eq}})$ of (23.15) for which $z = r$. This corresponds to an LQR criterion of the form

$$J_{\text{LQR}} := \int_0^\infty \tilde{z}(t)' \bar{Q} \tilde{z}(t) + \rho\, \tilde{u}(t)' \bar{R} \tilde{u}(t)\ dt, \qquad (23.16)$$

where $\tilde{z} := z - r,\ \tilde{u} := u - u_{\text{eq}}$.

Such equilibrium point $(x_{\text{eq}}, u_{\text{eq}})$ must satisfy the equation

Note. For $\ell = 1$, we can take $u_{\text{eq}} = 0$ when the matrix A has an eigenvalue at the origin, and this mode is observable through z (cf. Exercise 23.6)

$$\begin{cases} Ax_{\text{eq}} + Bu_{\text{eq}} = 0 \\ r = Gx_{\text{eq}} + Hu_{\text{eq}} \end{cases} \quad \Leftrightarrow \quad \begin{bmatrix} -A & B \\ -G & H \end{bmatrix}_{(n+\ell)\times(n+k)} \begin{bmatrix} -x_{\text{eq}} \\ u_{\text{eq}} \end{bmatrix} = \begin{bmatrix} 0 \\ r \end{bmatrix}. \quad (23.17)$$

To understand when these equations have a solution, three distinct cases should be considered.

1. When the number of inputs k is strictly smaller than the number of controlled outputs ℓ, we have an *underactuated system.* In this case, the system of equations (23.17) generally does not have a solution, because it presents more equations than unknowns.

2. When the number of inputs k is equal to the number of controlled outputs ℓ, (23.17) always has a solution as long as Rosenbrock's system matrix

Attention! This Rosenbrock's matrix is obtained by regarding the controlled output z as the only output of the system.

$$P(s) := \begin{bmatrix} sI - A & B \\ -G & H \end{bmatrix}$$

is nonsingular for $s = 0$. This means that $s = 0$ should not be an invariant zero of the system, and therefore it cannot also be a transmission zero of the transfer matrix $G(sI - A)^{-1}B + H$.

Note. Recall that a transmission zero of a transfer matrix is always an invariant zero of its state-space realizations (cf. Theorem 19.2).

Intuitively, one should expect problems when $s = 0$ is an invariant zero of the system, because when the state converges to an equilibrium point, the control input $u(t)$ must converge to a constant. By the zero-blocking property, one should then expect the controlled output $z(t)$ to converge to zero and not to r.

3. When the number of inputs k is strictly larger than the number of controlled outputs ℓ, we have an *overactuated system*, and (23.17) generally has multiple solutions.

Note. We shall
confirm in
Exercise 23.4 that
(23.18) is indeed a
solution to (23.17).

When $P(0)$ is full row-rank, i.e., when it has $n + \ell$ linearly independent rows, the $(n + \ell) \times (n + \ell)$ matrix $P(0)P(0)'$ is nonsingular, and one solution to (23.17) is given by

Note.
$P(0)'\big(P(0)P(0)'\big)^{-1}$
is called the
pseudoinverse of $P(0)$
(cf. Definition 17.2).

$$\begin{bmatrix} -x_{\text{eq}} \\ u_{\text{eq}} \end{bmatrix} = P(0)'\big(P(0)P(0)'\big)^{-1}\begin{bmatrix} 0 \\ r \end{bmatrix}. \tag{23.18}$$

Also in this case, $s = 0$ should not be an invariant zero of the system, because otherwise $P(0)$ cannot be full rank.

23.6.1 STATE FEEDBACK: REDUCTION TO OPTIMAL REGULATION

The optimal set point problem can be reduced to that of optimal regulation by considering an auxiliary system with state $\tilde{x} := x - x_{\text{eq}}$, whose dynamics are

$$\dot{\tilde{x}} = Ax + Bu = A(x - x_{\text{eq}}) + B(u - u_{\text{eq}}) + (Ax_{\text{eq}} + Bu_{\text{eq}})$$
$$\tilde{z} = Gx + Hu - r = G(x - x_{\text{eq}}) + H(u - u_{\text{eq}}) + (Gx_{\text{eq}} + Hu_{\text{eq}} - r).$$

The last terms on each equation cancel because of (23.17), and we obtain

$$\dot{\tilde{x}} = A\tilde{x} + B\tilde{u}, \qquad\qquad \tilde{z} = G\tilde{x} + H\tilde{u}. \tag{23.19}$$

We can then regard (23.16) and (23.19) as an *optimal regulation problem* for which the optimal solution is given by

$$\tilde{u}(t) = -K\tilde{x}(t),$$

as in Theorem 20.1. Going back to the original input and state variables u and x, we conclude that the optimal control for the set point problem defined by (23.15) and (23.16) is given by

$$u(t) = -K\big(x(t) - x_{\text{eq}}\big) + u_{\text{eq}}, \qquad\qquad t \geq 0. \tag{23.20}$$

Since the solution to (23.17) can be written in the form

Note. As seen in
Exercise 23.6, the
feed-forward term Nr
is absent when the
process has an
integrator.

$$x_{\text{eq}} = Fr, \qquad\qquad u_{\text{eq}} = Nr,$$

for appropriately defined matrices F and N, this corresponds to the control architecture in Figure 23.2.

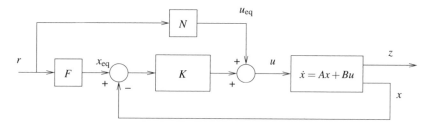

Figure 23.2. Linear quadratic set point control with state feedback.

CLOSED-LOOP TRANSFER MATRICES. To determine the transfer matrix from the reference r to the control input u, we use the diagram in Figure 23.2 to conclude that

$$\hat{u} = N\hat{r} + KF\hat{r} - K(sI - A)^{-1}B\hat{u} \quad \Leftrightarrow \quad \hat{u} = \big(I + \hat{L}(s)\big)^{-1}(N + KF)\hat{r},$$

where $\hat{L}(s) := K(sI - A)^{-1}B$ is the open-loop gain of the LQR state feedback controller. We therefore conclude the following:

1. When the open-loop gain $\hat{L}(s)$ is small, we essentially have

$$\hat{u} \approx (N + KF)\hat{r}.$$

Since at high frequencies $\hat{L}(s)$ falls at -20 dB/decade, the transfer matrix from r to u will always converge to $N + KF \neq 0$ at high frequencies.

Note. $N + KF$ is always nonzero, since otherwise the reference would not affect the control input.

2. When the open-loop gain $\hat{L}(s)$ is large, we essentially have

$$\hat{u} \approx \hat{L}(s)^{-1}(N + KF)\hat{r}.$$

To make this transfer matrix small, we need to increase the open-loop gain $\hat{L}(s)$.

The transfer matrix from r to the controlled output z can be obtained by composing the transfer matrix from r to u just computed with the transfer matrix from u to z,

$$\hat{z} = \hat{T}(s)\big(I + \hat{L}(s)\big)^{-1}(N + KF)\hat{r},$$

where $\hat{T}(s) := G(sI - A)^{-1}B + H$. We therefore conclude the following:

1. When the open-loop gain $\hat{L}(s)$ is small, we essentially have

$$\hat{z} \approx \hat{T}(s)(N + KF)\hat{r},$$

and therefore the closed-loop transfer matrix mimics that of the process.

2. When the open-loop gain $\hat{L}(s)$ is large, we essentially have

$$\hat{z} \approx \hat{T}(s)\hat{L}(s)^{-1}(N + KF)\hat{r}.$$

Note. Since z converges to a constant r, we must have $\|\hat{z}(0)\| = \|\hat{r}(0)\|$. Therefore when $\|\hat{L}(0)\| \gg 1$, we must have $\|N + KF\| \approx \sqrt{\rho}$.

Moreover, from Kalman's equality, we also have $\|\hat{L}(j\omega)\| \approx \frac{1}{\sqrt{\rho}}\|\hat{T}(j\omega)\|$ when $\|\hat{L}(j\omega)\| \gg 1$, $R := \rho I$, and $H = 0$ (cf. Section 22.3). In this case we obtain

$$\|\hat{z}(j\omega)\| \approx \frac{\|N + KF\|}{\sqrt{\rho}}\|\hat{r}(j\omega)\|,$$

which shows a flat Bode plot from r to z.

23.6.2 OUTPUT FEEDBACK

When the state is not accessible, we need to replace (23.20) by

$$u(t) = -K\big(\hat{x}(t) - x_{\text{eq}}\big) + u_{\text{eq}}, \qquad t \geq 0,$$

where $\hat{x}$ is the state estimate produced by an LQG/MEE state estimator

$$\dot{\hat{x}} = (A - LC)\hat{x} + Bu + Ly = (A - LC - BK)\hat{x} + BKx_{\text{eq}} + Bu_{\text{eq}} + Ly.$$

Defining $\bar{x} := x_{\text{eq}} - \hat{x}$ and using the fact that $Ax_{\text{eq}} + Bu_{\text{eq}} = 0$, we conclude that

$$\dot{\bar{x}} = -(A - LC - BK)\hat{x} + (A - BK)x_{\text{eq}} - Ly$$
$$= (A - LC - BK)\bar{x} - L(y - Cx_{\text{eq}}).$$

Note. When $z = y$, we have $G = C$, $H = 0$, and in this case $Cx_{\text{eq}} = r$. This corresponds to $CF = 1$ in Figure 23.3. When the process has an integrator, we get $N = 0$ and obtain the usual unity-feedback configuration.

This allows us to rewrite the equations for the *LQG/LQR set point controller* as

$$\dot{\bar{x}} = (A - LC - BK)\bar{x} - L(y - Cx_{\text{eq}}), \qquad u = K\bar{x} + u_{\text{eq}}, \qquad (23.21)$$

which corresponds to the control architecture shown in Figure 23.3.

CLOSED-LOOP TRANSFER MATRICES.   The closed-loop transfer matrices from the reference r to the control input u and controlled output z are now given by

$$\hat{u} = \big(I + \hat{C}(s)\hat{P}(s)\big)^{-1}(N + \hat{C}(s)CF)\hat{r},$$
$$\hat{y} = \hat{T}(s)\big(I + \hat{C}(s)\hat{P}(s)\big)^{-1}(N + \hat{C}(s)CF)\hat{r},$$

where

$$\hat{C}(s) := K(sI - A + LC + BK)^{-1}L, \qquad \hat{P}(s) := C(sI - A)^{-1}B.$$

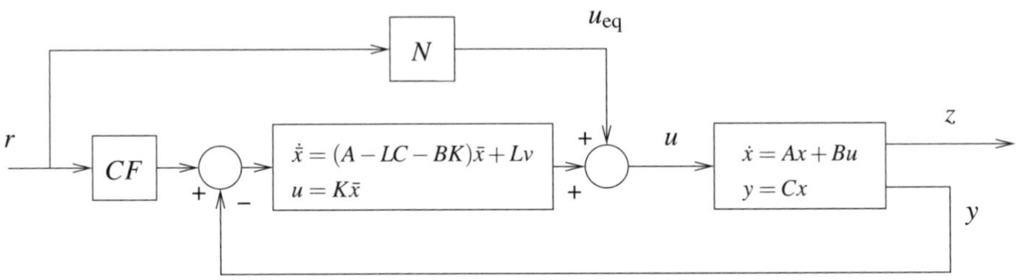

Figure 23.3. LQG/LQR set point control.

When LTR succeeds, i.e., when

$$\hat{C}(j\omega)\hat{P}(j\omega) \approx \hat{L}(j\omega), \qquad \forall\omega \in [0, \omega_{\max}],$$

the main difference between these and the formulas seen before for state feedback is that the matrix $N + KF$ multiplying by $\hat{r}$ has been replaced by the transfer matrix $N + \hat{C}(s)CF$.

When $N = 0$, this generally leads to smaller transfer matrices when the loop gain is low, because we now have

$$\hat{u} \approx \hat{C}(s)CF\hat{r}, \qquad\qquad \hat{y} \approx \hat{T}(s)\hat{C}(s)CF\hat{r},$$

and $\hat{C}(s)$ falls at least at -20 dB/decade.

23.7 LQR/LQG WITH MATLAB®

MATLAB® Hint 46 (`kalman`). The command

```
[est,L,P]=kalman(sys,QN,RN)
```

computes the optimal LQG estimator gain for the process

$$\dot{x} = \mathrm{A}x + \mathrm{B}u + \mathrm{BB}d, \qquad\qquad y = \mathrm{C}x + n,$$

where $d(t)$ and $n(t)$ are uncorrelated zero-mean Gaussian noise processes with covariance matrices

$$\mathrm{E}\big[d(t)\,d'(\tau)\big] = \delta(t - \tau)\mathrm{QN}, \qquad \mathrm{E}\big[n(t)\,n'(\tau)\big] = \delta(t - \tau)\mathrm{RN}.$$

The variable `sys` should be a state-space model created using

```
sys=ss(A,[B BB],C,0).
```

This command returns the optimal estimator gain `L`, the solution `P` to the corresponding algebraic Riccati equation, and a state-space model `est` for the estimator. The inputs to `est` are $[u;\ y]$, and its outputs are $[\hat{y};\ \hat{x}]$.

For loop transfer recovery (LTR), one should set

$$\mathrm{BB} = \mathrm{B}, \qquad\qquad \mathrm{QN} = I, \qquad\qquad \mathrm{RN} = \sigma I, \ \ \sigma \to 0. \qquad\qquad \square$$

Note. As discussed in Section 23.3, this LQG estimator is also an MEE estimator with cost (23.4), where $Q = \mathrm{RN}^{-1}$ and $R = \mathrm{QN}^{-1}$ (pay attention to the inverses and the exchange between Qs and Rs).

Note. See Example 23.1.

MATLAB® Hint 47 (`reg`). The function `reg(sys,K,L)` computes a state-space model for a *positive* output feedback LQG/LQR controller for the process with state-space model `sys` with regulator gain `K` and estimator gain `L`. $\qquad\qquad \square$

23.8 LTR DESIGN EXAMPLE

Example 23.1 (Aircraft roll dynamics, continued). Figure 23.4(a) shows Bode plots of the open-loop gain for the state feedback LQR state feedback controller versus the open-loop gain for several output feedback LQG/LQR controllers obtained for the aircraft roll dynamics in Example 22.1. The LQR controller was designed using the controlled output $z := \begin{bmatrix} \theta & \gamma & \dot{\theta} \end{bmatrix}'$, with $\gamma = 0.1$ and $\rho = 0.01$ (see Example 22.1). For the LQG state estimators, we used the parameters for the loop transfer recovery theorem (Theorem 23.3): $\bar{B} = B$, $R = 1$, and $Q = \sigma$ for several values of σ in the MEE cost (23.4) [or the corresponding LQG disturbance and noise (23.11)]. The matrix gain L, the LQG/LQR output feedback controller, and the corresponding Bode plot of the open-loop gain can be computed using the following sequence of MATLAB® commands:

MATLAB® Hint 49. See MATLAB® Hints 46 (p. 235) and 47 (p. 235).

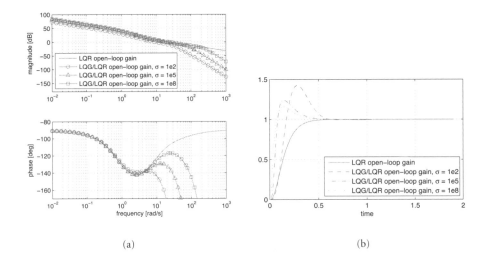

Figure 23.4. Bode plots of the open-loop gain and closed-loop step response for the LQR controllers in Example 23.1.

```
R=1;  Q=sigma;                    % weight matrices
Pkal=ss(A,[B B],C,0);             % process for the kalman()
                                  % command
[est,L]=kalman(Pkal,inv(R),inv(Q));  % compute LQG gain
P=ss(A,B,C,0);                    % process for the reg()
                                  % command
Cs=-reg(P,K,L);                   % LQG/LQR controller
                                  % (negative feedback)
bode(Cs*P);                       % bode plot of the
                                  % open-loop gain
```

We can see that, as σ increases, the range of frequencies over which the open-loop gain of the output feedback LQG/LQR controller matches that of the state feedback LQR state feedback increases. Moreover, at high frequencies the output feedback controllers exhibit much faster (and better!) decays of the gain's magnitude. $\quad\square$

23.9 EXERCISES

Note. This result is less interesting than Theorem 23.2, because often (A, C) is not observable, just detectable. This can happen when we augmented the state of the system to construct a "good" controlled output z, but these augmented states are not observable through y.

23.1 (Solution to the dual ARE). Assume that the pair $(-A, \bar{B})$ is stabilizable and that the pair (A, C) is observable. Prove the following.

(a) There exists a symmetric positive-definite solution P to the ARE (23.5), for which $-A - \bar{B}R^{-1}\bar{B}'P$ is a stability matrix.

(b) There exists a symmetric positive-definite solution $S := P^{-1}$ to the dual ARE (23.7), for which $A - LC$ is a stability matrix. $\quad\square$

23.2. Show that if the pair $(A, \bar{B})$ is controllable, then the pair

$$(-A - \bar{B}R^{-1}\bar{B}'S^{-1}, Y), \qquad Y := SC'QCS + \bar{B}R^{-1}\bar{B}'.$$

is also controllable for Q and R symmetric and positive-definite.

Hint: Use the eigenvector test.   □

23.3 (Separation principle). Verify that the LQG/LQR controller (23.14) makes the closed-loop system asymptotically stable.

Hint: Write the state of the closed-loop system in terms of x and $e := x - \hat{x}$.   □

23.4. Verify that a solution to (23.17) is given by (23.18).

Hint: Use direct substitution of the "candidate" solution into (23.17).   □

23.5. Verify that the LQG/LQR set point controller (23.21) makes the closed-loop system asymptotically stable.

Hint: Write the state of the closed loop in terms of $x - x_{eq}$ and $e := x - \hat{x}$.   □

23.6 (Set point control with integrator). Show that for a single controlled output ($\ell = 1$), we can take $u_{eq} = 0$ in (23.17) when the matrix A has an eigenvalue at the origin and this mode is observable through z. Note that in this case the process has an integrator.   □

LECTURE 24

LQG/LQR and the Q Parameterization

CONTENTS

This lecture shows how a given LQG/LQR controller can be used to parameterize all feedback controllers capable of stabilizing a given LTI system. This parameterization is subsequently used as the basis for a control design method based on numerical optimization.

1. *Q*-augmented LQG/LQR Controller
2. Properties
3. *Q* Parameterization
4. Exercises

24.1 *Q*-AUGMENTED LQG/LQR CONTROLLER

Consider a continuous-time LTI system of the form

$$\dot{x} = Ax + Bu, \qquad y = Cx, \qquad x \in \mathbb{R}^n,\ u \in \mathbb{R}^k,\ y \in \mathbb{R}^m, \qquad \text{(CLTI)}$$

where u is the control signal and y is the measured output. We saw in Lecture 23 that an LQG/LQR output feedback controller is of the form

$$\dot{\hat{x}} = (A - LC)\hat{x} + Bu + Ly, \qquad\qquad u = -K\hat{x}, \qquad (24.1)$$

where $A - LC$ and $A - BK$ are both stability matrices.

Suppose, however, that instead of (24.1) we use

$$\dot{\hat{x}} = (A - LC)\hat{x} + Bu + Ly, \qquad\qquad u = -K\hat{x} + v, \qquad (24.2)$$

where $v \in \mathbb{R}^k$ is the output of an asymptotically stable system driven by the *output estimation error* $\tilde{y} := y - C\hat{x} \in \mathbb{R}^m$:

Notation. The system (24.3) is often called the *Q* system and the overall controller is called a *Q-augmented LQG/LQR controller*.

$$\dot{x}_Q = A_Q x_Q + B_Q \tilde{y}, \qquad v = C_Q x_Q + D_Q \tilde{y}, \qquad \tilde{y} \in \mathbb{R}^m,\ v \in \mathbb{R}^k, \qquad (24.3)$$

with A_Q a stability matrix. We can rewrite (24.2) and the output estimation error as

$$\dot{\hat{x}} = (A - LC - BK)\hat{x} + Ly + Bv, \qquad\qquad (24.4\text{a})$$

$$u = -K\hat{x} + v, \qquad\qquad (24.4\text{b})$$

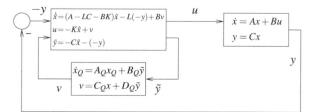

Figure 24.1. Q-augmented LQG/LQR controller.

Note. When the transfer matrix of (24.3) is equal to zero, we recover the original LQG/LQR controller.

$$\tilde{y} = -C\hat{x} + y, \tag{24.4c}$$

which corresponds to the negative-feedback control architecture shown in Figure 24.1. We shall see shortly that the resulting *Q-augmented LQG/LQR controller* defined by (24.3)–(24.4) still stabilizes the original process (CLTI).

24.2   PROPERTIES

Attention! In this interconnection we are excluding the subsystem (24.3).

Note. When one includes noise and disturbance in the process equations, then $e := x - \hat{x}$ and $\tilde{y}$ do not converge to zero, but both remain bounded as long as the noise and disturbance are bounded (cf. Section 23.2.3).

Attention! In the absence of a reference signal, measurement noise, and disturbances, the Q-augmented controller results in the same asymptotic closed-loop behavior as the original one. However, the two may lead to completely different transients, as well as different closed-loop transfer matrices from noise/disturbances to the output.

The Q-augmented LQG/LQR controller (24.3)–(24.4) has several important properties, which are explored below.

Properties (Q-augmented LQG/LQR controller). Assume that the matrices $A - LC$, $A - BK$, and A_Q are all stability matrices.

P24.1 Consider the interconnection of the process (CLTI) with the system (24.4), taking v as the input and $\tilde{y}$ as the output . The transfer matrix from v to $\tilde{y}$ is equal to zero.

Proof. The output estimation error can be written as

$$\tilde{y} := y - C\hat{x} = C(x - \hat{x}).$$

In the absence of noise and disturbances , and because $A - LC$ is a stability matrix, we saw in Lecture 23 that the state estimation error $e := x - \hat{x}$ converges to zero for any process input u. This means that, for *every input signal v* to the interconnection of (CLTI) with (24.4), its output signal $\tilde{y}$ converges to zero. This is only possible if the transfer matrix from v to $\tilde{y}$ is equal to zero.   ∎

P24.2 The controller (24.3)–(24.4) makes the closed-loop system asymptotically stable.

Proof. We have just seen that $\tilde{y}$ always converges to zero when one interconnects (CLTI) with (24.4). Since $\tilde{y}$ is the input to (24.3) and A_Q is a stability matrix, we conclude that the output v to this system also converges to zero. We thus conclude that the Q-augmented controller (24.3)–(24.4) has the same asymptotic behavior as the original LQG/LQR controller (24.1). In particular, all signals converge to zero, which means that the closed-loop system must be asymptotically stable.   ∎

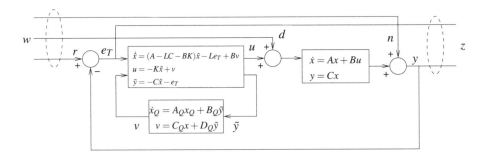

Figure 24.2. Q-augmented LQG/LQR controller with an exogenous input vector $w(t)$ containing a reference signal $r(t)$, an additive disturbance $d(t)$ to the control input, and additive measurement noise $n(t)$. The controlled output $z(t)$ contains the process output $y(t)$ and the tracking error $e_T(t)$.

To study the closed-loop transfer matrices obtained with the Q-augmented controller, we now add inputs and outputs to the feedback control loop in Figure 24.1:

1. a vector $w(t)$ of *exogenous inputs* that may include, e.g., a reference signal $r(t)$, measurement noise $n(t)$, and/or a disturbance signal $d(t)$, and

2. a vector $z(t)$ of *controlled outputs* that may include, e.g., the process output, a tracking error, and/or the control input.

Attention! This vector of controlled outputs need not be the same used in the LQR criterion.

Figure 24.2 contains an example of the resulting closed-loop system.

Stacking the states x and $\hat{x}$ of the process and state estimator, respectively, into a single column vector $\bar{x}$, we can write the resulting closed-loop dynamics in state-space form as

$$\dot{\bar{x}} = \bar{A}\bar{x} + \bar{B}\begin{bmatrix} w \\ v \end{bmatrix}, \qquad \begin{bmatrix} z \\ \tilde{y} \end{bmatrix} = \bar{C}\bar{x} + \bar{D}\begin{bmatrix} w \\ v \end{bmatrix}, \qquad (24.5)$$

$$\dot{x}_Q = A_Q x_Q + B_Q \tilde{y}, \qquad v = C_Q x_Q + D_Q \tilde{y}. \qquad (24.6)$$

We can view (24.5) as an LTI system with input vector $[w'\ v']'$ and output vector $[z'\ \tilde{y}']'$, whose transfer matrix can be partitioned as

$$\begin{bmatrix} \hat{z} \\ \hat{\tilde{y}} \end{bmatrix} = \begin{bmatrix} \hat{P}_{zw}(s) & \hat{P}_{zv}(s) \\ \hat{P}_{\tilde{y}w}(s) & 0 \end{bmatrix} \begin{bmatrix} \hat{w} \\ \hat{v} \end{bmatrix}, \qquad \begin{bmatrix} \hat{P}_{zw}(s) & \hat{P}_{zv}(s) \\ \hat{P}_{\tilde{y}w}(s) & 0 \end{bmatrix} := \bar{C}(sI - \bar{A})^{-1}\bar{B} + \bar{D},$$

$$(24.7)$$

where $\hat{z}$, $\hat{y}$, $\hat{w}$, and $\hat{v}$ denote the Laplace transforms of z, $\tilde{y}$, w, and v, respectively. Because of P24.1, the transfer matrix from v to $\tilde{y}$ is equal to zero, which explains the zero in the bottom right corner of the transfer matrices in (24.7).

Properties (Q-augmented LQG/LQR controller, continued).

P24.3 The transfer matrices $\hat{P}_{zw}(s)$, $\hat{P}_{zv}(s)$, and $\hat{P}_{\tilde{y}w}(s)$ are all BIBO stable. Moreover, $\hat{P}_{zw}(s)$ is equal to the transfer matrix from w to z obtained with the original LQG/LQR controller.

Proof. When $v = 0$, the Q-augmented controller reverts to the original LQG/LQR controller, and therefore (24.5) with $v = 0$ corresponds precisely to the closed-loop system obtained with the original controller.

Since the original controller asymptotically stabilizes the process, this means that $\bar{A}$ must be a stability matrix. Therefore all the transfer matrices $\hat{P}_{zw}(s)$, $\hat{P}_{zv}(s)$, and $\hat{P}_{\bar{y}w}(s)$ are BIBO stable.

Moreover, when $v = 0$ we have $\hat{z} = \hat{P}_{zw}(s)\hat{w}$, and therefore the transfer matrix $\hat{P}_{zw}(s)$ must be equal to the transfer matrix from w to z for the original controller.   ■

P24.4 With the Q-augmented controller, the closed-loop transfer matrix $\hat{T}(s)$ from the exogenous input w to the controlled output z is given by

$$\hat{T}(s) = \hat{P}_{zw}(s) + \hat{P}_{zv}(s)\hat{Q}(s)\hat{P}_{\bar{y}w}(s), \tag{24.8}$$

where $\hat{Q}(s) := C_Q(sI - A_Q)^{-1}B_Q + D_Q$ is the (BIBO stable) transfer matrix of (24.6).

Proof. From (24.6) and (24.7) we conclude that

$$\hat{z} = \hat{P}_{zw}(s)\hat{w} + \hat{P}_{zv}(s)\hat{v}, \qquad \hat{y} = \hat{P}_{\bar{y}w}(s)\hat{w}, \qquad \hat{v} = \hat{Q}(s)\hat{y}.$$

Therefore

$$\hat{z} = \hat{P}_{zw}(s)\hat{w} + \hat{P}_{zv}(s)\hat{Q}(s)\hat{P}_{\bar{y}w}(s)\hat{w},$$

which confirms that the transfer matrix from w to z is indeed given by (24.8).   ■

24.3 Q PARAMETERIZATION

We just saw in P24.2 that the controller (24.3)–(24.4) stabilizes the closed loop *for every LTI asymptotically stable LTI system* (24.3). It turns out that this controller architecture has several other important properties, summarized in the following result.

Theorem 24.1 (Q parameterization). *Assume that the matrices $A - LC$ and $A - BK$ are stability matrices.*

1. *The controller (24.3)–(24.4) makes the closed-loop system asymptotically stable for every stability matrix A_Q.*

Note. Different inputs and outputs will correspond to different transfer matrices $\hat{T}_0(s)$, $\hat{L}(s)$, and $\hat{R}(s)$ but the closed-loop transfer matrices will always be affine in $\hat{Q}(s)$.

2. *The closed-loop transfer function $\hat{T}(s)$ from any given exogenous input signal to any given controlled output can always be written as*

$$\hat{T}(s) = \hat{T}_0(s) + \hat{L}(s)\hat{Q}(s)\hat{R}(s), \tag{24.9}$$

where $\hat{Q}(s) := C_Q(sI - A_Q)^{-1}B_Q + D_Q$ is the (BIBO stable) transfer function of (24.3) and $\hat{T}_0(s)$, $\hat{L}(s)$, and $\hat{R}(s)$ are BIBO stable transfer matrices.

Attention! This
realization will
generally not be
minimal, but it is
always stabilizable and
detectable.

3. *For every controller transfer function $\hat{C}(s)$ that asymptotically stabilizes* (CLTI),
   there exist matrices A_Q, B_Q, C_Q, and D_Q, with A_Q a stability matrix, such that
   (24.3)–(24.4) *is a realization of $\hat{C}(s)$.*   □

Items 1 and 2 are just restatements of the properties P24.2 and P24.4, respectively,
but the new item 3 is nontrivial. The proof of this result can be found, e.g., in [5,
Chapter 5].

Theorem 24.1 states that one can construct *every controller that stabilizes an LTI
process* and *every stable closed-loop transfer matrix* by Q-augmenting any given LQG/
LQR controller, by allowing $\hat{Q}(s)$ to range over the set of all BIBO stable transfer
matrices with m inputs and k outputs. Because of this, we say that (24.3)–(24.4)
provides a *parameterization of all stabilizing controllers for the process* (CLTI) and also
that (24.9) is a *parameterization of the set of all stable closed-loop transfer matrices for
the process* (CLTI).

24.4 EXERCISE

24.1 (Q-augmented LQG/LQR controller). Show that the controller (24.3)–(24.4)
can be realized as

$$\begin{bmatrix} \dot{\hat{x}} \\ \dot{x}_Q \end{bmatrix} = \begin{bmatrix} A - LC - BK - BD_QC & BC_Q \\ -B_QC & A_Q \end{bmatrix} \begin{bmatrix} \hat{x} \\ x_Q \end{bmatrix} + \begin{bmatrix} L + BD_Q \\ B_Q \end{bmatrix} y$$

$$u = \begin{bmatrix} -K - D_QC & C_Q \end{bmatrix} \begin{bmatrix} \hat{x} \\ x_Q \end{bmatrix} + D_Q y.$$ □

LECTURE 25

Q Design

CONTENTS

This lecture describes a control design method based on the Q parameterization and numerical optimization.

1. Control Specifications for Q Design
2. The Q Design Feasibility Problem
3. Finite-dimensional Optimization: Ritz Approximation
4. Q Design using MATLAB® and CVX
5. Q Design Example
6. Exercise

25.1 CONTROL SPECIFICATIONS FOR Q DESIGN

The idea behind Q design is to take a controller that does not meet all the required specifications for the closed-loop system and augment it so as to improve its performance. The original controller is typically designed using LQG/LQR as discussed in Lecture 23, and it is then Q-augmented using the architecture described in Lecture 24. The search for the Q parameter is done numerically.

Q design can address a wide range of *closed-loop* specifications , i.e., specifications that can be expressed in terms of closed-loop responses or transfer functions. The basic feedback control architecture is the one considered in Lecture 24 and depicted in Figure 24.2. It consists of

Note. Specifications on gain/phase margins and open-loop gain cannot be directly addressed by Q design.

Note. For simplicity, we consider in Figure 24.2 an input disturbance that is additive to the control input, but more general disturbances are also allowed [e.g., as in (23.2)].

1. a continuous-time LTI system of the form

$$\dot{x} = Ax + Bu, \qquad y = Cx, \qquad x \in \mathbb{R}^n,\ u \in \mathbb{R}^k,\ y \in \mathbb{R}^m, \qquad \text{(CLTI)}$$

where u is the control signal, y is the measured output, d is an input disturbance, and n is the measurement noise, and

2. a Q-augmented LQG/LQR controller

$$\dot{\hat{x}} = (A - LC - BK)\hat{x} - Le_T + Bv, \quad u = -K\hat{x} + v, \qquad \tilde{y} = -C\hat{x} - e_T,$$
$$(25.1a)$$

$$\dot{x}_Q = A_Q x_Q + B_Q \tilde{y}, \qquad\qquad v = C_Q x_Q + D_Q \tilde{y}, \qquad (25.1b)$$

driven by the tracking error

$$e_T := r - y,$$

where r is a reference input.

Note. A stable prefilter that would be applied to the reference prior to the subtraction by y would also be allowed.

The matrix gains K and L should make the matrices $A - BK$ and $A - LC$ both stability matrices and can be designed using the LQG/LQR control design methodology previously discussed. Before addressing the design of the Q system, we review a few typical closed-loop specifications addressed by Q design.

To express these specifications, it is convenient to stack all the exogenous inputs in a vector w and all the controlled outputs of interest in a vector z, e.g.,

$$w := \begin{bmatrix} r \\ d \\ n \end{bmatrix}, \qquad\qquad z := \begin{bmatrix} y \\ e_Y \\ u \end{bmatrix}.$$

TIME DOMAIN SPECIFICATIONS. *Time domain* specifications refer to desired properties of the closed-loop response to specific exogenous inputs. Typical time domain specifications include the following.

1. *Norm bounds.* For a given test exogenous input $w_{\text{test}}(t)$, $t \geq 0$, a vector $\bar{z}(t)$ containing one or more entries of the controlled output $z(t)$ should satisfy

Note. Typically, one needs to consider multiple specifications, corresponding to different test inputs and different entries of the controlled output.

$$\int_0^\infty \|\bar{z}(t)\| dt \leq c \qquad (25.2)$$

or

$$\int_0^\infty \|\bar{z}(t)\|^2 dt \leq c \qquad (25.3)$$

or

$$\|\bar{z}(t)\| \leq c, \quad \forall t \geq 0 \qquad (25.4)$$

for a given constant $c > 0$. The specification given by (25.2) is generally called an L_1 *norm bound*, the one given by (25.3) is called an L_2 *norm bound*, and the one given by (25.4) is called an L_∞ *norm bound*.

2. *Interval bounds.* For a given test exogenous input signal $w_{\text{test}}(t)$, $t \geq 0$, the ith entry $z_i(t)$ of the controlled output $z(t)$ should satisfy

$$s_{\min}(t) \leq z_i(t) \leq s_{\max}(t), \quad \forall t \geq 0 \qquad (25.5)$$

for two given time functions $s_{\min}(t)$ and $s_{\max}(t)$.

Interval-bound specifications can be used to impose undershoot, overshoot, and settling times for a step response. In particular, if one selects $w_{\text{test}}(t)$ to be a unit step at the reference input, sets $z_i(t)$ to be the process output, and sets

$$s_{\min}(t) = \begin{cases} -p_{\text{under}} & t \leq t_{\text{settle}} \\ 1 - p_{\text{settling}} & t > t_{\text{settle}} \end{cases}, \quad s_{\max}(t) = \begin{cases} 1 + p_{\text{over}} & t \leq t_{\text{settle}} \\ 1 + p_{\text{settling}} & t > t_{\text{settle}} \end{cases},$$

then the interval-bound specification (25.5) guarantees an overshoot no larger than p_{over}, an undershoot no larger than p_{under}, and a settling time around $1 \pm p_{settling}$ no larger than t_{settle}.

Interval-bound specifications are also typically used to guarantee that the control signal does not exceed safe ranges for typical exogenous inputs. In this case, $w_{test}(t)$ should be the typical exogenous input, $z_i(t)$ should be the control signal, and the functions $s_{min}(t)$ and $s_{max}(t)$ should be constants that define the safe range.

INPUT-OUTPUT SPECIFICATIONS.   *Input-output* specifications refer to desired properties of the closed-loop transfer function from a given exogenous input to a given controlled output. Typical input-output specifications include the following.

1. *Frequency domain.* The transfer function $\hat{T}(s)$ from a vector $\bar{w}$ containing one or more entries of the exogenous input w to a vector $\bar{z}$ containing one or more entries of the controlled output z should satisfy

Note. Typically, one needs to consider specifications for different transfer functions between different entries of w and z.

$$\|\hat{T}(j\omega)\| \leq \ell(\omega), \qquad\qquad \forall \omega \in [\omega_{min}, \omega_{max}], \qquad (25.6)$$

for a given function $\ell(\omega)$.

When $\omega_{min} = 0$, $\omega_{max} = \infty$, and $\ell(\omega) = \gamma$, $\forall \omega \geq 0$, this specification guarantees that

$$\left(\int_0^\infty \|\bar{z}(t)\|^2 dt \right)^{\frac{1}{2}} \leq \gamma \left(\int_0^\infty \|\bar{w}(t)\|^2 dt \right)^{\frac{1}{2}},$$

for every $\bar{w}(t), t \geq 0$ when all other entries of the exogenous input w are equal to zero and the closed-loop system has zero initial conditions (see, e.g., [5]). In this case, it is said that the closed-loop system has *root mean square (RMS) gain* from $\bar{w}$ to $\bar{z}$ no larger than γ.

Notation. The root mean square (RMS) gain is often also called the *H-infinity norm* of the system.

2. *Impulse response.* The impulse response $\bar{h}(t)$ from a vector $\bar{w}$ containing one or more entries of the exogenous input w to a vector $\bar{z}$ containing one or more entries of the controlled output z should satisfy

$$\int_0^\infty \|\bar{h}(t)\| dt \leq \rho,$$

for a given constant ρ.

This specification guarantees that

$$|\bar{z}(t)| \leq \rho \sup_{\tau \geq 0} \|\bar{w}(\tau)\|, \quad \forall t \geq 0$$

for every $\bar{w}(t), t \geq 0$ when all other entries of the exogenous input w are equal to zero and the closed-loop system has zero initial conditions (see, e.g., [5]). In this case, it is said that the closed-loop system has *peak gain* from $\bar{w}$ to $\bar{z}$ no larger than ρ.

Notation. The peak gain is often also called the L_1 *norm* of the system.

Attention! The Q design method can accommodate many control specifications not considered here, e.g., specifications related to the closed-loop response to stochastic inputs. The reader is referred to [2, Part III] for additional specifications. $\square$

25.2   THE Q DESIGN FEASIBILITY PROBLEM

As we saw in Lecture 24, any closed-loop transfer function for the closed-loop (CLTI), (25.1) can be written as

$$\hat{T}(s) = \hat{T}_0(s) + \hat{L}(s)\hat{Q}(s)\hat{R}(s), \tag{25.7}$$

where

$$\hat{Q}(s) := C_Q(sI - A_Q)^{-1}B_Q + D_Q$$

is the $k \times m$ BIBO stable transfer function of the Q system (25.1b), and $\hat{T}_0(s)$, $\hat{L}(s)$, and $\hat{R}(s)$ are BIBO stable transfer matrices.

Given a family $\mathcal{D}_1, \mathcal{D}_2, \ldots, \mathcal{D}_K$ of time domain and input-output closed-loop specifications, the Q *design method* consists of finding a Q system that meets all the specifications, i.e.,

$$\text{find} \qquad \hat{Q}(s) \text{ BIBO stable} \tag{25.8a}$$

$$\text{such that} \qquad \hat{T}_0(s) + \hat{L}(s)\hat{Q}(s)\hat{R}(s) \text{ satisfies } \mathcal{D}_1, \mathcal{D}_2, \ldots, \mathcal{D}_K, \tag{25.8b}$$

and then using this Q system to construct the Q-augmented LQG/LQR controller (25.1).

Notation. The statement in (25.8) is called a *feasibility problem*, since its goal is to determine whether the given set of constraints is feasible, in the sense that it can be satisfied by some controller.

To devise efficient numerical procedures to solve the feasibility problem (25.8), it is convenient for the specifications to be convex. A closed-loop *control specification is said to be convex* if, given any two closed-loop transfer functions $\hat{T}_1(s)$ and $\hat{T}_2(s)$ that satisfy the constraint, the closed-loop transfer functions

$$\lambda \hat{T}_1(s) + (1 - \lambda)T_2(s), \quad \forall \lambda \in [0, 1]$$

also satisfy the constraint.

25.3   FINITE-DIMENSIONAL OPTIMIZATION: RITZ APPROXIMATION

The key difficulty in solving the feasibility problem (25.8) is that it involves a search over the set of all BIBO stable transfer matrices. However, this problem can be converted into a numerical search over a finite number of scalars, suitable for numerical optimization.

The *Ritz approximation* allows one to convert the infinite-dimensional search over the BIBO stable transfer matrices into a search over a finite-dimensional space. To achieve this, one starts by selecting a sequence of $k \times m$ BIBO stable transfer functions

$$\hat{Q}_1(s), \hat{Q}_2(s), \ldots, \hat{Q}_i(s), \ldots,$$

Note 15. A possible sequence with this property is obtained by selecting all entries of each $\hat{Q}_i(s)$ equal to zero, except for one entry that is set to be of the form $\left(\frac{\alpha}{s+\alpha}\right)^\ell$, $\ell \geq 0$ for some fixed constant $\alpha > 0$.
▶ p. 247

which should be *complete* in the sense that for every BIBO stable transfer function $\hat{Q}(s)$ there should be a *finite* linear combination of the $\hat{Q}_i(s)$ arbitrarily close to $\hat{Q}(s)$. One then restricts the search to linear combinations of the first N matrices in the sequence; i.e., one restricts $\hat{Q}(s)$ to be of the form

$$\hat{Q}(s) := \sum_{i=1}^N \alpha_i \, \hat{Q}_i(s), \qquad \alpha_i \in \mathbb{R}. \tag{25.9}$$

For this choice of $\hat{Q}(s)$, the closed-loop transfer function (25.7) can be written as

$$\hat{T}(s) = \hat{T}_0(s) + \sum_{i=1}^{N} \alpha_i \, \hat{T}_i(s), \qquad \hat{T}_i(s) := \hat{L}(s)\hat{Q}_i(s)\hat{R}(s), \qquad (25.10)$$

and the feasibility problem (25.8) becomes to

$$\text{find} \qquad \alpha_1, \alpha_2, \ldots, \alpha_N \in \mathbb{R} \qquad (25.11a)$$

$$\text{such that} \qquad \hat{T}_0(s) + \sum_{i=1}^{N} \alpha_i \, \hat{T}_i(s) \text{ satisfies } \mathcal{D}_1, \mathcal{D}_2, \ldots, \mathcal{D}_K. \qquad (25.11b)$$

If this problem is feasible, then one uses the corresponding Q system to construct the Q-augmented LQG/LQR controller (25.1). Otherwise, the problem (25.11) may not be feasible for two reasons.

1. The original feasibility problem (25.8) has a solution, but not of the form (25.9). In this case, one should increase N to enlarge the search space.

2. The original feasibility problem (25.8) has no solution. In this case one needs to relax one or more specifications.

Unfortunately, in general it is not possible to determine the cause for (25.11) to be infeasible, and one simply tries to increase N until the resulting numerical optimization becomes computationally intractable or until the order of the resulting Q-augmented controller would be unacceptable.

Note 15 (Complete Q sequence). A sequence that is *complete* in the sense that for every BIBO stable transfer function $\hat{Q}(s)$ there exists a *finite* linear combination of the $\hat{Q}_i(s)$ arbitrarily close to $\hat{Q}(s)$ that can be obtained by selecting all entries of each $\hat{Q}_i(s)$ equal to zero, except for one entry that is set to be of the form

$$\left(\frac{\alpha}{s+\alpha}\right)^{\ell}, \qquad \ell \geq 0,$$

for some fixed constant $\alpha > 0$. This leads to the following sequence for the $\hat{Q}_i(s)$.

Note. In general, one chooses the pole α to fall within the range of frequencies for which the closed-loop response is expected to have an "interesting behavior" (e.g., the closed-loop bandwidth from reference to output or the inverse of the dominant time constant of the step response).

$$
\begin{bmatrix} 1 & 0 & \cdots & 0 \\ 0 & 0 & \cdots & 0 \\ \vdots & \vdots & \ddots & 0 \\ 0 & 0 & \cdots & 0 \end{bmatrix}, \quad
\begin{bmatrix} 0 & 1 & \cdots & 0 \\ 0 & 0 & \cdots & 0 \\ \vdots & \vdots & \ddots & 0 \\ 0 & 0 & \cdots & 0 \end{bmatrix}, \cdots,
$$

$$
\begin{bmatrix} 0 & 0 & \cdots & 0 \\ 1 & 0 & \cdots & 0 \\ \vdots & \vdots & \ddots & 0 \\ 0 & 0 & \cdots & 0 \end{bmatrix}, \quad
\begin{bmatrix} 0 & 0 & \cdots & 0 \\ 0 & 1 & \cdots & 0 \\ \vdots & \vdots & \ddots & 0 \\ 0 & 0 & \cdots & 0 \end{bmatrix}, \cdots,
$$

$$
\begin{bmatrix} 0 & 0 & \cdots & 0 \\ 0 & 0 & \cdots & 0 \\ \vdots & \vdots & \ddots & 0 \\ 0 & 0 & \cdots & 1 \end{bmatrix}, \quad
\begin{bmatrix} \frac{\alpha}{s+\alpha} & 0 & \cdots & 0 \\ 0 & 0 & \cdots & 0 \\ \vdots & \vdots & \ddots & 0 \\ 0 & 0 & \cdots & 0 \end{bmatrix}, \quad
\begin{bmatrix} 0 & \frac{\alpha}{s+\alpha} & \cdots & 0 \\ 0 & 0 & \cdots & 0 \\ \vdots & \vdots & \ddots & 0 \\ 0 & 0 & \cdots & 0 \end{bmatrix}, \cdots,
$$

$$
\begin{bmatrix} 0 & 0 & \cdots & 0 \\ \frac{\alpha}{s+\alpha} & 0 & \cdots & 0 \\ \vdots & \vdots & \ddots & 0 \\ 0 & 0 & \cdots & 0 \end{bmatrix}, \quad
\begin{bmatrix} 0 & 0 & \cdots & 0 \\ 0 & \frac{\alpha}{s+\alpha} & \cdots & 0 \\ \vdots & \vdots & \ddots & 0 \\ 0 & 0 & \cdots & 0 \end{bmatrix}, \cdots,
$$

$$\begin{bmatrix} 0 & 0 & \cdots & 0 \\ 0 & 0 & \cdots & 0 \\ \vdots & \vdots & \ddots & 0 \\ 0 & 0 & \cdots & \frac{\alpha}{s+\alpha} \end{bmatrix}, \quad \begin{bmatrix} \left(\frac{\alpha}{s+\alpha}\right)^2 & 0 & \cdots & 0 \\ 0 & 0 & \cdots & 0 \\ \vdots & \vdots & \ddots & 0 \\ 0 & 0 & \cdots & 0 \end{bmatrix}, \quad \begin{bmatrix} 0 & \left(\frac{\alpha}{s+\alpha}\right)^2 & \cdots & 0 \\ 0 & 0 & \cdots & 0 \\ \vdots & \vdots & \ddots & 0 \\ 0 & 0 & \cdots & 0 \end{bmatrix}, \cdots ,$$

$$\begin{bmatrix} 0 & 0 & \cdots & 0 \\ \left(\frac{\alpha}{s+\alpha}\right)^2 & 0 & \cdots & 0 \\ \vdots & \vdots & \ddots & 0 \\ 0 & 0 & \cdots & 0 \end{bmatrix}, \quad \begin{bmatrix} 0 & 0 & \cdots & 0 \\ 0 & \left(\frac{\alpha}{s+\alpha}\right)^2 & \cdots & 0 \\ \vdots & \vdots & \ddots & 0 \\ 0 & 0 & \cdots & 0 \end{bmatrix}, \cdots ,$$

$$\begin{bmatrix} 0 & 0 & \cdots & 0 \\ 0 & 0 & \cdots & 0 \\ \vdots & \vdots & \ddots & 0 \\ 0 & 0 & \cdots & \left(\frac{\alpha}{s+\alpha}\right)^2 \end{bmatrix}, \quad \begin{bmatrix} \left(\frac{\alpha}{s+\alpha}\right)^i & 0 & \cdots & 0 \\ 0 & 0 & \cdots & 0 \\ \vdots & \vdots & \ddots & 0 \\ 0 & 0 & \cdots & 0 \end{bmatrix}, \quad \begin{bmatrix} 0 & \left(\frac{\alpha}{s+\alpha}\right)^i & \cdots & 0 \\ 0 & 0 & \cdots & 0 \\ \vdots & \vdots & \ddots & 0 \\ 0 & 0 & \cdots & 0 \end{bmatrix}, \cdots ,$$

$$\begin{bmatrix} 0 & 0 & \cdots & 0 \\ \left(\frac{\alpha}{s+\alpha}\right)^i & 0 & \cdots & 0 \\ \vdots & \vdots & \ddots & 0 \\ 0 & 0 & \cdots & 0 \end{bmatrix}, \quad \begin{bmatrix} 0 & 0 & \cdots & 0 \\ 0 & \left(\frac{\alpha}{s+\alpha}\right)^i & \cdots & 0 \\ \vdots & \vdots & \ddots & 0 \\ 0 & 0 & \cdots & 0 \end{bmatrix}, \cdots ,$$

$$\begin{bmatrix} 0 & 0 & \cdots & 0 \\ 0 & 0 & \cdots & 0 \\ \vdots & \vdots & \ddots & 0 \\ 0 & 0 & \cdots & \left(\frac{\alpha}{s+\alpha}\right)^i \end{bmatrix}, \cdots \qquad\blacksquare$$

25.4 Q DESIGN USING MATLAB® AND CVX

The Q design procedure can be implemented in MATLAB® with the CVX toolbox using the following four steps, to be discussed next.

1. Construct the sequence $\hat{Q}_1(s), \hat{Q}_2(s), \cdots , \hat{Q}_N(s)$ of $k \times m$ BIBO stable transfer functions for the Ritz approximation.

2. Determine the sequence $\hat{T}_0(s), \hat{T}_1(s), \cdots , \hat{T}_N(s)$ needed to compute the closed-loop transfer function using (25.10).

3. Numerically solve the finite-dimensional convex feasibility problem (25.11).

4. Construct the Q-augmented LQG/LQR controller (25.1).

25.4.1 $\hat{Q}_i(s)$ SEQUENCE

The sequence $\hat{Q}_1(s), \hat{Q}_2(s), \cdots , \hat{Q}_N(s)$ for the Ritz approximation considered in Note 15 (p. 247) can be constructed using the following sequence of MATLAB® commands:

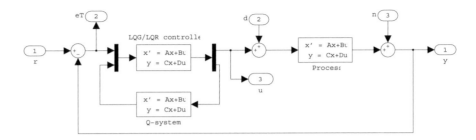

Figure 25.1. Simulink® diagram to compute the transfer functions $\hat{T}_i(s)$.

```
% Inputs:
%    k      - number of rows for the Qi(s)
%    m      - number of columns for the Qi(s)
%    alpha  - pole for the Qi(s) in the Ritz expansion
%    q      - maximum multiplicity for the Qi(s) poles in the
%             Ritz expansion
% Output:
%    Qi     - cell array with Q1(s), Q2(s), ..., QN(s)
Qi = cell(1,k*m*(q+1));
N  = 0;
for ell = 0:q
    for i = 1:k
        for j = 1:m
            N = N+1;
            Qi{N}        = ss(zeros(k,m));
            Qi{N}(i,j)   = (alpha/(tf('s')+alpha))^ell;
        end
    end
end
```

The output to this script is a cell array `Qi` with `N` elements, containing the `k×m` transfer functions $\hat{Q}_i(s)$, $i \geq 1$ in the Note 15 (p. 247). The sequence contains transfer matrices with $\frac{\alpha}{s+\alpha}$ raised to powers from 0 to `q`.

25.4.2 $\hat{T}_i(s)$ TRANSFER FUNCTIONS

MATLAB® Hint 50.
When using MATLAB®/Simulink® to compute these transfer functions, one generally does not obtain minimal realization, so one should use `minreal(sys)` to remove unnecessary states (cf. MATLAB® Hint 36, p. 163).

The transfer functions $\hat{T}_i(s)$ in (25.11) can be computed directly using Simulink®. To do this, one starts by constructing a Simulink® diagram with the process connected to the Q-augmented LQG/LQR controller as in Figure 24.2. In this diagram, the exogenous inputs should come from input ports, and the controlled outputs should be connected to output ports (see Figure 25.1). The Q system should be a Simulink® state-space block with matrices `AQ`, `BQ`, `CQ`, `DQ` *left as MATLAB® variables to be taken from the workspace*. The $\hat{T}_i(s)$ can then be constructed using the following sequence of MATLAB® command:

```
% Inputs:
%   k      - number of rows for the Qi(s)
%   m      - number of columns for the Qi(s)
%   Qi     - cell array with Q1(s), Q2(s), ..., QN(s)
% Output:
%   Ti     - cell array with T0(s), T1(s), T2(s), ..., TN(s)
Ti = cell(length(Qi)+1,1);              % initialize cell array
% compute T0(s)
[AQ,BQ,CQ,DQ] = ssdata(tf(zeros(k,m)));
                                        % set Q = 0 in simulink
[a,b,c,d]     = linmod('augmented_closedloop');  % get T0(s)
Ti{1} = minreal(ss(a,b,c,d));                        % store T0
% compute T1(s), T2(s), ...
for i = 1:length(Qi)
    [AQ,BQ,CQ,DQ]=ssdata(Qi{i});        % set Q = Qi in simulink
    [a,b,c,d]=linmod('augmented_closedloop');
                                        % get T0(s) + Ti(s)
    Ti{i+1}=minreal(ss(a,b,c,d)-Ti{1});          % store Ti(s)
end
```

The output to this script is a cell array Ti with N+1 elements, containing the $\hat{T}_i(s)$, $i \geq 0$ in (25.10).

25.4.3 NUMERICAL SOLUTION TO THE FEASIBILITY PROBLEM

The feasibility problem (25.11) can be solved numerically using the MATLAB® toolbox CVX [7]. Using CVX, the computation of a Q-augmented LQG/LQR controller can be accomplished with the following sequence of MATLAB®/CVX commands:

MATLAB® Hint 51.
The CVX minimize directive can be used to select an optimal controller that minimizes a particular criterion among those that satisfy all the specifications.   ▶ p. 253

```
%% create cell array cell array with Q1(s), Q2(s), ..., QN(s)
  {... code in previous sections ...}

%% create cell array with T0(s), T1(s), T2(s), ..., TN(s)
  {... code in previous sections ...}

%% CVX problem specification
  N=length(Qi);
  cvx_begin
    variable alpha(N)       % declare variables to be optimize

    % declare closed-loop specifications through inequalities
    % on alpha(1:N)
    {... code in the remainder of this section ...}

  cvx_end

%% compute Q system
  Q = ss(zeros(size(Qi{1})));
  for i=1:N
    Q=Q+alpha(i)*Qi{i};
  end
  Q=minreal(Q);
```

In the remainder of this section, we discuss how to convert some of the most common closed-loop control specifications $\mathcal{D}_1, \mathcal{D}_2, \ldots, \mathcal{D}_K$ into explicit constraints on the parameters $\alpha_1, \alpha_2, \ldots, \alpha_N$, to be used by CVX.

INTERVAL BOUND TIME-DOMAIN SPECIFICATIONS.   For a given test exogenous input $w_{\text{test}}(t)$, $t \geq 0$, the resulting forced controlled output is given by

$$z(t) = \left(\mathcal{L}^{-1}\left[\hat{T}_0(s) + \sum_{i=1}^{N} \alpha_i\, \hat{T}_i(s) \right] \star w_{\text{test}} \right)(t) = \zeta_0(t) + \sum_{i=1}^{N} \alpha_i \zeta_i(t),$$

where

$$\zeta_0(t) := \left(\mathcal{L}^{-1}[\hat{T}_0(s)] \star w_{\text{test}} \right)(t), \quad \zeta_i(t) := \left(\mathcal{L}^{-1}[\hat{T}_i(s)] \star w_{\text{test}} \right)(t), \quad \forall t \geq 0.$$

An interval bound on the lth entry of the forced response $z(t)$ to $w_{\text{test}}(t)$, $t \geq 0$ can then be expressed as

$$s_{\min}(t) \leq \zeta_{0,l}(t) + \sum_{i=1}^{N} \alpha_i\, \zeta_{i,l}(t) \leq s_{\max}(t), \quad \forall t \geq 0, \tag{25.12}$$

where $\zeta_{i,l}(t)$ denotes the lth entry of $\zeta_i(t)$, $i \geq 0$. The following sequence of MATLAB®/CVX commands declares the constraint (25.12) for an exogenous input $w_{\text{test}}(t)$, $t \geq 0$ corresponding to a step at the jth entry of the exogenous input $w(t)$:

Note 16. In practice, we do not test (25.12) at every point in the interval $[0, \infty)$. Instead, (25.12) will be enforced on a grid of sample times (variable ts in the MATLAB® code). This grid should be sufficiently fine and long that (25.12) is not violated between samples or after the last point in the grid. However, a large grid also increases the computation time required to solve the feasibility problem.

```
% Inputs:
%    Ti     - cell array with T0(s), T1(s), T2(s), ..., TN(s)
%    alpha  - vector with the alpha_i
%    ts     - sample times at which the interval bound will be
%             tested
%    j      - entry of w(t) where step is applied
%    l      - entry of z(t) that appears in the interval bound
%    step_response_min - lower bound sampled at times ts
%    step_response_max - upper bound sampled at times ts
step_response=step(Ti{1}(l,j),ts);
for i=1:N
   step_response=step_response+alpha(i)*step(Ti{i+1}(l,j),ts);
end
% interval bound constraint
step_response_min <= step_response;
step_response <= step_response_max;
```

To enforce a given settling time, overshoot, and undershoot, one would construct the variables step_response_min and step_response_max, which appear in the code above, with the following MATLAB® commands:

```
% Inputs:
%    ts          - sample times at which the interval bounds
%                  will be tested
%    overshoot   - desired maximum overshoot
%    undershoot  - desired maximum undershoot
%    tsettle     - desired settling time
%    psettle     - percentage defining the settling time
```

```
step_response_min=-undershoot*ones(size(ts));    % undershoot
step_response_max=(1+overshoot)*ones(size(ts));  % overshoot
step_response_min(ts>=tsettle)=1-psettle;        % settling
step_response_max(ts>=tsettle)=1+psettle;        % time
```

FREQUENCY DOMAIN INPUT-OUTPUT SPECIFICATIONS. A frequency domain specification on the SISO closed-loop transfer function from the jth entry of w to the lth entry of z of the form

$$|\hat{T}_{lj}(j\omega)| \le \ell(\omega), \qquad\qquad \forall \omega \in [\omega_{\min}, \omega_{\max}],$$

can be expressed in terms of the α_i as

Note 17. In practice, we do not test (25.13) at every point in the interval $[\omega_{\min}, \omega_{\max}]$. Instead, (25.13) is enforced only on a grid of sample frequencies (variable ws in the MATLAB® code). This grid should be sufficiently fine so that (25.13) is not violated between samples. However, a large grid also increases the computation time required to solve the feasibility problem.

$$\left| \hat{T}_{0,lj}(s) + \sum_{i=1}^{N} \alpha_i \, \hat{T}_{i,lj}(s) \right| \le \ell(\omega), \qquad \forall \omega \in [\omega_{\min}, \omega_{\max}], \qquad (25.13)$$

where $\hat{T}_{i,lj}(s)$ denotes the ljth entry of the closed-loop transfer matrix $\hat{T}_i(s)$ from w to z. The following sequence of MATLAB®/CVX commands declares the constraint (25.13):

```
% Inputs:
%   Ti    - cell array with T0(s), T1(s), T2(s), ..., TN(s)
%   alpha - vector with the alpha_i
%   ws    - sample frequencies at which the bound will be
%           tested
%   j     - entry(s) of w(t) for the input(s)
%   l     - entry(s) of z(t) for the output(s)
%   freq_response_max - norm upper bound ell(ws) sampled
%    at frequencies ws
freq_response=reshape(freqresp(Ti{1}(l,j),ws),length(ws),1);
for i=1:N
   freq_response=freq_response ...
                +alpha(i)*reshape(freqresp(Ti{i+1}(l,j),ws),
                                  length(ws),1);
end
abs(freq_response) <= freq_response_max;
```

IMPULSE RESPONSE INPUT-OUTPUT SPECIFICATIONS. An impulse response specification on the SISO closed-loop transfer function from the jth entry of w to the lth entry of z of the form

$$\int_0^\infty |h_{lj}(t)| dt \le \rho$$

can be expressed in terms of the α_i as

Note 18. In practice, the integral in (25.14) is computed numerically using a fine grid of sample times (variable ti in the MATLAB® code). This grid should be sufficiently *fine* and *long* that the integral in (25.14) is well approximated by a finite sum. However, a large grid also increases the computation time required to solve the feasibility problem.

$$\int_0^\infty \left| h_{0,lj}(s) + \sum_{i=1}^{N} \alpha_i \, h_{i,lj}(s) \right| dt \le \rho, \qquad (25.14)$$

where $h_{i,lj}(t)$ denotes the ljth entry of the closed-loop impulse response $h_i(t)$ from w to z. The following sequence of MATLAB®/CVX commands declares the constraint (25.14):

```
% Inputs:
%   Ti    - cell array with T0(s), T1(s), T2(s), ..., TN(s)
%   alpha - vector with the alpha_i
%   ti    - equally spaced sample times at which the impulse
%           response will be computed
%   j     - entry(s) of w(t) for the input(s)
%   l     - entry(s) of z(t) for the output(s)
%   rho   - upper bound on the L1-norm
impulse_response=impulse(Ti{1}(l,j),ti);
for i=1:N
  impulse_response=impulse_response+alpha(i)
                  *impulse(Ti{i+1}(l,j),ti);
end
sum(abs(impulse_response))*(ti(2)-ti(1)) <= rho;
```

MATLAB® Hint 51 (`minimize`). The CVX `minimize` directive can be used to select an optimal controller that minimizes a particular criterion from among those that satisfy all the specifications. For example:

1. To minimize the overshoot, one would add the following MATLAB®/CVX commands before the `cvx_end` directive in the code on page 250:

    ```
    variable J
    step_response <= J
    minimize J
    ```

2. To minimize the root mean square (RMS) gain, one would add the following MATLAB®/CVX commands before the `cvx_end` directive in the code on page 250:

    ```
    variable J
    freq_response <= J
    minimize J
    ```

3. To minimize the peak gain, one would add the following MATLAB®/CVX commands before the `cvx_end` directive in the code on page 250:

    ```
    minimize sum(abs(impulse_response))
    ```

However, only a single `minimize` directive is allowed in a CVX program. □

25.5 *Q* DESIGN EXAMPLE

Example 25.1 (Aircraft roll dynamics, continued). Consider the LQG/LQR controller designed in Example 23.1 for $\sigma = 10^8$. Our goal is to Q-augment this controller to satisfy the following control specifications.

1. Decrease the peak gain from reference to control input to at most 20 (from about 1476 for the nonaugmented controller).

2. Decrease the overshoot to at most 10% (from about 16% for the nonaugmented controller).

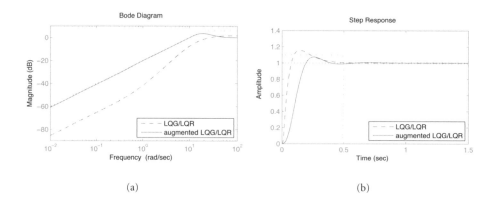

(a)   (b)

Figure 25.2. Bode plots of the closed-loop transfer function $\hat{T}_{e_T,r}(s)$ from reference to tracking error and step response for the LQG/LQR controller designed in Example 23.1 for $\sigma = 10^8$ and the Q-augmented controller in Example 25.1. The Q-augmented controller has a peak gain equal to 17.2 and overshoot of 7.5%, whereas the original LQG/LQR controller has peak gain of 1476 and overshoot of 16%.

3. Keep the 1% settling time below 0.5 s (already satisfied for the nonaugmented controller).

4. Keep the magnitude of the transfer function $\hat{T}_{e_T,r}(s)$ from reference to tracking error below the value of

Note. Specification 4 guarantees good tracking for low frequencies, especially for frequencies below 5 rad/s, and also zero steady-state error.

$$|\hat{T}_{e_T,r}(j\omega)| \leq \ell(\omega) := \begin{cases} \frac{\omega}{10} & \omega \leq 20 \text{ rad/s} \\ 2 & \omega \geq 20 \text{ rad/s} \end{cases} \qquad \forall \omega \geq 0$$

(already satisfied for the nonaugmented controller).

For the numerical optimization, the impulse response was sampled from 0 to 2 s with a sampling time of 0.001 s (see Note 18, p. 252, the step response was sampled from 0 to 4 s with a sampling time of 0.01 s (see Note 16, p. 251), and the frequency response was sampled from 0.01 to 100 rad/s with 200 points logarithmically spaced produced by the MATLAB® command `logspace` (see Note 17, p. 252).

To augment the original LQG/LQR controller, we used the following sequence for the Ritz approximation

$$\hat{Q}_i(s) = \left(\frac{5}{s+5}\right)^i, \quad i \in \{0, 1, 2, \ldots, 10\}.$$

Among all the controllers that satisfied the specifications above, we selected the one with the smallest peak gain (see MATLAB® Hint 51, p. 253), which led to a 10-dimensional Q system and 13-dimensional controller. Figure 25.2 shows the closed-loop step responses and Bode plots from reference to tracking error of the original and the augmented controller.   □

25.6 EXERCISE

25.1 (Convex specification). Show that all the control specifications considered in Section 25.1 are convex.   □

BIBLIOGRAPHY

[1] Antsaklis, P. J., and A. N. Michel. *Linear systems*. McGraw-Hill Series in Electrical and Computer Engineering. New York: McGraw-Hill, 1997.

[2] Boyd, S. P., and C. H. Barratt. *Linear Controller design: Limits of Performance*. Englewood Cliffs, NJ: Prentice Hall, 1991.

[3] Craig, J. J. Introduction to robotics Mechanics and Control. Reading, MA: Addison Wesley, 1986.

[4] Cremean, L., W. Dumbar, D. vanGogh, J. Hickey, E. Klavins, J. Meltzer, and R. Murray. The Caltech Multi-Vehicle Wireless Testbed. In *Proc. of the 41st Conf. on Decision and Contr.*, Dec. 2002.

[5] Dullerud, G. E., and F. Paganini. *A Course in Robust Control Theory*. Texts in Applied Mathematics 36. New York: Springer, 1999.

[6] Franklin, G. F., J. D. Powell, and A. Emami-Naeini. *Feedback Control of Dynamic Systems*. 4th ed. Upper Saddle River, NJ: Prentice Hall, 2002.

[7] Grant, M., S. Boyd, and Y. Ye. *CVX: Matlab Software for Disciplined Convex Programming*. Palo Alto, CA: Stanford University, June 2008. Available at http://www. stanford.edu/~boyd/cvx/.

[8] Khalil, H. K. *Nonlinear Systems*. 3rd ed. Englewood Cliffs, NJ: Prentice Hall, 2002.

[9] Kwakernaak, H., and R. Sivan. *Linear Optimal Control Systems*. New York: Wiley, 1972.

[10] Laub, A. J. *Matrix Analysis for Scientists and Engineers*. Philadelphia: SIAM, 2005.

[11] Maciejowski, J. M. *Multivariable Feedback Design*. Electronic Systems Engineering Series. Workingham, UK: Addison Wesley, 1989.

[12] Oppenheim, A. V., A. S. Willsky, and S. H. Nawad. *Signals and Systems*. Signal Series. 2nd ed. London: Prentice Hall, 1996.

[13] Sánchez-Peña, R. S., and M. Sznaier. *Robust Systems: Theory and applications*. Adaptive and Learning Systems for Signal Processing, Communications, and Control. New York: Wiley, 1998.

[14] Strang, G. *Linear Algebra and Its Applications*. 3rd ed. Fort Worth: Harcourt Brace Jovanovich, 1988.

[15] Vegte, J. V. *Feedback Control Systems*. 3rd ed. Englewood Cliffs, NJ: Prentice Hall, 1994.

INDEX